S0-BXA-016

Difference Equations with Public Health Applications

Biostatistics: A Series of References and Textbooks

Series Editor

Shein-Chung Chow

President, U.S. Operations
StatPlus, Inc.
Yardley, Pennsylvania

Adjunct Professor
Temple University
Philadelphia, Pennsylvania

1. *Design and Analysis of Animal Studies in Pharmaceutical Development*, edited by Shein-Chung Chow and Jen-pei Liu
2. *Basic Statistics and Pharmaceutical Statistical Applications*, James E. De Muth
3. *Design and Analysis of Bioavailability and Bioequivalence Studies, Second Edition, Revised and Expanded*, Shein-Chung Chow and Jen-pei Liu
4. *Meta-Analysis in Medicine and Health Policy*, edited by Dalene K. Stangl and Donald A. Berry
5. *Generalized Linear Models: A Bayesian Perspective*, edited by Dipak K. Dey, Sujit K. Ghosh, and Bani K. Mallick
6. *Difference Equations with Public Health Applications*, Lemuel A. Moyé and Asha Seth Kapadia

ADDITIONAL VOLUMES IN PREPARATION

Medical Biostatistics, Abhaya Indrayan and Sanjeev B. Sarmukaddam

Statistical Methods for Clinical Trials, Mark X. Norleans

Difference Equations with Public Health Applications

Lemuel A. Moyé
Asha Seth Kapadia
The University of Texas–Houston School of Public Health
Houston, Texas

Marcel Dekker, Inc. New York • Basel

Library of Congress Cataloging-in-Publication Data

Moyé, Lemuel A.
Difference equations with public health applications / Lemuel A. Moyé, Asha Seth Kapadia.
p. cm.— (Biostatistics ; vol 6)
Includes index.
ISBN 0-8247-0447-9 (alk. paper)
1. Public Health—Research—Methodology. 2. Difference equations. I. Kapadia, Asha Seth. II. Title. III. Biostatistics (New York, N.Y.) ; 6.

RA440.85 .M693 2000
614.4'07'27—dc21 00-040473

This book is printed on acid-free paper.

Headquarters
Marcel Dekker, Inc.
270 Madison Avenue, New York, NY 10016
tel: 212-696-9000; fax: 212-685-4540

Eastern Hemisphere Distribution
Marcel Dekker AG
Hutgasse 4, Postfach 812, CH-4001 Basel, Switzerland
tel: 41-61-261-8482; fax: 41-61-261-8896

World Wide Web
http://www.dekker.com

The publisher offers discounts on this book when ordered in bulk quantities. For more information, write to Special Sales/Professional Marketing at the headquarters address above.

Current printing (last digit):
10 9 8 7 6 5 4 3 2 1

PRINTED IN THE UNITED STATES OF AMERICA

To

My mother Florence Moyé, my wife Dixie, my daughter Flora, my brother Eric, and the DELTS

LAM

To

My mother Sushila Seth, my sisters Mira, Gita, Shobha, and Sweety, my son Dev, and my friend, Jürgen

ASK

Series Introduction

The primary objectives of the Biostatistics Series are to provide useful reference books for researchers and scientists in academia, industry, and government, and also to offer textbooks for undergraduate and/or graduate courses in the area of biostatistics. This series provides a comprehensive and unified presentation of statistical designs and analyses of important applications in biostatistics, such as those in biopharmaceuticals. A well-balanced summary is given of current and recently developed statistical methods and interpretations for both statisticians and researchers/scientists with minimal statistical knowledge who are engaged in the field of applied biostatistics. The series is committed to providing easy-to-understand, state-of-the-art references and textbooks. In each volume, statistical concepts and methodologies are illustrated through real world examples.

The difference equation is a powerful tool for providing analytical solutions to probabilistic models of dynamic systems in health related research. The health related applications include, but are not limited to, issues commonly encountered in stochastic processes, clinical research, and epidemiology. In practice, many important applications, such as the occurrence of a clinical event and patterns of missed clinic visits in randomized clinical trials, can be described in terms of recursive elements. As a result, difference equations are motivated and are solved using the generating function approach to address issues of interest. This volume not only introduces the important concepts and methodology of difference equations, but also provides applications in public health research through practical examples. This volume serves as a bridge among biostatisticians, public health related researchers/scientists, and regulatory agents by providing a comprehensive understanding of the applications of difference equations regarding design, analyses, and interpretations in public health research.

Shein-Chung Chow

Preface

The authors have worked together since 1984 on the development and application of difference equations. Sometimes our work was episodic, at other times it was continuous. Sometimes we published in the peer reviewed literature while at other times we wrote only for ourselves. Sometimes we taught, and at other times, we merely talked. Difference equations have provided a consistent cadence for us, the background beat to which our careers unconsciously moved.

For each of us, difference equations have held a fascination in and of themselves, and it has always been natural for the two of us to immerse ourselves in their solution to the exclusion of all else. However, working in a school of public health means involvement in public health issues. We have committed to continually explore the possibility of linking difference equations to public health topics. Issues in public health, whether environmental, economic, or patient-care oriented, involve a host of interdependencies which we suspected might be clearly reflected in families of difference equations.

A familiar medical school adage is that physicians cannot diagnose a disease they know nothing about. Similarly, quantitative students in public health cannot consider utilizing difference equations to formulate a public health problem unless they are comfortable with their formulations and solutions. Difference equations have been the focus of attention by several authors including Samuel Goldberg's classic text, *Introduction to Difference Equations*, published in 1958. However, none of these have focused on the use of difference equations in public health. Although generating functions cannot be used to solve *every* difference equation, we have chosen to elevate them to the primary tool. Mastering these tools provides a path to the solution of complicated difference equations.

We would like to express our appreciation and thanks to Ms. Kathleen Baldonado for her valuable suggestions, extraordinary editing efforts, and keeping these turbulent authors calm.

Lemuel A. Moyé
Asha Seth Kapadia

Introduction

Difference equations are an integral part of the undergraduate and graduate level courses in stochastic processes. Instruction in difference equations provides solid grounding for strengthening the mathematical skills of the students by providing them with the ability to develop analytical solutions to mathematical and probabilistic prediction models of dynamic systems. *Difference Equations with Public Health Applications* is a new contribution to didactic textbooks in mathematics, offering a contemporary introduction and exploration of difference equations in public health. The approach we have chosen for this book is an incremental one–we start with the very basic definitions and operations for difference equations and build up sequentially to more advanced topics. The text begins with an elementary discussion of difference equations with definitions of the basic types of equations. Since intuition is best built incrementally, development of the solution of these equations begins with the simplest equations, moving on to the more complicated solutions of more complex equations.

In addition, many examples for the complete solutions of families of difference equations in their complete detail are offered. The solution methods of induction and of partial fractions are discussed and provided. Solutions from partial fractions are contrasted to other forms of the solution in terms of the effort in obtaining the solution (identification and the relative comprehensive ease of the solutions). Expansive exercises are provided at the end of each chapter.

The major contribution of *Difference Equations with Public Health Applications* is its scope of coverage. We begin with a simple, elementary discussion of the foundations of difference equations and move carefully from there through an examination of their structures and solutions. Solutions for homogeneous and nonhomogeneous equations are provided. A major emphasis of the text is obtaining solutions to difference equations using generating functions. Complete solutions are provided for difference equations through the fourth order. An approach is offered that will in general solve the kth order difference equation, for $k > 1$.

In addition, this textbook makes two additional contributions. The first is the consistent emphasis on public health applications of the difference equation approach. Issues in hydrology, health services, cardiology, and vital status ascertainment in clinical trials present themselves as problems in difference equations. We recognize this natural formulation and provide their solution using difference equation techniques. Once the solution is obtained, it is discussed and interpreted in a public health context.

The second contribution of this book is the focus on the generating function approach to difference equation solutions. This book provides a thorough examination of the role of generating functions to understand and solve difference equations. The generating function approach to difference equation solutions is presented throughout the text with applications in the most simple to the most complex equations.

Although mention is made of computer-based solutions to difference equations, minimal emphasis is placed on computing. Software use is not emphasized and, in general, computing is used not as a tool to solve the equations, but only to explore the solution identified through the generating function approach. This computing de-emphasis supports our purpose to provide the opportunity for students to improve their own mathematical skills through use of generating function arguments. In addition, computing software, like most software, is ephemeral in nature. A book focused on the application of one class of software ties its utility to the short life expectancy of the software. To the contrary, the time-tested generating function approach to difference equations is an elegant, stable, revealing approach to their solution. An analytic solution to a system of difference equations as identified through generating functions is a full revelation of the character of the equation.

Students in undergraduate or graduate programs in statistics, biostatistics, operations research, stochastic processes, and epidemiology compose the target audience for this book, as well as researchers whose work involves solutions to difference equations. It is envisioned that this book will be the foundation text for a course on difference equations in biostatistics, mathematics, engineering, and statistics. Difference equations are introduced at a very introductory level. A course in calculus is strongly recommended, and students so equipped will be able to practically take the tools discussed and developed in this text to solve complicated difference equations.

Chapter 1 provides a general introduction to the difference equation concept, establishes notation, and develops the iterative solution useful for first-order equations. Chapters 2 and 3 discuss the generating function concept and develop the tools necessary to successfully implement this approach. Chapter 2 provides an introduction to generating functions, and extensively develops techniques in manipulating generating functions, sharpening the mathematical skills and powers of observation of the students. Chapter 2 also provides the complete derivation for generating functions from commonly used discrete probability distribution functions. Chapter 3 provides a complete comprehensive discussion of the use of generating function inversion procedures when the denominator of the generating function is a polynomial. Chapter 4 combines the results of the first three chapters, demonstrating through discussion and numerous examples, the utility of the generating function approach in solving homogeneous and nonhomogenous difference equations. Chapter 5 addresses the complicated topic of difference equations with variable coefficients.

Chapter 6 introduces a collection of difference equation systems developed by the authors which are useful in public health research. Many of these applications are based in run theory, which the authors have found is a natural setting for several important issues in public health. The public health problem is often naturally expressed using run theory and a family of difference equations is provided to describe the recursive elements of the public health problem. These difference equations are motivated and solved in their entirety using the generating function approach. Chapters 7–10 applies these systems to problems in public health. Chapter 7 applies difference equations to drought predictions. Chapter 8 explores the applicability of difference equations to predict the occurrence of life-saving cardiac arrhythmias. In Chapter 9, difference equations are provided to describe patterns of missed visits in a randomized controlled clinical trial. Chapter 10 applies difference equations to health services organization plans for cash flow management at small ambulatory clinics. In each application, the solutions of the equations are formulated in a way to provide additional insight into the public health aspect of the problem. Finally, the application of difference equations to epidemiologic models is provided through the treatment of immigration, birth, death, and emigration models in Chapters 11 and 12. Solutions for combinations of these models are also provided.

Difference Equations with Public Health Applications provides solid grounding for students in stochastic processes with direct meaningful applications to current public health issues.

Lemuel A. Moyé
Asha Seth Kapadia

Contents

1

Difference Equations: Structure and Function

1.1 Sequences and Difference Equations

1.1.1 Sequence definitions and embedded relationships

To understand the concept and to appreciate the utility of difference equations, we must begin with the idea of a sequence of numbers. A sequence is merely a collection of numbers that are indexed by integers. An example of a finite sequence is {5, 7, -0.3, 0.37, –9 }, which is a sequence representing a collection of five numbers. By describing this sequence as being indexed by the integers, we are only saying that it is easy to identify the members of the sequence. Thus, we can identify the 2nd member in the sequence or, as another example, the 5th member. This is made even more explicit by identifying the members as y_1, y_2, y_3, y_4, and y_5 respectively. Of course, sequences can be infinite as well. Consider for example the sequence $\left\{1, \frac{1}{2}, \frac{1}{4}, \frac{1}{8}, \frac{1}{16}, \frac{1}{32}, \ldots\right\}$ This sequence contains an infinite number of objects or elements. However, even though there are an infinite number of these objects, they are indexed by integers, and can therefore have a counting order applied to them. Thus, although we cannot say exactly how many elements there are in this sequence, we can just as easily find the 8th element or the 1093rd element.* Since we will be working with sequences, it will be useful to refer to their elements in a general way. The manner we will use to denote a sequence in general will be by using the variable y with an integer subscript. In this case the notation y_1, y_2, y_3, ... ,y_k, ... represents a general sequence of numbers. We are often interested in discovering the values of the individual members of the sequence, and therefore must use whatever tools we have to allow us to discover their values. If the individual elements in the sequence are independent of each other, i.e., knowledge of one number (or a collection of elements) tells us nothing about the value of the element in question, it is very difficult to predict the value of the sequence element. However, many sequences are such that there exists embedded relationships between the elements; the elements are not independent, but linked together by an underlying structure. Difference equations are equations that describe the underlying structure or relationship between the sequence element; solving this family of equations means using the information about the sequence element interrelationship that is contained in the family to reveal the identity of members of the sequence.

1.1.2 General definitions and terminology

In its most general form a difference equation can be written as

$$p_0(k)y_{k+n} + p_1(k)y_{k+n-1} + p_2(k)y_{k+n-2} + \cdots + p_n(k)y_k = R(k) \tag{1.1}$$

which is the most general form of difference equations. It consists of terms involving members of the $\{y_k\}$ sequence, and, in addition, coefficients such as $p_j(k)$, of the elements of the $\{y_k\}$ sequence in the equation. These coefficients may or may not be a function of k. When the coefficients are not functions of k, the difference equation has constant coefficients. Difference equations with coefficients that are functions of k are described as difference equations with variable coefficients. In general, difference equations with constant coefficients are easier to solve then those difference equations that have nonconstant coefficients.

---

* Sequences which are infinite but not countable would be, for example, all of the numbers between 0 and 1.

Begin with a brief examination of the right side of equation (1.1) to introduce additional terminology. Note that every term on the left side of the equation involves a member of the $\{y_k\}$ sequence. If the term R(k) on the right side of the equation is equal to zero, then the difference equation is *homogeneous*. If the right side of the equation is not zero, then equation (1.1) becomes a *nonhomogeneous* difference equation. For example, the family of difference equations

$$y_{k+2} = 6y_{k+1} - 3y_k \tag{1.2}$$

for k an integer from zero to infinity is homogeneous since each term in the equation is a function of a member of the sequence $\{y_k\}$. The equation

$$y_{k+2} = 6y_{k+1} - 3y_k + 12 \tag{1.3}$$

for $k = 1,2,\ldots,\infty$ is a nonhomogeneous one because of the inclusion of the term 12.

Finally, the order of a family of difference equations is the difference in sequence location between the term with the greatest index and the term with the smallest index of the $\{y_1, y_2, y_3, \ldots, y_k, \ldots\}$ sequence represented in the equation. In equation (1.1) the member of the sequence $\{y_k\}$ with the largest subscript is y_{k+n} and the member of the sequence with the smallest subscript in equation (1.1) is y_k. Thus the order of this difference equation is $k + n - k = n$, and equation (1.1) is characterized as an nth order difference equation. If $R(k) = 0$, we describe equation (1.1) as an nth order homogenous difference equation, and if $R(k) \neq 0$ equation (1.1) is an nth order nonhomogeneous difference equation.

As an example, the family of difference equations described by

$$y_{k+2} = 6y_{k+1} - 3y_k \tag{1.4}$$

for $k = 0, 1, 2, \ldots, \infty$ where y_0 and y_1 are known constants is a family of second-order homogeneous difference equation with constant coefficients, while

$$3y_{k+4} + (k+3)y_{k+3} + 2^{k+2}y_{k+2} + (k+1)y_{k+1} + 4ky_k = (k+2)(k+1) \tag{1.5}$$

would be designated as a fourth-order, nonhomogeneous family of difference equations with variable coefficients.

1.2 Families of Difference Equations

An important distinction between difference equations and other equations encountered earlier in mathematics is that difference equations represent a persistent relationship among the sequence members. Since these relationships continue as we move deeper into the sequence, it is clear that the "difference equation" is not just one equation that must be solved for a small number of unknown variables. Difference equations are in fact better described as families of equations. An example of a fairly simple family of difference equations is given in the following equation

$$y_{k+1} = 2y_k - 2, \quad \text{k an integer} > 0, \, y_0 \text{ known} \tag{1.6}$$

Recognize this as a first order, nonhomogeneous difference equation with constant coefficients. We must now expand our view to see that this formulation has three components, each of which is required to uniquely identify this family of difference equations. The first is the mathematical relationship that links one member of the sequence to other sequence members–this is provided by the essence of the mathematical equation. However, the second component is the range over the sequence for which the difference equations apply. The relationship $y_{k+1} = 2y_k - 2$ serves to summarize the following set of equations:

$$\begin{aligned} y_1 &= 2y_0 - 2 \\ y_2 &= 2y_1 - 2 \\ y_3 &= 2y_2 - 2 \\ y_4 &= 2y_3 - 2 \\ y_5 &= 2y_4 - 2 \\ y_6 &= 2y_5 - 2 \end{aligned} \tag{1.7}$$

The key to the family of equations is that the same equation is used to link one member of the sequence (here y_{k+1}) to at least one other member of the sequence. It is this embedded relationship among the sequence members that allows one to find the value for each member of the sequence.

Rather than be forced to write this sequence of 100 equations out one by one, we can write them succinctly by using the index k, and consideration of this index brings us to the second component of equation (1.6)'s formulation: namely, the range of sequence members for which the relationship is true. For equation (1.6) the sequence is only true for $y_0, y_1, y_2, \ldots, y_{100}$. Some other relationship may hold for sequence members with indices greater than 100. If the sequence of integers is finite, the number of equations that need to be solved is finite. An infinite sequence will produce an infinite number of equations that require solution.

The third and last component of the equation for a family of difference equations is the initial or boundary condition. For example, for k=0, the first member in the family of difference equations (1.6) reduces to $y_1 = 2y_0 - 2$. Since there is no equation relating y_0 to another sequence member, it is assumed that y_0 is a known constant. This is called a boundary condition of the difference equation family. The boundary conditions are necessary to anchor the sequence for a particular solution. For example, if $y_0 = 3$, the sequence governed by the set of difference equations $y_{k+1} = 2y_k - 2$ is {3, 4, 6, 10, 18,...} If, on the other hand, the sequence began with a different boundary condition $y_0 = -1$, the sequence becomes {–1, -4, -10, -22, -46, ...}. Thus, the same difference equation, with different boundary conditions, can produce very different sequences. Each of these three components is required to uniquely specify the difference equation.

1.2.1 Examples

There are infinite number of families of difference equations. Some of the examples are given below.

Example 1:

$$y_{k+3} = ay_{k+2} + by_{k+1} + cy_k \tag{1.8}$$

for k = any integer from 0 to 20: a, b, and c are known constants and y_0, y_1, and y_2 are known.

Example 2:

$$y_{k+1} = 8y_k + 7 \tag{1.9}$$

for k = 0, 1, 2,..., 10: y_0 a known constant

Example 3:

$$3ky_{k+2} = y_k - 0.952^k y_{k+1} \tag{1.10}$$

for k = 0,...,∞ : y_0 and y_1 are known.

Each of these three examples represents a difference equation family with its own unique solution depending on the relationship, the range of the index, and the boundary conditions. Thus, these families of equations represent unique sequences.

1.3 Solutions to Difference Equations

Being able to first formulate a public health problem in terms of a family of difference equations, and then find the solution to the family, is a useful skill for which we will develop powerful tools from first principles. There are several approaches to the solutions of difference equations. Chapters 2 and 3 will develop some analytical tools useful in their solution. However, two other rather intuitive approaches should be considered as well, and can themselves be useful and constructive: The first is the iterative approach to the solution of difference equations, and the second is the use of mathematical induction. Both are briefly discussed below.

1.3.1 Iterative solutions

The temptation is almost irresistible to solve difference equations iteratively. To begin, consider the following simple set of relationships in which the task is to explicitly find all values of the sequence $\{y_k\}$ where a is a constant

$$y_{k+1} = ay_k \tag{1.11}$$

for $k = 0,1,2,\ldots,\infty$: y_0 a known constant. This first-order, homogeneous family of difference equations represents a relatively easy set of relationships, and one can envision starting with the first elements of the sequence for which the values are known, then using the difference equations to move further into the sequence, utilizing the recently discovered value of the kth element to help identify the $k+1^{st}$ value.

$$\begin{aligned} y_1 &= ay_0 \\ y_2 &= ay_1 = a(ay_0) = a^2y_0 \\ y_3 &= ay_2 = a(a^2y_0) = a^3y_0 \\ y_4 &= ay_3 = a(a^3y_0) = a^4y_0 \\ &\vdots \end{aligned} \tag{1.12}$$

If we continue to develop the sequence elements, it can be seen that the general solution for this family of difference equations is $y_k = a^k y_0$. The logic of the underlying solution is embedded in the family of difference equations themselves, and by moving through the recursive equations one at a time, we allow the difference equation family essentially to solve itself.

As another example of the ease of this approach, consider the family of difference equations

$$y_{k+1} = ay_k + b \tag{1.13}$$

for all $k > 0$; y_0 known. Proceeding as in the earlier example, the examination of the values of the elements of the sequence of $\{y_k\}$ can proceed in an analogous fashion. From equation (1.13), the next value in the $\{y_k\}$ sequence y_{k+1} is obtained by multiplying the previous value in the sequence y_k by the known constant a, and then adding to this product the known constant b. Equation (1.13) provides the necessary information to evaluate each of the y_1, y_2, y_3,... .

$$\begin{aligned} y_1 &= ay_0 + b \\ y_2 &= ay_1 + b = a(ay_0 + b) + b = a^2y_0 + ab + b \\ y_3 &= ay_2 + b = a(a^2y_0 + ab + b) + b = a^3y_0 + a^2b + ab + b \\ y_4 &= ay_3 + b = a(a^3y_0 + a^2b + ab + b) + b = a^4y_0 + a^3b + a^2b + ab + b \end{aligned} \tag{1.14}$$

and so on. Again the solution is obtained through an iterative approach.

Unfortunately, this intuitive approach to the solution, which is very useful for first-order difference equations that were described above, becomes somewhat complicated if the order is increased. Consider, for example, the second-order, nonhomogeneous family of difference equations

$$y_{k+2} = ay_{k+1} + by_k + d \tag{1.15}$$

for $k = 0,1,2,\ldots,\infty$; y_0, y_1 are known. The iterative approach begins easily enough but rapidly becomes complicated.

$$\begin{aligned} y_2 &= ay_1 + by_0 + d \\ y_3 &= ay_2 + by_1 + d = a(ay_1 + by_0 + d) + by_1 + d \\ &= a^2y_1 + aby_0 + ad + by_1 + d \\ y_4 &= ay_3 + by_2 + d = a(a^2y_1 + aby_0 + ad + by_1 + d) \\ &+ b(ay_1 + by_0 + d) + d \\ &= a^3y_1 + a^2by_0 + a^2d + aby_1 + ad + aby_1 + b^2y_0 + bd + d \\ &= (a^3 + 2ab)y_1 + (a^2b + b^2)y_0 + (a^2 + a + b + 1)d \end{aligned} \tag{1.16}$$

From the developing pattern it can be seen, that, in general, y_k will always be written as a function of the constant coefficients a, b, and d, as well as in terms of the boundary conditions (i.e., the values of y_0 and y_1). Recognizing these terms is by itself not sufficient for attaining the actual solution. It is difficult to see a pattern emerging since y_k becomes an increasingly complicated function of a and b as k increases. The situation becomes even more complicated for third-order families of difference equations. Although only algebra is required for the iterative solutions of these equations, this algebra itself becomes more complicated as solutions are found for values deep in the $\{y_k\}$ sequence. Thus the iterative approach will always work, but it becomes difficult in complicated difference equations to envision the general solution for any sequence member y_k.

1.3.2 The induction method

One way to solve these simple equations is to guess the solution and then prove that the guessed solution is correct through the use of induction. Briefly, the induction argument outlines a simple sequence of three steps in the proof of an assertion about the non-negative integers. The steps are

(1) Demonstrate the assertion is true for $k = 1$
(2) Assume the solution is true for k
(3) Use (1) and (2) to prove that the assertion is true for k+1

This time-tested tool has been very useful in proving assertions in algebra for integers. We will use as a simple example an examination of a possible solution to the nonhomogeneous, first-order equation

$$y_{k+1} = ay_k + b \tag{1.17}$$

for all $k > 0$; y_0 a known constant. Assume the solution

$$y_k = a^k y_0 + b\sum_{j=0}^{k-1} a^j \tag{1.18}$$

To begin the proof by induction, first check if this solution is true for k=1

$$y_1 = ay_0 + b = ay_0 + b\sum_{j=0}^{0} a^j \tag{1.19}$$

so the assertion is true for $k = 1$.
The next step in the induction argument is to assume that equation (1.17) is true for k (i.e., assume that the assertion is true for some non-negative integer k), and

then, using this assumption, prove that the proposed solution is true for $k = k+1$. The task is to prove that

$$y_{k+1} = a^{k+1}y_0 + b\sum_{j=0}^{k} a^j \tag{1.20}$$

This is accomplished easily with simple algebra

$$\begin{aligned} y_{k+1} &= ay_k + b = a\left[a^k y_0 + b\sum_{j=0}^{k-1} a^j\right] + b = \\ &= a^{k+1}y_0 + ba\sum_{j=0}^{k-1} a^j + b = a^{k+1}y_0 + b\sum_{j=0}^{k} a^j \end{aligned} \tag{1.21}$$

so the assertion is true for k = k+1 and we have the solution for the difference equation family. Of course, an obstacle to this approach is the requirement of first having an idea of the solution of the family of equations. Induction does not work without a good guess at the solution. Looking back at our work on the second-order, nonhomogeneous family of difference equations with constant coefficients given by

$$y_{k+2} = ay_{k+1} + by_k + d \tag{1.22}$$

for $k = 0,1,2,\ldots,\infty$; y_0, y_1 are known it is difficult to guess at the solution. In general guessing will not serve us well, and we will use the generating function argument to be developed in the next two chapters to guide us to the solutions.

1.4 Some Applications

Difference equation families can be elegant mathematical expressions of relationships that extend themselves through a sequence. There are many examples of their applications. Some of the more classic ones occur in Goldberg [1], Feller [2], Saaty [3], Chiang [4], and Bailey [5]. They do in fact appear in public health. Here are three illustrations.

1.4.1 Health services research

Consider a situation in which you are hired as a manager of a medical complex. This organization includes physicians, nurses, physician assistants, and administrators. One of the new manager's first responsibilities is to help resolve an issue that has become an important problem for the employees of the company who have been employed the longest, and have the company's best and greatest experience. The standing company policy has been to award each company employee an increase in their hourly wages. The increase comes from a table of hourly rates. The table has ten rows, each row representing the number of years of service. For each row the new hourly wage is supplied. In this company, several of the employees have been in the organization for more than ten years (Table 1.1).

Table 1.1 Observed Clinic Hourly Wage by Years of Service

Year of Service	Wage
1	6.00
2	6.48
3	7.00
4	7.56
5	8.16
6	8.82
7	9.52
8	10.28
9	11.11
10	11.99

However, since the table only contains ten entries, those employees who have been in the organization for more than ten years cease to get yearly increments. The assignment is to compute the correct adjustment in wages for these longstanding employees.*

A review of Table 1.1 quickly reveals that the increase in hourly wage represents about an 8% increase over the previous year's hourly wage. Your task is to represent the worker's hourly wage for this year in terms of last year's hourly wage. Choose y_k as the hourly wage for the kth year. Then the hourly wage an individual makes is indexed by the number of years he/she has worked, and a worker who has worked for k years has the sequence of yearly hourly wages $y_1, y_2, y_3, y_4, y_5, y_6, \ldots$ depicting their history of hourly wages. What is

* The employees in this friendly group did not ask for back wages, or for interest associated with the lost wages, only for the immediate adjustment of their wages to the correct rate.

required is a relationship between the hourly wage for the k+1th year, and the hourly wage for a salary for the previous year. You write

$$y_{k+1} = 1.08y_k \tag{1.23}$$

for k = 1 to ∞ and y_1 a fixed, known, constant. Note that this equation is a first-order homogeneous equation with constant coefficients, and is the same as the first difference equation that was solved using the iterative approach. The solution for that equation was $y_k = a^k y_0$ for k = 1, 2, 3, 4, 5, ...

which in the circumstances of this clinic becomes

Table 1.2 Predicted Clinic Hourly Wage by Years of Service

Year of Service	Wage
1	6.00
2	6.48
3	7.00
4	7.56
5	8.16
6	8.82
7	9.52
8	10.28
9	11.11
10	11.99
11	12.95
12	13.99
13	15.11
14	16.32
15	17.62
16	19.03
17	20.56
18	22.20
19	23.98
20	25.89

$$y_k = (1.08)^{k-1} y_1 \tag{1.24}$$

$k = 1$ to ∞ and you can complete the table for the senior employees.

Obtaining y_1 from the first row of Table 1.2, compute the hourly wage for an employee who is in their twelfth year of service is $y_{12} = (1.08)^{11}(6.00) =$ \$13.99 per hour. Of course, employees who have been there for many years, would have exhausted this company's payroll. For example, an employee who has been working for forty years would have an hourly wage of $y_{40} = (1.08)^{39}(6.00) =$ \$120.69. Fortunately for the organization, there was no employee with more than fifteen years of service.

1.4.2 Screening clinic payments

Another useful example of the derivation of a popular (indeed, infamous) family of difference equations involves debt repayment. Consider the construction of a women's clinic in an underserved area, which will play a major role in screening women for breast and cervical cancer. The clinic owners arranged for a business loan of \$500,000 for equipment and operations costs to be paid back in 30 years. If the clinic has arranged for an annual interest rate of r, then what must the monthly payments be over the next thirty years?

As before, we do not start with the difference equation to solve. Instead, we begin with the issue and attempt to formulate it as a difference equation. We expect that there is a recursive relationship built into the sequence of monthly interest payments and monthly principal payments that needs to be discovered. Let's start our approach to this problem by introducing some helpful notation. Let m_k be the total monthly payment that is due. It is assembled from two components: The first is the interest payment for that month, i_k; the second component is the principle that is due for the kth month,* p_k. Then

$$m_k = i_k + p_k \tag{1.25}$$

for $k = 1, 2, 3, \ldots, 360$.

Begin by finding i_1, the interest payment for the first month. Recall that the amount of interest due in any month is based on the unpaid principle. Since in the first month, no principle has yet been paid,

$$i_1 = 500{,}000\left(\frac{r}{1200}\right) \tag{1.26}$$

The term $\frac{r}{1200}$ is the annual interest rate divided by 100 (to get the annual rate in percentage points) and then divided by 12 (to obtain the monthly rate). If r =

---

* For simplicity, we will ignore premium mortgage insurance, property insurance, and taxes which would increase the monthly payment even more

12, then it is $\frac{12}{(100)\cdot 12}$. Compute the interest payment due in the second payment month as

$$i_2 = (500{,}000 - p_1)\left(\frac{r}{1200}\right) \tag{1.27}$$

Another critical feature of this system of interrelationships is that the monthly payments are constant. Even though the interest payments change each month and the amount of the principle changes with each payment, the monthly payment itself (which is the sum of these payments) remains constant. That is, $m_1 = m_2 = m_3 = \ldots = m_{360}$, as there are 360 monthly payments during the term of the thirty year loan. Thus

$$\begin{aligned} m_1 &= m_2 \\ i_1 + p_1 &= i_2 + p_2 \end{aligned} \tag{1.28}$$

and therefore

$$500{,}000\left(\frac{r}{1200}\right) + p_1 = (500{,}000 - p_1)\frac{r}{1200} + p_2 \tag{1.29}$$

This suggests that we can solve for the 2nd month's principle, p_2 in terms of p_1 to find

$$p_2 = \frac{1200 + r}{1200} p_1 \tag{1.30}$$

We can proceed to the next step in the sequence of payments, now evaluating the relationship between p_3 and p_2.

$$\begin{aligned} m_2 &= m_3 \\ i_2 + p_2 &= i_3 + p_3 \end{aligned} \tag{1.31}$$

which becomes

$$(500000 - p_1)\frac{r}{1200} + p_2 = (500000 - p_1 - p_2)\frac{r}{1200} + p_3 \tag{1.32}$$

Solving p_3 in terms of p_2

$$p_3 = \frac{1200 + r}{1200} p_2 \tag{0.32}$$

and, in general

$$p_{k+1} = \frac{1200 + r}{1200} p_k \tag{0.33}$$

for k = 1, 2, ..., 360. So, the sequence of 360 principle payments has embedded in them a relationship reflected by a difference equation. Recognize this equation as a first-order, homogeneous equation with constant coefficients. Furthermore, letting $\frac{1200+r}{1200} = a$, it can be seen that equation (0.33) is another form of equation $y_{k+1} = ay_k$, for which it is known the solution is $y_k = a^k y_1$. Applying this principle here, we find

$$p_{k+1} = \left(\frac{1200 + r}{1200}\right)^k p_1 \tag{0.34}$$

However, p_1 is not yet known. Since we will need a tool from Chapter 2 to find this, we must postpone our solution until later. As it turns out, knowing p_1 is the one piece of the puzzle that is needed. Since i_1 is already known the interest payment for the first month, compute m_1 from $m_1 = i_1 + p_1$ to compute the monthly payment*. We will return to a complete solution for this problem in chapter three. The point here is that it can take some work and ingenuity to identify the appropriate family of difference equations.

1.4.3 Chronobiology and mixed difference equations

It is sometimes of interest to combine information for a system that includes two families of difference equations. The solutions provided for this procedure will demonstrate the adaptability of the difference equation perspective. Consider a system that describes the likelihood that a woman is most likely to ovulate on a given day. There is important variability in the day a woman in the population will ovulate. Assume that the ovarian cycle is regular and of 28 days duration (this is different than a calendar month). The women's likelihood of ovulation increases as the day comes closer to the middle of the cycle. Let y_k be the likelihood that a woman will ovulate on the kth day, k = 1, 2, 3, ..., 28. Suppose as a first approximation, the relationship is

---

* A general solution for this equation will be presented in Chapter 3, after the concept of generating functions is introduced in Chapter 2.

$$y_k = 1.46y_{k-1} \tag{0.36}$$

for k = 2, 3,..., 28 (let y_1=1.46). Note, that equation (0.36) states that the likelihood of ovulation increases, and, is in fact a multiple of the previous day's ovulation likelihood. However, this family of equations has two problems related to the unbounded nature of y_k in this relationship. The first is that these likelihoods are not probabilities and do not sum to one across the 28 day cycle period. The second problem is that it is known that the likelihood of ovulation does not increase monotonically–rather, it peaks at the middle of the cycle and decreases thereafter until the cycle ends. Thus, the system requires a model that allows the ovulation likelihood to increase to a particular day in the month, and then begin to decrease. One way to build this into the difference equation family would be to let the form of the difference equation relationship be a function of the day. The difference equation will be

$$\begin{aligned} y_k &= 1.46y_{k-1} \text{ for } 2 \le k \le 14 \\ &= 0.47y_{k-1} \text{ for } 15 \le k \le 28 \end{aligned} \tag{0.37}$$

where y_1 is a known constant. This system is actually really a hybrid of two difference equation families. For $k \le 14$, the likelihood monotonically increases. However, for $k > 14$, the relationship between y_k and y_{k-1} changes, with a decreasing value of the likelihood of ovulation as the days progress to those later in the 28-day cycle.

Using the iterative approach to identify a solution to this family of difference equations.

$$y_2 = (1.46)y_1 : y_3 = (1.46)^2 y_1 : L \;\; y_{14} = (1.46)^{13} y_1 \tag{0.38}$$

When k = 15, use the second summand of equation (0.37) to find $y_{15} = 0.47y_{14} = 0.47(1.46^{13}y_1)$ and proceed on to the end of the 28-day cycle

$$\begin{aligned} y_{16} &= (0.47)^2 (1.46)^{13} y_1 \\ y_{17} &= (0.47)^3 (1.46)^{13} y_1 \\ y_{18} &= (0.47)^4 (1.46)^{13} y_1 \end{aligned} \tag{0.39}$$

This evaluation reveals that the solution for y_k for $k \ge 15$ has two components. The first reflects the influence of the first 14 days of the month, as measured by $(1.46)^{13}y_1$. The second component reflects the influence of the sequence members $\{y_k\}$ for $k \geq 15$. This component is calculated to be $(0.47)^{k-13}$. We can see that the general solution for this problem demonstrates that

$$\begin{aligned} y_k &= \left(1.46^{k-1}\right) y_1 \text{ for } 1 \leq k \leq 14 \\ &= \left(1.46^{13}\right)\left(0.47^{k-13}\right) y_1 \text{ for } 15 \leq k \leq 28 \end{aligned} \tag{1.40}$$

Figure 1.1 is a plot of this relationship as a function of k , the day in the cycle. The values of y_k are normed so that $\sum_{k=1}^{28} y_k = 1$. In Figure 1.1 the relationship between the probability of ovulation (suitably normalized) as given by equation (1.40) and k, the day of the cycle is displayed. By mixing the two difference equation families, the appropriate features of the probability of ovulation were obtained. Asymmetry can be introduced, by changing the decline parameter from 0.47 to 0.35, as denoted in Figure 1.2.

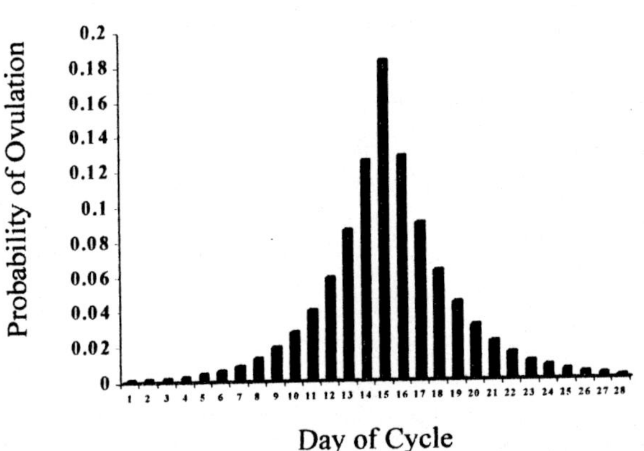

Figure 1.1 The application of a difference equation family to depict a symmetric relationship between the probability of ovulation and the day in the 28-day menstrual cycle.

The degree of asymmetry is completely in the control of the investigator (and the patient!) by choosing different values of a and b in the family of difference equations provided by $y_k = a^{k-1} y_1$ for $k \leq 14$ and $y_k = a^{13} b^{k-14} y_1$ for $15 \leq k \leq 28$.

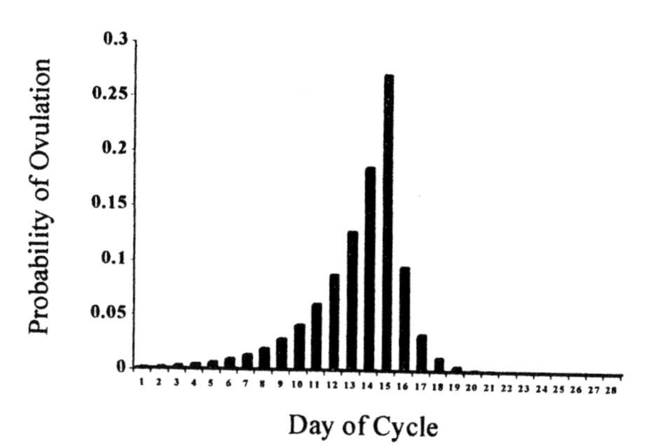

Figure 1.2 The application of a difference equation family to depict an asymmetric relationship between the probability of ovulation and the day in the 28-day menstrual cycle.

1.5 Summary

This first chapter has served as an introduction to the concept of difference equation applications and has provided a glimpse of how they may be applied in public health settings. It is necessary, in order to implement the tools, not only to understand how these equations may be solved, and to develop a fundamental understanding of the public health problem to which the family of difference equations will ultimately be applied, but also to develop some intuition for public health scenarios. It is first necessary to be able to envision the relationship between the variables of interest as having a recursive character; once this is done, exploration of the data may suggest the nature of the relationship. As the relationship is uncovered, we can then begin to see the formulation as a difference equation.

The great sculptor, Michelangelo, was asked how he is able to sculpt so well. He replied that to him, the slab of stone facing him was not a slab of stone but a statue incased in stone. All he did was hammer and brush away the stone that blocked his view of the statue. The key here is vision. One cannot bring mathematics, such as difference equations, to bear in a public health problem if he or she doesn't really understand the fundamental nature of the problem. It is crucial to examine the public health problem from many directions and perspectives, clearing away that which obstructs our view–we can then see if, at its core, the problem has a recursive relationship that may be represented by a difference equation.

Once the equation is identified, it must be solved. The goal of the next four chapters is to equip one for this task.

Problems

1. For each of the following five families of difference equations, give the order and classify it as homogeneous or nonhomogenous.
   a) $2y_{k+1} = 7y_k$: $k = 0$ to ∞, y_0 a known constant
   b) $3y_{k+2} - 11y_{k+1} = 2y_k - 1$: $k = 0$ to ∞, y_0, y_1 are known constants
   c) $ay_{k+3} = 2y_{k+1}$: $k = 0$ to ∞, y_0, y_1, y_2 are known constants
   d) $y_{k+2}=(-1)^k y_k - 1$: $k = 0$ to ∞, y_0, y_1 are known constants
   e) $y_{k+3} - y_{k+2} + y_{k+1} - y_k = k$: $k = 0$ to ∞, y_0, y_1, y_2 are known constants

2. Solve the following families of difference equations for y_k, $k = 0\ldots10$ using the iterative procedure
   a) $5y_{k+1} = y_k$: $k = 0$ to ∞, y_0 a known constant
   b) $13y_{k+2} - y_{k+1} = 5y_k - 5$: $k = 0$ to ∞, y_0, y_1 are known constants
   c) $7y_{k+3} = 8y_{k+1}$: $k = 0$ to ∞, y_0, y_1, y_2 are known constants
   d) $11y_{k+2}=(-0.5)^k y_k - 1$: $k = 0$ to ∞, y_0, y_1 are known constants
   e) $2y_{k+3} - 2y_{k+2} + y_{k+1} - 2y_k = k$: $k = 0$ to ∞, y_0, y_1, y_2 are known constants.

3. You are in charge of providing adequate nursing staff for a small urban hospital. Nursing staff is in great demand, which means that new nurses are always available for hire, but nurses often leave your hospital after several years because of fatigue or other career opportunities. You wish to model the number of nurses you will have at your clinic in the kth year. Consider the number of nurses you have at the clinic's inception is $y_0 = 14$. You expect that in any given year you will add new nurses who have never worked at your hospital before. However, in that same year, you will lose 30% of the nurses who worked in your system last year, but will gain back 5% of the nurses who worked in your system two years ago.

   A. Write a difference equation for the number of nurses in your clinic in the kth year, based on the changes in nursing availability in the $k-1$ and $k-2$ years. Compute iteratively the number of nurses you will have for the next fifteen years.

   B. How large must d be to sustain at least fourteen nurses available to work in your hospital.

4. Consider the screening clinic payment problem of section 1.4.2. In this generalization, assume that the clinic administration will be able to "pay

ahead, " i.e., the clinic will pay an additional amount on the principle of the loan, each month paying an additional v dollars for the principle. With this complication, compute the recursive relationship for i_k the monthly interest to be

$$i_k = \left[500,000 - \sum_{j=1}^{k-1} p_k - (k-1)v \right] \frac{r}{1200} \tag{0.40}$$

and p_k the monthly principle can be expressed as

$$p_k = \left[\frac{1200 - r}{1200} \right] p_{k-1} + \frac{rv}{1200} \tag{0.41}$$

References

1. Goldberg S. Introduction to Difference Equations. New York. John Wiley and Sons. 1958.
2. Feller W. An Introduction to Probability Theory and Its Applications. New York. John Wiley and Sons. 1968.
3. Saaty T.L. Elements of Queuing Theory with Applications. New York. McGraw-Hill. 1961.
4. Chiang C.L. An Introduction to Stochastic Processes and Their Applications. New York. Robert E. Krieger Publishing Company. 1980.
5. Bailey N. The Elements of Stochastic Processes. New York. John Wiley and Sons. 1964.

2

Generating Functions I: Inversion Principles

2.1 Introduction

In this chapter we will devote our efforts to introducing and building up experience with using generating functions. It would first be useful, however, to show why any such experience is necessary. A relevant question the reader may ask at the conclusion of Chapter 1 is "Why is another approach, such as generating functions, necessary to solve difference equations?" In Chapter 1, some useful tools in identifying solutions to difference equations were demonstrated. One of these tools was to solve the equations one at a time, starting with $k = 0$ and then working up through the sequence, using the solution for $(k - 1)^{st}$ difference equation to obtain the solution for the kth equation. This is a very natural approach, as it is very helpful in many circumstances. The other approach is to try to guess the difference equation solution and to prove that answer for the entire difference equation family. As attractive as these approaches are, their limitations become clear when one considers a second-order nonhomogenous family of difference equations, e.g.,

$$y_{k+2} = 3y_{k+1} + 2y_k + 7 \tag{2.1}$$

for $k = 0,1,2,\ldots,\infty$; y_0, y_1 are known constants. If we try to solve this family of difference equations iteratively, the process begins as follows. For $k = 0$

$$y_2 = 3y_1 + 2y_0 + 7 \tag{2.2}$$

Proceeding to the next equation ($k = 1$)

$$\begin{aligned} y_3 &= 3y_2 + 2y_1 + 7 = 3(3y_1 + 2y_0 + 7) + 2y_1 + 7 \\ &= 9y_1 + 6y_0 + 21 + 2y_1 + 7 \\ &= 11y_1 + 6y_0 + 28 \end{aligned} \tag{2.3}$$

The third equation yields

$$\begin{aligned} y_4 &= 3y_3 + 2y_2 + 7 = 3(11y_1 + 6y_0 + 28) + 2(3y_1 + 2y_0 + 7) + 7 \\ y_4 &= 33y_1 + 18y_0 + 84 + 6y_1 + 4y_0 + 14 + 7 \\ &= 39y_1 + 22y_0 + 105 \end{aligned} \tag{2.4}$$

So, after solving the first three equations,

$$\begin{aligned} y_2 &= 3y_1 + 2y_0 + 7 \\ y_3 &= 11y_1 + 6y_0 + 28 \\ y_4 &= 39y_1 + 22y_0 + 105 \end{aligned} \tag{2.5}$$

Only an algebraic genius would be able to see a pattern in the solutions of y_2, y_3, and y_4 that would help to identify the values for each of y_6, y_7, and y_8. The absence of this intuition is keenly felt here. Since the general solution to the family of difference equations as expressed in equation (2.1) cannot be intuited, neither of the methods discussed in Chapter 1 will be very helpful. It is because of these difficulties that arise from a very simple difference equation family that we turn to generating functions as a useful tool.

2.2 Introduction to Generating Functions

2.2.1 Motivation for using generating functions

In this section we define generating functions and show how they can be applied to the solution of a family of difference equations. A fine introduction to this development is provided by Goldberg [1]. A generating function is a function that contains the information necessary to produce or generate the sequence of y_k's. G(s) is defined as

$$G(s) = \sum_{k=0}^{\infty} y_k s^k \tag{2.6}$$

G(s) provides the content of the sequence $\{y_k\}$ which will be the solution to the family of difference equations. First, we will demonstrate the role G(s) plays in the solution of a difference equation, providing a view of the skills the reader will require and gain in the use of this tool. Let's begin with a simple first-order, homogeneous family of difference equations.

$$y_{k+1} = 3y_k \tag{2.7}$$

for $k = 0,1,2,\ldots,\infty$; y_0 is a known constant. Know that this is an infinite collection of equations ordered by the index k. Consider the solutions to each of these equations as a member of an infinite sequence of elements y_0, y_1, y_2,... We begin the generating function approach by choosing a constant s, where $0 < s < 1$ and multiplying k^{th} equation in the difference equation family by s^k. Thus $s^1y_1 = 3s^1y_0$, $s^2y_2 = 3s^2y_1$, and $s^3y_3 = 3s^3y_2$. In general

$$s^k y_{k+1} = 3s^k y_k \tag{2.8}$$

for k = 0,1,2,... After this multiplication, the next step is to add these equations from k = zero to infinity, recognizing that the sum of left side of these equations is equal to the sum of the right side of all of these equations. This reveals

$$\sum_{k=0}^{\infty} s^k y_{k+1} = \sum_{k=0}^{\infty} 3s^k y_k \tag{2.9}$$

Note that this maneuver has converted the original, infinite number of difference equations into one equation. The equations have become consolidated.

Admittedly, this one equation is complicated with the use of a summation sign. Let's begin the simplification of equation (2.9) by recalling our definition

$$G(s) = \sum_{k=0}^{\infty} s^k y_k \tag{2.10}$$

G(s) is the generating function of the sequence $\{y_k\}$. Since the goal is to identify the values of each of the y_k's, we will use knowledge of G(s) to "generate" or produce the y_k's for us. In order to do this, we will attempt to formulate equation (2.7) in terms of G(s). This will require that we convert each of the terms in equation (2.9) into a term that involves G(s). For example, the term on the right side of equation (2.9) can be written as

$$\sum_{k=0}^{\infty} 3s^k y_k = 3\sum_{k=0}^{\infty} s^k y_k = 3G(s) \tag{2.11}$$

Movement of the constant 3 through the summation sign allows us recognition of G(s) in this term and to write the entire expression as a function of G(s).

Returning to the expression on the left side of equation (2.9), see that this term is not G(s); however, it is related to G(s). This relationship is illuminated by taking two steps: the first is to adjust the exponent of s so that it matches the index of k in the infinite sum.

$$\sum_{k=0}^{\infty} s^k y_{k+1} = \sum_{k=0}^{\infty} s^{-1}s^1 s^k y_{k+1} = \sum_{k=0}^{\infty} s^{-1}s^{k+1} y_{k+1} = s^{-1}\sum_{k=0}^{\infty} s^{k+1} y_{k+1} \tag{2.12}$$

The summation in the far right expression of equation (2.12) is almost G(s). However, after reviewing equation (2.12), we see that we still do not have G(s). In fact, we are missing the term $s^0 y_0$. This can be taken care of by adding and then subtracting terms to preserve the equality as follows:

$$s^{-1}\sum_{k=0}^{\infty} s^{k+1} y_{k+1} = s^{-1}\left[\sum_{k=0}^{\infty} s^{k+1} y_{k+1} + s^0 y_0 - s^0 y_0\right] \tag{2.13}$$

Further simplification reveals

$$
\begin{aligned}
&= s^{-1}\left[\sum_{k=0}^{\infty}\left(s^{k+1}y_{k+1} + s^0 y_0\right) - s^0 y_0\right] \\
&= s^{-1}\left[\sum_{k=0}^{\infty} s^k y_k - y_0\right] \qquad (2.14) \\
&= s^{-1}\left[G(s) - y_0\right]
\end{aligned}
$$

This entire process deserves some comment. We began with a summation which had the term $s^k y_{k+1}$. Our destination was a summation with the summand of $s^k y_k$. The manipulation began by first multiplying $\sum_{k=0}^{\infty} s^k y_{k+1}$ by one in the form $s^{-1}s^1$. This manipulation permits us to rewrite the summand in a form in which the power of s matches the index of y (i.e., $s^{k+1}y_{k+1}$). However, when the summation process is carried out, the term $s^0 y_0$ is missing. The missing term can easily be replaced by adding and subtracting the $s^0 y_0$ term. We have gained by being able to now substitute G(s), but the price we now pay is to carry "$-y_0$" term in our computations.

We have now converted each of the terms in equation (2.9) that had a summation involving s and may now write equation (2.9) as

$$
s^{-1}\left[G(s) - y_0\right] = 3G(s) \qquad (2.15)
$$

which is easily solved for G(s) by multiplying throughout by s^{-1}, obtaining

$$
\begin{aligned}
&G(s) - y_0 = 3sG(s) \\
&G(s) = \frac{y_0}{1-3s} \qquad (2.16)
\end{aligned}
$$

This has been quite a bit of work, and a relevant question at this point would certainly be "What have we gained?" How is equation (2.16) related to our original family of difference equations ($y_{k+1} = 3y_k$ $k = 0,1,2,\ldots,\infty$; y_0 is a known constant)? The first reward from this effort is consolidation. The process begin with an infinite number of equations, each with its own solution in terms of a particular y_k. This set of infinite number of equations has been reduced to one equation in terms of G(s). The remaining job now is to transform or convert the information about the y_k's contained in G(s) in equation (2.16) into a solution for each of the y_k's. This final transformation step we will call the inversion of G(s). The conclusion of this step will reveal the values of each of the y_k's, k = 1, 2, 3, … . Experience with the inversion process will allow us to recognize at once the value of members of the sequence $\{y_k\}$ upon direct inspection of G(s).

2.2.2 Conversion, consolidation, and inversion

Inversion is the process by which each of the y_k's is identified from an examination of the generating function G(s). Thus, the process of using generating functions to solve difference equations will be a three step process: (1) conversion, (2) consolidation, and (3) inversion. Conversion involves transforming the family of difference equations from functions of the y_k's to a function of G(s). This conversion results in collapsing the infinite number of equations down to one equation which can be written in terms of G(s). Consolidation involves solving this one final, condensed equation for G(s). Inversion, which we will now explore, requires us to reformulate the one equation for G(s) into the solution for each of the y_k's. We will use G(s) to generate the individual y_k's.

2.3 Generating Functions as Infinite Series

2.3.1 The geometric series and the first inversion

We have already defined the generating function mathematically, so it is now time to define it conceptually. A generating function is a function that contains the information necessary to produce or generate the sequence of y_k's. In general, inverting G(s) is based on the recognition that, since G(s) is defined as

$$G(s) = \sum_{k=0}^{\infty} y_k s^k \tag{2.17}$$

We will take apart this sum in such a way as to identify each of its infinite component-summands. It is quite an amazing concept, that a function such as this contains explicit information to identify each and every y_k in an infinite sequence. We will now show how this is possible, and how straightforward and direct a task this really is.

To begin, start with the most elementary example, and then move to increasingly complicated ones. Let s be a number less than one in absolute value and begin with the finite sequence.

$$S_k = 1 + s + s^2 + s^3 + s^4 + \ldots + s^k \tag{2.18}$$

Then multiply each side by s to find

$$sS_k = s + s^2 + s^3 + s^4 + s^5 + \ldots + s^{k+1} \tag{2.19}$$

Let's subtract equation (2.19) from equation (2.18) to show

$$(1-s)S_k = 1 - s^{k+1} \tag{2.20}$$

or

$$S_k = \frac{1-s^{k+1}}{1-s} \tag{2.21}$$

Now, examine how each side of equation (2.21) behaves as k approaches infinity. Since the right side is a continuous function, the examination is a direct one

$$\lim_{k\to\infty} S_k = \sum_{k=0}^{\infty} s^k = \lim_{k\to\infty} \frac{1-s^{k+1}}{1-s} = \frac{1}{1-s} \tag{2.22}$$

This is a straightforward evaluation, since $|s| < 1$. For this range of s, as k gets larger s^{k+1} approaches zero.

Now, if $\frac{1}{1-s}$ is to be a generating function, then there must be an infinite series such that $\sum_{k=0}^{\infty} y_k s^k = \frac{1}{1-s}$. Is there such a series? Yes! Let $y_k=1$ for all k

$$\sum_{k=0}^{\infty} y_k s^k = \sum_{k=0}^{\infty} 1s^k = \sum_{k=0}^{\infty} s^k = \frac{1}{1-s} \tag{2.23}$$

Thus, $\frac{1}{1-s}$ is a generating function, and its inversion is $y_k = 1$ for all integers $k \geq 0$. Using a straightforward summation of the simple geometric series, it was possible to find a one-to-one correspondence between the sum of the series and the coefficients of each terms in that series. It is this one-to-one correspondence that is the heart of the generating function inversion process.

2.3.2 The meaning of $\triangleright$

This is our first generating function inversion. Since we will have many more generating functions to explore and invert, it would be helpful to have a shorthand notion for the statement "is inverted to the sequence." We will use the symbol $\triangleright$ to portray this, and write

$$G(s) = \frac{1}{1-s} \triangleright \{1\} \tag{2.24}$$

Expression (2.24) means that $\frac{1}{1-s}$ is a generating function, and the inversion of the generating function $\frac{1}{1-s}$ is the infinite sequence of y_k, for which $y_k = 1$ for all $k \geq 0$. The symbol $\triangleright$ simply means "is inverted to the sequence for which the coefficient of s^k is ... ."

Proceeding with the growth of our inversion ability, a development similar to the one which led to the conclusion that $\frac{1}{1-s} \triangleright \{1\}$ leads to the conclusion that $G(s) = \frac{1}{1-as} \triangleright \{a^k\}$. This time begin with $S_k = 1 + as + a^2s^2 + a^3s^3 + a^4s^4 + \ldots + a^ks^k$, and write

$$\begin{aligned} S_k &= 1 + as + a^2s^2 + a^3s^3 + \ldots + a^ks^k \\ (as)S_k &= \left(as + a^2s^2 + a^3s^3 + \ldots + a^ks^k + a^{k+1}s^{k+1}\right) \end{aligned} \tag{2.25}$$

The subtraction in equation (2.25) reveals

$$S_k = \frac{1 - a^{k+1}s^{k+1}}{1 - as} \tag{2.26}$$

and

$$\lim_{k \to \infty} S_k = \sum_{k=0}^{\infty} a^k s^k = \lim_{k \to \infty} \frac{1 - a^{k+1}s^{k+1}}{1 - as} = \frac{1}{1 - as} = G(s) \tag{2.27}$$

allowing the conclusion that $G(s) = \frac{1}{1-as} \triangleright \{a^k\}$. Again, as in the earlier case, the inversion of G(s) required the identification of a one-to-one correspondence between the coefficient of s^k and the sum of the series.

2.4 Recognizing Generating Functions – General Principles

Much of the effort in the inversion of a generating function involves recognition of key components of G(s). In many of the examples we will be working with in this text, G(s) will at first glance appear complicated. However, further inspection will reveal that G(s) is composed of component pieces, each of which provides a key to the inversion. Thus, although G(s) may be complicated, it will be disassembled into component features, each of which may be easily inverted. In this section some tools and maneuvers will be developed that will be useful in converting a complicated generating function to one which is more easily inverted.

2.4.1 Scaling a generating function – the scaling tool

As we work through this chapter, we will build a list of generating functions and their inversions, as well as develop the skills in manipulating generating functions. This first tool is very easy to apply. Consider the infinite series ($|as| < 1$)

$$1+as+a^2s^2+a^3s^3+L \quad = \sum_{k=0}^{\infty} a^k s^k = \frac{1}{1-as} \tag{2.28}$$

Now multiply both sides by a known constant c to find

$$c+cas+ca^2s^2+ca^3s^3+\ldots = \sum_{k=0}^{\infty} ca^k s^k = c\sum_{k=0}^{\infty} a^k s^k = \frac{c}{1-as} = G(s) \tag{2.29}$$

From this see that $\frac{c}{1-as} > \left\{ca^k\right\}$, or the generating function $\frac{c}{1-as}$ is the sum of the infinite series whose k^{th} term is $ca^k s^k$. Thus define

The scaling tool for generating function inversion

Let $G_1(s)$ be a generating function such that $G_1(s) > \{y_k\}$. Let $G_2(s) = cG_1(s)$. Then $G_2(s) > \{cy_k\}$.

The scaling tool will be a helpful device throughout this text. Some easy manipulations follow from the application of this simple scaling tool. For example, the scaling principle can be implemented at once to assess whether $G(s) = \frac{1}{a+bs}$ is a generating function.

$$G(s) = \frac{1}{a+bs} = \frac{1}{a\left[1+\frac{b}{a}s\right]} = \frac{\frac{1}{a}}{1+\frac{b}{a}s} = \frac{\frac{1}{a}}{1-\left(\frac{-b}{a}\right)s} \tag{2.30}$$

and now using the scaling tool to see that

$$G(s) = \frac{1}{a+bs} > \left\{\frac{1}{a}\left(\frac{-b}{a}\right)^k\right\} \tag{2.31}$$

2.4.2 Additive principle of generating function inversion

To continue increasing familiarity with generating functions, now consider two generating functions $G_1(s)$ and $G_2(s)$. Assume $G_1(s) > \{y_{1k}\}$ and $G_2(s) > \{y_{2k}\}$.

The question before us is whether $G_1(s) + G_2(s)$ is a generating function.

$$G_1(s)+G_2(s)=\sum_{k=0}^{\infty} y_{1k}s^k+\sum_{k=0}^{\infty} y_{2k}s^k=\sum_{k=0}^{\infty}(y_{1k}+y_{2k})s^k \quad (2.32)$$

and we find that $G_1(s) + G_2(s) > \{y_{1k} + y_{2k}\}$. We can now state the

The additive principle of generating functions

Assume $G_1(s) > \{y_{1k}\}$ and $G_2(s) > \{y_{2k}\}$. Then $G_1(s) + G_2(s) > \{y_{1k} + y_{2k}\}$

This principle will be most useful when dealing with a generating function that is a sum. If each of the summands can be inverted then we have the inversion of the original G(s) in our hands, since all that is needed is to obtain the kth term from each of the summands. In a completely analogous fashion, it can be seen that $G_1(s) - G_2(s) > \{y_{1k} - y_{2k}\}$.

2.4.3 Convolution principle of generating functions

Unfortunately, the product of generating functions is not the product of the kth terms of each of the infinite series. However, we do not have to look too much farther in order to identify the terms of the infinite series that represent the true value of the kth term of the series when the series is the product of at least two other series. The solution to this problem will simplify the inversion of complicated forms of G(s), when G(s) can be written as the product of two other generating functions. Begin with two generating functions, $G_1(s) > \{a_k\}$ and $G_2(s) > \{b_k\}$ as follows

$$\begin{aligned} G_1(s)G_2(s) &= \left(a_0+a_1s+a_2s^2+a_3s^3+L\right)\left(b_0+b_1s+b_2s^2+b_3s^3+L\right) \\ &\neq> \left\{a_ib_j\right\} \end{aligned} \quad (2.33)$$

Here, we only need evaluate this product term by term and gather together those coefficients which are associated with like powers of s. A table which relates the powers of s to the coefficient s to that power quickly reveals the pattern.

In general write the coefficient of s^k, y_k as $y_k = \sum_{j=0}^{k} a_j b_{k-j}$ and so observe

Table 2.1 Powers and Coefficients of s for $G_1(s)G_2(s)$

Power	Coefficient
0	a_0b_0
1	$a_0b_1 + a_1b_0$
2	$a_0b_2 + a_1b_1 + a_2b_0$
3	$a_0b_3 + a_1b_2 + a_2b_1 + a_3b_0$
4	$a_0b_4 + a_1b_3 + a_2b_2 + a_3b_1 + a_4b_0$

$$G_1(s)G_2(s) \triangleright \left\{ \sum_{j=0}^{k} a_j b_{k-j} \right\} \tag{2.34}$$

Note that for each term in the summand, the sum of the subscripts is equal to k. This constant sum of subscripts, as j varies from 0 to k, is what makes this particular sum of terms a convolution. Thus, we may write

The Convolution Principle of Generating Function Inversion
Begin with two generating functions, $G_1(s) \triangleright \{a_k\}$ and $G_2(s) \triangleright \{b_k\}$.Then

$$G_1(s)G_2(s) \triangleright \left\{ \sum_{j=0}^{k} a_j b_{k-j} \right\} \tag{2.35}$$

The convolution principle will be extremely useful throughout our experience of generating function inversion. For example, consider the generating function

$$G(s) = \frac{1}{(1-5s)(1-2s)} \tag{2.36}$$

We know $\frac{1}{1-5s} > \{5^k\}$ and $\frac{1}{1-2s} > \{2^k\}$. This is all that we need to apply the convolution principle to see that

$$G(s) = \frac{1}{(1-5s)(1-2s)} > \left\{\sum_{j=0}^{k} 5^j 2^{k-j}\right\} \tag{2.37}$$

We can see the solution immediately upon visualization of G(s) as a convolution. Proceeding, we can compute $y_0 = 1$, $y_1 = (1)(2) + (5)(1) = 7$, $y_2 = (1)(4) + (5)(2) + (25)(1) = 39$, $y_3 = (1)(8) + (5)(4) + (25)(2) + (125)(1) = 203$. The entire sequence can be generated in this fashion. Also, one need not build up the sequence of solutions for $y_0, y_1, y_2, y_3, \ldots, y_{k-1}$ to find the solution to y_k. Instead, using the generating function inversion principles, we can "jump" directly to the solution for y_k. This is an attractive feature of generating functions. We can, for example, compute the value of y_6, without computing y_0 through y_5.

The convolution principle propagates nicely. For example, if we have three generating functions $G_1(s)$, $G_2(s)$, and $G_3(s)$ where $G_1(s) > \{a_k\}$, $G_2(s) > \{b_k\}$, $G_3(s) > \{c_k\}$ then we can demonstrate that

$$G_1(s)G_2(s)G_3(s) > \left\{\sum_{i=0}^{k}\sum_{j=0}^{k-i} a_i b_j c_{k-i-j}\right\} \tag{2.38}$$

To begin this demonstration, first observe that

$$\begin{aligned} &G_1(s)G_2(s)G_3(s) \\ &= \left(a_0 + a_1 s + a_2 s^2 + a_3 s^3 + L\right)\left(b_0 + b_1 s + b_2 s^2 + b_3 s^3 + L\right) \\ &\quad \left(c_0 + c_1 s + c_2 s^2 + c_3 s^3 + L\right) \end{aligned} \tag{2.39}$$

We can write the right side of equation (2.39) one product at a time.

$$
\begin{aligned}
&\left(a_0+a_1s+a_2s^2+a_3s^3+\cdots\right)\left(b_0+b_1s+b_2s^2+b_3s^3+\cdots\right)\\
&=a_0b_0+\left(a_0b_1+a_1b_0\right)s+\left(a_0b_2+a_1b_1+a_2b_0\right)s^2\\
&+\left(a_0b_3+a_1b_2+a_2b_1+a_3b_0\right)s^3\cdots
\end{aligned}
\tag{2.40}
$$

resulting in

$$
\left[\sum_{j=0}^{0}a_jb_{0-j}\right]+\left[\sum_{j=0}^{1}a_jb_{1-j}\right]s+\left[\sum_{j=0}^{2}a_jb_{2-j}\right]s^2+\left[\sum_{j=0}^{3}a_jb_{3-j}\right]s^3+\ldots+ \tag{2.41}
$$

The coefficients of s^k in equation (2.41) represents the inversion of $G_1(s)G_2(s)$. The triple generating function product can be written as

$$
\begin{aligned}
&G_1(s)G_2(s)G_3(s)\\
&=\left(\left[\sum_{j=0}^{0}a_jb_{0-j}\right]+\left[\sum_{j=0}^{1}a_jb_{1-j}\right]s+\left[\sum_{j=0}^{2}a_jb_{2-j}\right]s^2+\left[\sum_{j=0}^{3}a_jb_{3-j}\right]s^3+\cdots\right)\\
&\qquad\cdot\left(c_0+c_1s+c_2s^2+c_3s^3+\cdots\right)
\end{aligned}
\tag{2.42}
$$

and we just take another convolution of these remaining two sequences in this product to show

$$
G_1(s)G_2(s)G_3(s) \;\triangleright\; \left\{\sum_{i=0}^{k}\sum_{j=0}^{k-i}a_ib_jc_{k-i-j}\right\} \tag{2.43}
$$

As another example of the productivity of this principle, when confronted with a generating function of the form

$$
G(s)=\frac{1}{(1-as)(1-bs)(1-cs)} \tag{2.44}
$$

First write

$$
\begin{aligned}
G(s)&=\frac{1}{(1-as)(1-bs)(1-cs)}=\left(\frac{1}{1-as}\right)\left(\frac{1}{1-bs}\right)\left(\frac{1}{1-cs}\right)\\
&=G_1(s)G_2(s)G_3(s)
\end{aligned}
\tag{2.45}
$$

and then use the convolution principle to see (note that $a_0 = b_0 = c_0 = 1$)

$$G(s) = \frac{1}{(1-as)(1-bs)(1-cs)} \triangleright \left\{ \sum_{i=0}^{k} \sum_{j=0}^{k-i} a_i b_j c_{k-i-j} \right\} \tag{2.46}$$

2.4.4 Sliding tool of generating functions

Thus far we have considered generating functions in which the denominator is a function of s but not the numerator. In this short section, we will see how to handle generating functions where the numerator also contains terms involving s. Begin with $G(s) = \frac{1}{1-bs}$ and find

$$\frac{1}{1-bs} = 1 + bs + b^2s^2 + b^3s^3 + b^4s^4 + \ldots \tag{2.47}$$

and we know that $G(s) = \frac{1}{1-bs} \triangleright \{b^k\}$. We can multiply each side of equation (2.47) by s to obtain

$$\frac{s}{1-bs} = s + bs^2 + b^2s^3 + b^3s^4 + b^4s^5 + \ldots \tag{2.48}$$

here the coefficient of s^k is not b^k, but b^{k-1}. Having the term involving s in the numerator has essentially slid the necessary coefficient in the series one term to the left. Thus, to invert

$$G(s) = \frac{s^3}{1-4s} \tag{2.49}$$

first recognize a subcomponent of G(s), $\frac{1}{1-4s} \triangleright \{4^k\}$. The numerator of G(s) however reveals that the kth term of the sequence $\{4^k\}$ is three terms to the right of the correct term. Thus, sliding three terms to the left reveals that $G(s) \triangleright \left\{4^{k-3}\right\}$. As another example, consider the inversion of

$$G(s) = \frac{s^2}{6-s} + \frac{s}{6-s} \tag{2.50}$$

Inversion of the generating function in equation (2.50) is eased by the common denominator in each of the terms on the right side of the equality. The inversion itself will involve the application of the scaling, summation, and sliding principles. Begin with the recognition that

$$\frac{1}{6-s} = \frac{1/6}{1/6(6-s)} = \frac{1/6}{1-1/6\,s} \triangleright \left\{\left(\frac{1}{6}\right)^{k+1}\right\} \tag{2.51}$$

The inversion of the factor $(6\text{-}s)^{-1}$ common to each term of equation (2.50) is complete. Now use the sliding tool and summation principle to write

$$G(s) = \frac{s^2}{6-s} + \frac{s}{6-s} \triangleright \left\{\left(\frac{1}{6}\right)^{k-1} + \left(\frac{1}{6}\right)^{k}\right\} \triangleright \left\{\frac{7}{6}\left(\frac{1}{6}\right)^{k-1}\right\} \tag{2.52}$$

The sliding tool works in the reverse direction as well. For example consider

$$G(s) = \frac{1}{s^3 - 2s^4} = \frac{1}{s^3(1-2s)} = \frac{s^{-3}}{1-2s} \triangleright \left\{2^{k+3}\right\} \tag{2.53}$$

In this case the s^{-3} in the denominator forced a slide to the right, increasing the required term in the $\{y_k\}$ sequence by 3.

2.5 Examples of Generating Function Inversion

In this section we demonstrate by example the inversion of several generating functions using the tools and principles developed thus far. These examples will be provided in increasing order of complexity. It is important that the reader understand each step of the process for every example.

2.5.1 Example 1

$$G(s) = \frac{1}{(s-2)(s-1)} \tag{2.54}$$

The convolution principle allows us to decompose this generating function as

$$G(s) = \frac{1}{(s-2)(s-1)} = G_1(s)G_2(s) \tag{2.55}$$

If each of $G_1(s)$ and $G_2(s)$ can be inverted, the convolution principle will permit us to construct the sequence $\{y_k\}$ associated with the inversion of G(s). Beginning with $G_1(s)$

$$G_1(s) = \frac{1}{s-2} = \frac{-1}{2-s} = \frac{-\frac{1}{2}}{\frac{1}{2}(2-s)} = \frac{-\frac{1}{2}}{1-\frac{1}{2}s} \triangleright \left\{\left(\frac{-1}{2}\right)\left(\frac{1}{2}\right)^k\right\} \triangleright \left\{\frac{-1}{2^{k+1}}\right\} \tag{2.56}$$

inverting the second generating function of equation (2.55) similarly

$$G_2(s) = \frac{1}{s-1} = \frac{-1}{1-s} \triangleright \{-1\} \tag{2.57}$$

We can now apply the convolution principle, the coefficient of s^k in the product of the sequences $\left\{\frac{-1}{2^{k+1}}\right\}$ and {-1} is

$$G_1(s)G_2(s) \triangleright \left\{\sum_{j=0}^{k}\left(\frac{-1}{2^{j+1}}\right)(-1) = \sum_{j=0}^{k}\frac{1}{2^{j+1}} = \frac{1}{2}\sum_{j=0}^{k}\left(\frac{1}{2}\right)^j = 1-\left(\frac{1}{2}\right)^{k+1}\right\} \tag{2.58}$$

So

$$G(s) \triangleright \left\{1-\left(\frac{1}{2}\right)^{k+1}\right\} \tag{2.59}$$

2.5.2 Example 2

To invert

$$G(s) = \frac{s^2 - s^4}{(1-5s)(1-9s)} \tag{2.60}$$

first proceed by recognizing that the numerator will utilize the sliding tool.

$$G(s) = \frac{s^2 - s^4}{(1-5s)(1-9s)} = \left(s^2 - s^4\right)\left[\frac{1}{(1-5s)(1-9s)}\right] \tag{2.61}$$

Begin by inverting the last term using the convolution principle, observing

$$\left[\frac{1}{(1-5s)(1-9s)}\right] \triangleright \left\{\sum_{j=0}^{k} 5^j 9^{k-j}\right\} \tag{2.62}$$

Now use the sliding tool and the summation principle to compute

$$\left[\frac{s^2 - s^4}{(1-5s)(1-9s)}\right] \triangleright \left\{\sum_{j=0}^{k-2} 5^j 9^{k-2-j} - \sum_{j=0}^{k-4} 5^j 9^{k-4-j}\right\} \tag{2.63}$$

2.5.3 Example 3

Consider

$$G(s) = \frac{s^3}{(s-3)(s-4)(s-5)} = s^3\left(\frac{1}{s-3}\right)\left(\frac{1}{s-4}\right)\left(\frac{1}{s-5}\right) \tag{2.64}$$

It can be seen from the denominator of G(s) that the convolution principle will be required not once, but twice. Begin by considering each of the terms in the denominator of equation (2.64). A simple application of the scaling principle reveals

$$\frac{1}{s-3}=\frac{-1}{3\ -\ (s)}=\frac{\left(\frac{-1}{3}\right)}{1-\left(\frac{1}{3}\right)s}\triangleright\left\{\left(\frac{-1}{3}\right)\left(\frac{1}{3}\right)^{k}\right\}=\left\{-\left(\frac{1}{3}\right)^{k+1}\right\} \tag{2.65}$$

Thus we see that

$$\frac{1}{s-3}\triangleright\left\{-\left(\frac{1}{3}\right)^{k+1}\right\},\ \frac{1}{s-4}\triangleright\left\{-\left(\frac{1}{4}\right)^{k+1}\right\},\text{ and }\ \frac{1}{s-5}\triangleright\left\{-\left(\frac{1}{5}\right)^{k+1}\right\} \tag{2.66}$$

What remains is to reassemble G(s) from the product of these terms, while keeping track of the coefficients of s^k. Apply the convolution principle for the first time to see

$$\frac{1}{(s-4)(s-5)}\ \triangleright\ \left\{\sum_{i=0}^{k}\left(\frac{1}{4}\right)^{i+1}\left(\frac{1}{5}\right)^{k+1-i}\right\} \tag{2.67}$$

and again to find

$$\frac{1}{(s-3)(s-4)(s-5)}\ \triangleright\ \left\{\sum_{i=0}^{k}\sum_{j=0}^{k-i}(-1)\left(\frac{1}{3}\right)^{i+1}\left(\frac{1}{4}\right)^{j+1}\left(\frac{1}{5}\right)^{k-i-j+1}\right\} \tag{2.68}$$

and invoking the sliding tool

$$G(s)=\frac{s^3}{(s-3)(s-4)(s-5)}\ \triangleright\ \left\{\sum_{i=0}^{k-3}\sum_{j=0}^{k-3-i}(-1)\left(\frac{1}{3}\right)^{i+1}\left(\frac{1}{4}\right)^{j+1}\left(\frac{1}{5}\right)^{k-i-j-2}\right\} \tag{2.69}$$

2.6 Inverting Functions of $(1-s^n)^{-1}$

One form of G(s) which will be of particular importance to us is $G(s)=\frac{1}{1-s^n}$. In this section, we will invert generating functions that are of this form, beginning with smaller values of the positive integer k and then building up to larger values of k.

2.6.1 Inversion of G(s) for n = 2

The solution for n = 1 has already been provided. To see the solution for n = 2, only a small amount of algebra is required. Begin by writing

$$\frac{1}{1-s^2}=\frac{1}{1-s\cdot s} \tag{2.70}$$

If the variable was a constant s, such as the constant a, the inversion procedure would be straightforward as depicted in equation (2.70)

$$\frac{1}{1-as}=1+as+a^2s^2+a^3s^3+a^4s^4+\ldots \rhd \left\{a^k\right\} \tag{2.71}$$

Now just substituting s for a in both the left side and the right side of equation (2.71) provides an interesting solution for G(s)

$$\begin{aligned}\frac{1}{1-s^2}&=1+s\cdot s+s^2s^2+s^3s^3+s^4s^4+\ldots\\&=1+s^2+s^4+s^6+s^8+\ldots\end{aligned} \tag{2.72}$$

and see what the coefficients are. If k is odd, the coefficient of s^k is zero. On the other hand, if k is even, then the coefficient of s^k is one. We may write

$$\frac{1}{1-s^2} \rhd \left\{I_{k \bmod 2=0}\right\} \tag{2.73}$$

The expression $I_{k\bmod 2=0}$ is a combination of two useful functions. Let $I_{x=A}$ be the indicator function, equal to one when x = A and equal to zero otherwise. For example, $I_{x=1,2,3}$ is a function of x, equal to 1, when x is equal to 1, 2 or 3, and equal to zero for all other values of x. The function k mod 2 is short for k modulus 2, which is the integer remainder of k/2. For even values of k, k mod 2 is equal to 0. For odd values of k, k mod 2 is equal to one.

We have seen that the inversion of $G(s) = \frac{1}{1-s^2}$ requires us to include a one for each even power s, and a zero for an odd power of s. The function $I_{k \bmod 2=0}$ is precisely the function we need to carry out this task.

2.6.2 Inversion for n = 3

The inversion process for $G(s) = \frac{1}{1-s^3}$, proceeds analogously. Once we write G(s) as

$$\begin{aligned} G(s) &= \frac{1}{1-s^3} = \frac{1}{1-s^2 \cdot s} = 1 + s^2 s + s^4 s^2 + s^6 s^3 \\ &= 1 + s^3 + s^6 + s^9 + \ldots \end{aligned} \tag{2.74}$$

Using the indicator function and the modulus function to write

$$G(s) = \frac{1}{1-s^3} \triangleright \{I_{k \bmod 3=0}\} \tag{2.75}$$

2.6.3 The general solution for positive integer values of n

We can proceed with the general solution for $\frac{1}{1-s^n}$. as

$$\begin{aligned} G(s) &= \frac{1}{1-s^n} = \frac{1}{1-s^{n-1} \cdot s} = 1 + s^{n-1}s + s^{2(n-1)}s^2 + s^{3(n-1)}s^3 \\ &= 1 + s^n + s^{2n} + s^{3n} + \ldots \end{aligned} \tag{2.76}$$

and we can use the indicator function and the modulus function to write

$$G(s) = \frac{1}{1-s^n} \triangleright \{I_{k \bmod n=0}\} \tag{2.77}$$

2.6.4 Additional examples

In this section, we will apply the experience developed thus far to related generating functions. Consider the generating function

$$G(s)=\frac{1}{\left(1-s^3\right)\left(1-s^4\right)}=\frac{1}{\left(1-s^3\right)}\cdot\frac{1}{\left(1-s^4\right)} \tag{2.78}$$

We see that we can use the convolution principle for inversion of G(s).

$$\frac{1}{1-s^3}\triangleright\left\{I_{k \bmod 3=0}\right\}:\frac{1}{1-s^4}\triangleright\left\{I_{k \bmod 4=0}\right\} \tag{2.79}$$

and invoking the convolution principle to write

$$G(s)=\frac{1}{\left(1-s^3\right)\left(1-s^4\right)}\triangleright\left\{\sum_{j=0}^{k} I_{j \bmod 3=0} I_{(k-j) \bmod 4=0}\right\} \tag{2.80}$$

As another example, consider the generating function

$$G(s)=\frac{a}{b-cs^5} \tag{2.81}$$

Begin by writing

$$\frac{a}{b-cs^5}=\frac{\frac{a}{b}}{1-\frac{c}{b}s^5}=\frac{a/b}{1-\left(\frac{c}{b}s^4\right)s} \tag{2.82}$$

and proceed by writing

$$\begin{aligned}\frac{1}{1-\left(\frac{c}{b}s^4\right)s}&=1+\frac{c}{b}s^4s+\left(\frac{c}{b}s^4\right)^2 s^2+\left(\frac{c}{b}s^4\right)^3 s^3+\ldots\\&=1+\frac{c}{b}s^5+\left(\frac{c}{b}\right)^2 s^{10}+\left(\frac{c}{b}\right)^3 s^{15}+\ldots\end{aligned} \tag{2.83}$$

and

$$\frac{1}{1-\frac{c}{b}s^5} \triangleright \left\{\left[\frac{c}{b}\right]^{k/5} I_{k \bmod 5=0}\right\} \tag{2.84}$$

Continue by invoking the scale principle to write

$$G(s) = \frac{a}{b-cs^5} \triangleright \left\{\left[\frac{a}{b}\right]\left[\frac{c}{b}\right]^{k/5} I_{k \bmod 5=0}\right\} \tag{2.85}$$

For an additional example consider

$$G(s) = \frac{3s-4}{(3s-2)(2-s^3)} \tag{2.86}$$

The work in this inversion is in the denominator. Once the denominator is inverted, use the sliding and scaling tools. Proceed by writing

$$\frac{1}{3s-2} = \frac{-1}{2-3s} = \frac{-1/2}{1-\frac{3}{2}s} \triangleright \left\{\left[\frac{-1}{2}\right]\left[\frac{3}{2}\right]^k\right\} \tag{2.87}$$

noting

$$\frac{1}{2-s^3} = \frac{1/2}{1-\frac{1}{2}s^3} \triangleright \left\{\left[\frac{1}{2}\right]^{\frac{k}{3}+1} I_{k \bmod 3=0}\right\} \tag{2.88}$$

Now invoking the convolution principle,

$$\frac{1}{(3s-2)(2-s^3)} \triangleright \left\{\sum_{j=0}^{k}\left[\frac{-1}{2}\right]\left[\frac{3}{2}\right]^j\left[\frac{1}{2}\right]^{\frac{k-j}{3}+1} I_{(k-j) \bmod 3=0}\right\} \tag{2.89}$$

And now invoking the scaling and addition principles complete the inversion

$$G(s) = \frac{3s-4}{(3s-2)(2-s^3)} \triangleright$$
$$\left\{ 3\sum_{j=0}^{k-1} \left[\frac{-1}{2}\right]\left[\frac{3}{2}\right]^j \left[\frac{1}{2}\right]^{\frac{k-j-1}{3}+1} I_{(k-j-1)\bmod 3=0} - 4\sum_{j=0}^{k} \left[\frac{-1}{2}\right]\left[\frac{3}{2}\right]^j \left[\frac{1}{2}\right]^{\frac{k-j}{3}+1} I_{(k-j)\bmod 3=0} \right\} \quad (2.90)$$

2.7 Generating Functions and Derivatives

Thus far as we have explored combinations of generating functions, each maneuver has broadened our ability to invert G(s), identifying the coefficients y_k of the sequence of interest. Continue this examination through the use of derivatives. If G(s) is differentiable, then the derivative of this function with respect to s is written as

$$\frac{dG(s)}{ds} = G'(s) \quad (2.91)$$

Where G'(s) will imply that the derivative of G(s) is to be taken with respect to s. If G(s) = cs where c is a known constant, then G'(s)=c. If G(s) = s^k, then G'(s)=ks^{k-1}. The only function of calculus used here is that of the chain rule, which the reader will recall, can be used to compute G'(s) when G(s) = g(f(s)). In this case G'(s) = G'(f(s))f'(s). With these simple rules, we can further expand the generating functions of interest for us to evaluate. Begin with the simplest generating function.

$$G(s) = \frac{1}{1-s} = 1 + s + s^2 + s^3 + s^4 + \cdots + s^k + \cdots \quad (2.92)$$

Taking a derivative on each side of the equality reveals

$$G'(s) = \frac{1}{(1-s)^2} = 1 + 2s + 3s^2 + 4s^3 + 5s^4 + \cdots + (k+1)s^k + \cdots \quad (2.93)$$

and we see at once that $\frac{1}{(1-s)^2} \triangleright \{k+1\}$. Note this could be verified by using the product principle of generating functions on G(s) when written in the form $\left(\frac{1}{1-s}\right)\left(\frac{1}{1-s}\right)$. We can multiply equation (2.93) by s to see

$$\frac{s}{(1-s)^2} = s + 2s^2 + 3s^3 + 4s^4 + 5s^5 + \cdots + ks^k + \cdots \tag{2.94}$$

or $\frac{s}{(1-s)^2} \triangleright \{k\}$. Of course, additional uses of the translation principle at this point leads to $\frac{s^2}{(1-s)^2} \triangleright \{k-1\}$, $\frac{s^3}{(1-s)^2} \triangleright \{k-2\}$, $\frac{s^4}{(1-s)^2} \triangleright \{k-3\}$ and so on. Similarly, $\frac{1}{s^3(1-s)^2} \triangleright \{k+4\}$ through an application of the translation process in the opposite direction.

We can continue with this development, taking another derivative from equation (2.93) to find

$$G''(s) = \frac{2}{(1-s)^3} = (2\cdot 1) + (3\cdot 2)s^1 + (4\cdot 3)s^2 + (5\cdot 4)s^3 + \cdots + (k+2)\cdot(k+1)s^k + \cdots \tag{2.95}$$

which demonstrates that $\frac{2}{(1-s)^3} \triangleright \{(k+2)\cdot(k+1)\}$. Following the development above, one can see through the translation principle that

$$\frac{2s^2}{(1-s)^3} \triangleright \{k\cdot(k-1)\} \tag{2.96}$$

and also that $\frac{1}{s^4(1-s)^3} \triangleright \{2^{-1}(k+6)\cdot(k+5)\}$. This process can proceed indefinitely. We will examine one more example.

$$G'''(s) = \frac{3\cdot 2}{(1-s)^4} = 3\cdot 2\cdot 1 + 4\cdot 3\cdot 2s + 5\cdot 4\cdot 3s^2 + \cdots + (k+3)\cdot(k+2)\cdot(k+1)s^k + \cdots \tag{2.97}$$

and $\frac{3\cdot 2}{(1-s)^4} \triangleright \{(k+3)\cdot(k+2)\cdot(k+1)\}$. In fact, it is easy to see a pattern here. In general

$$\frac{r!}{(1-s)^{r+1}} \triangleright \left\{\frac{(k+r)!}{k!}\right\} \tag{2.98}$$

Taking advantage of the relationship that $\frac{(k+r)!}{k!r!} = \binom{k+r}{r}$, we can write

$$\left(\frac{1}{1-s}\right)^r \triangleright \binom{k+r-1}{k} \tag{2.99}$$

Finally, we can change the base series on which we apply the derivative. For example, it has been demonstrated that if $G(s) = \frac{s}{(1-s)^2}$, then $G(s) \triangleright \{k\}$. Following the pattern of taking a derivative on each side, $G'(s) = \frac{1+s}{(1-s)^3} \triangleright \{(k+1)^2\}$. We only need invoke the translation principle to see that

$$G'(s) = \frac{s(1+s)}{(1-s)^3} \triangleright \{k^2\} \tag{2.100}$$

This process involves taking a derivative followed by the use of the scaling tool. This may be continued one additional time. Continuing in this vein, a straightforward demonstration reveals that $\frac{s^2+4s+1}{(1-s)^4} \triangleright \left\{(k+1)^3\right\}$ and a translation reveals that $\frac{s(s^2+4s+1)}{(1-s)^4} \triangleright \left\{k^3\right\}$.

The use of these principles applies to the notion of simplification and recognition. The more experience one can get, the better the recognition skills, and the less simplification will be required. To invert

$$G(s) = \frac{14(1-3s)(s-2)}{s^3(1-as)^3(4-bs)^2} \tag{2.101}$$

We recognize that the denominator will require the most work. The numerator merely suggests that the final inversion will involve repetitive applications of the sliding tool and the principle of addition.
Since the denominator of equation (2.101) contains the product of two polynomials, the multiplication principle can be applied. Rewrite G(s) as

$$\begin{aligned} G(s) &= \frac{-42s^{-1}+98s^{-2}-28s^{-3}}{(1-as)^3(4-bs)^2} \\ &= \left[-42s^{-1}+98s^{-2}-28s^{-3}\right]\left[\frac{1}{(1-as)^3}\right]\left[\frac{1}{(4-bs)^2}\right] \end{aligned} \tag{2.102}$$

Evaluating the quotients first,

$$\frac{1}{(1-as)^3} \triangleright \left\{(k+2)\cdot(k+1)\frac{a^k}{2}\right\} \tag{2.103}$$

by taking two successive derivatives of the sequence for which $G(s) = \frac{1}{(1-as)^3}$.

Proceeding in an analogous fashion for $\frac{1}{(4-bs)^2}$, write

$$\frac{1}{(4-bs)^2} = \frac{1/16}{\left(1-\frac{b}{4}s\right)^2} \triangleright \left\{\frac{1}{16}\left(\frac{b}{4}\right)^k (k+1)\right\} \tag{2.104}$$

Applying the convolution principle reveals

$$\left[\frac{1}{(1-as)^3}\right]\left[\frac{1}{(4-bs)^2}\right] \triangleright \sum_{j=0}^{k}\left[(j+2)(j+1)\frac{a^j}{2}\right]\left[\frac{1}{16}\left(\frac{b}{4}\right)^{k-j}(k-j+1)\right] \tag{2.105}$$

and

$$G(s) \triangleright \left\{\begin{array}{l} -42\sum_{j=0}^{k+1}\left[(j+2)(j+1)\frac{a^j}{2}\right]\left[\frac{1}{16}\left(\frac{b}{4}\right)^{k+1-j}(k-j+2)\right] \\ 98\sum_{j=0}^{k+2}\left[(j+2)(j+1)\frac{a^j}{2}\right]\left[\frac{1}{16}\left(\frac{b}{4}\right)^{k+2-j}(k-j+3)\right] \\ -28\sum_{j=0}^{k+3}\left[(j+2)(j+1)\frac{a^j}{2}\right]\left[\frac{1}{16}\left(\frac{b}{4}\right)^{k+3-j}(k-j+4)\right] \end{array}\right\} \tag{2.106}$$

A fine example of the derivative approach to the inversion of generating functions is to find the generating function G(s) such that

$$G(s) \triangleright \left\{k^2 a^k\right\} \tag{2.107}$$

Begin with $\frac{1}{1-as} \triangleright \{a^k\}$ and take a derivative of both sides with respect to s.

This reveals $\frac{a}{(1-as)^2} = \sum_{k=0}^{\infty} ka^k s^{k-1}$ and therefore $\frac{as}{(1-as)^2} = \sum_{k=0}^{\infty} ka^k s^k$.

Taking a second derivative with respect to s and simplifying gives

$$\frac{a + a^2 s}{(1-as)^3} = \sum_{k=0}^{\infty} k^2 a^k s^{k-1} \tag{2.108}$$

Use of the sliding tool reveals $\frac{as + a^2 s^2}{(1-as)^3} \triangleright \{k^2 a^k\}$.

Another example of the principles developed in this section is the evaluation of the generating function

$$G(s) = \left[\frac{as+b}{cs+d}\right]^n \tag{2.109}$$

for integer n (> 0). The inversion of this generating function will require the tools of derivatives, the binomial theorem, as well as the sliding and scaling tools. Begin by writing G(s) as the product of two generating functions

$$G(s) = \left[\frac{as+b}{cs+d}\right]^n = (as+b)^n \left[\frac{1}{cs+d}\right]^n = G_1(s)G_2(s) \tag{2.110}$$

Recognize $G_1(s)$ through a direct invocation of the binomial theorem. Write

$$G_1(s) = (as+b)^n = \sum_{k=0}^{n} \binom{n}{k} (as)^k b^{n-k} = \sum_{k=0}^{n} \binom{n}{k} a^k b^{n-k} s^k \tag{2.111}$$

The next task is to invert $G_2(s)$.

$$\left[\frac{1}{cs+d}\right]^n = \left[\frac{\frac{1}{d}}{1-\left(-\frac{c}{d}\right)s}\right]^n = \left(\frac{1}{d}\right)^n\left[\frac{1}{1-\left(-\frac{c}{d}\right)s}\right]^n \tag{2.112}$$

The last term in equation (2.112) can be considered $\left(\frac{1}{1-as}\right)^n$, which we will approach by taking consecutive derivatives.

$$\begin{aligned} G(s) &= \frac{1}{1-as} \triangleright \left\{a^k\right\} \\ G'(s) &= \frac{a}{(1-as)^2} \triangleright \left\{(k+1)a^{k+1}\right\} \\ \frac{1}{(1-as)^2} &\triangleright \left\{(k+1)a^k\right\} \end{aligned} \tag{2.113}$$

Taking a second derivative reveals

$$\begin{aligned} \frac{2a}{(1-as)^3} &\triangleright \left\{(k+2)(k+1)a^{k+1}\right\} \\ \frac{1}{(1-as)^3} &\triangleright \left\{\frac{(k+2)(k+1)}{2}a^k\right\} \end{aligned} \tag{2.114}$$

One additional derivative demonstrates

$$\begin{aligned} \frac{3a}{(1-as)^4} &\triangleright \left\{\frac{(k+3)(k+2)(k+1)}{2}a^{k+1}\right\} \\ \frac{1}{(1-as)^4} &\triangleright \left\{\frac{(k+3)(k+2)(k+1)}{3!}a^k\right\} \end{aligned} \tag{2.115}$$

In general

$$\left[\frac{1}{1-as}\right]^n \rhd \left\{\frac{(k+n-1)(k+n-2)\cdots(k+1)}{(n-1)!}a^k\right\}$$
$$\left[\frac{1}{1-as}\right]^n \rhd \left\{\binom{k+n-1}{n-1}a^k\right\} \tag{2.116}$$

Note that the source of the combinatoric term is the result of the sequence of successive derivatives applied to the sequence $\{s^k\}$ in equations (2.113), (2.114), and (2.115), each of which retrieves an exponent of s, making this exponent a coefficient of s. Now write

$$\left[\frac{1}{cs+d}\right]^n = \left[\frac{1/d}{1-\left(-c/d\right)s}\right]^n \tag{2.117}$$

Then,

$$\left[\frac{1/d}{1-\left(-c/d\right)s}\right]^n = \left\{\left[\frac{1}{d}\right]^n \frac{(k+n-1)(k+n-2)\cdots(k+1)}{(n-1)!}\left[-\frac{c}{d}\right]^k\right\}$$
$$= \left\{\left[\frac{1}{d}\right]^n \binom{k+n-1}{n-1}\left[-\frac{c}{d}\right]^k\right\} \rhd \{b_k\} \tag{2.118}$$

Note that

$$G_1(s) = (as+b)^n \rhd \left\{\binom{n}{k}a^k b^{n-k} I_{0\le k\le n}\right\} \tag{2.119}$$

As demonstrated in problem 47 of this chapter when the generating functions $G_1(s)$ and $G_2(s)$ are defined as $G_1(s) = \sum_{j=0}^{\infty} a_j s^j$ and $G_2(s) = \sum_{j=0}^{J} b_j s^j$ then

$$G_1(s)G_2(s) \rhd \left\{\sum_{j=0}^{\min(J,k)} a_{k-j} b_j\right\} \tag{2.120}$$

Applying this result to the original problem

$$G(s) = \left[\frac{as+b}{cs+d}\right]^n \\ > \left\{ \left[\frac{1}{d}\right]^n \sum_{j=0}^{\min(n,k)} \left(\binom{k-j+n-1}{n-1} \left[-\frac{c}{d}\right]^{k-j} \right) \binom{n}{j} a^j b^{n-j} \right\} \tag{2.121}$$

If $G(s) = \left[\frac{as+b}{c-ds}\right]^n$ then write

$$G(s) = \left[\frac{as+b}{c-ds}\right]^n = (as+b)^n \left(\frac{1}{c-ds}\right)^n = (as+b)^n \left[\frac{1}{c}\right]^n \frac{1}{\left(1-\frac{d}{c}s\right)^n} \tag{2.122}$$

and, since

$$\frac{1}{\left(1-\frac{d}{c}s\right)^n} > \left\{ \binom{n+k-1}{k} \left[\frac{d}{c}\right]^k \right\} \tag{2.123}$$

invert using a convolution and the scaling tool

$$\left[\frac{a+bs}{c-ds}\right]^n > \left\{ \sum_{j=0}^{\mathrm{Min}(n,k)} \binom{n+k-j-1}{n-1} \left[\frac{d}{c}\right]^{k-j} \binom{n}{j} a^j b^{n-j} \right\} \tag{2.124}$$

Finally, consider the inversion of $\frac{[as+b]^{n_1}}{[c+ds]^{n_2}}$ for $n_1 > 0$, $n_2 > 0$, and $n_1 \neq n_2$. By applying the binomial theorem to the numerator, the derivative process to the denominator, and a convolution, it can be shown that

$$\frac{[as+b]^{n_1}}{[c+ds]^{n_2}} > \left[\frac{1}{c}\right]^{n_2} \sum_{j=0}^{\min(k,n_1)} \binom{k-j+n_2-1}{n_2-1} \left[-\frac{d}{c}\right]^{k-j} \binom{n_1}{j} a^j b^{n_1-j} \tag{2.125}$$

2.8 Equivalence of Generating Function Arguments

At this point, it can be seen that for a given generating function G(s), we can find several different approaches to the inversion of G(s). The possibility of multiple approaches provides a useful demonstration that each of two approaches will lead to the same infinite series, and a check of our work. Consider the generating function $G(s) = \frac{1}{(a-bs)^2}$. There are two useful approaches to take for this inversion. The first might be inversion by the convolution principle, recognizing that the denominator of G(s) is the product of two polynomials. However, a second approach would involve taking a derivative. Since G(s) produces only one value of y_k for each k, the value of y_k obtained using the derivative approach is equivalent to that of the "derivative approach." First consider the solution using the convolution approach

$$\begin{aligned} G(s) &= \frac{1}{(a-bs)^2} = \left(\frac{1}{a-bs}\right)\left(\frac{1}{a-bs}\right) = \left(\frac{1/a}{\left(1-\frac{b}{a}s\right)}\right)\left(\frac{1/a}{\left(1-\frac{b}{a}s\right)}\right) \\ &= \frac{1/a^2}{\left(1-\frac{b}{a}s\right)\left(1-\frac{b}{a}s\right)} > \left\{\frac{1}{a^2}\sum_{j=0}^{k}\left(\frac{b}{a}\right)^j\left(\frac{b}{a}\right)^{k-j}\right\} \end{aligned} \tag{2.126}$$

Evaluating the kth term of this sequence reveals

$$\frac{1}{a^2}\sum_{j=0}^{k}\left(\frac{b}{a}\right)^{j}\left(\frac{b}{a}\right)^{k-j} = \frac{1}{a^2}\sum_{j=0}^{k}\left(\frac{b}{a}\right)^{k}$$
$$= \frac{1}{a^2}\left(\frac{b}{a}\right)^{k}\sum_{j=0}^{k}1 = \frac{1}{a^2}\left(\frac{b}{a}\right)^{k}(k+1) \tag{2.127}$$

and find that

$$G(s) = \frac{1}{(a-bs)^2} \triangleright \left[\frac{1}{a^2}\left(\frac{b}{a}\right)^{k}(k+1)\right] \tag{2.128}$$

Using the derivative approach

$$\left(\frac{1}{a-bs}\right) = \frac{\left(\frac{1}{a}\right)}{1-\frac{b}{a}s} \triangleright \left\{\left(\frac{1}{a}\right)\left(\frac{b}{a}\right)^{k}\right\} \tag{2.129}$$

and taking derivatives on both sides of equation (2.129) with respect to s,

$$\frac{b}{(a-bs)^2} \triangleright \left\{\frac{1}{a^2}\left(\frac{b}{a}\right)^{k}(k+1)\right\} \tag{2.130}$$

which is equivalent to our earlier result, obtained through the application of the product principle.

2.9 Probability Generating Functions

The skillful use of generating functions is invaluable in solving families of difference equations. However, generating functions have developed in certain specific circumstances. Consider the circumstance where the sequence $\{y_k\}$ are probabilities. Then, using our definition of G(s) we see that $G(s) = \sum_{k=0}^{\infty} p_k s^k$ where $\sum_{k=0}^{\infty} p_k = 1$. In probability it is often useful to describe the relative frequency of an event in terms of an outcome of an experiment. Among the most useful of these experiments are those for which outcomes are discrete (the integers or some subset of the integers). These discrete models have many applications, and it is often helpful to recognize the generating function associated with the models. The generating function associated with such a model is described as the probability generating function. We will continue to refer to these probability generating functions using the nomenclature G(s). This section will describe the use of probability generating functions and derive the probability generating function for the commonly used discrete distributions. Note that in this context of probability, the generating function can be considered an expectation, i.e., $G(s) = E[s^x]$

2.9.1 Bernoulli distribution

Consider one of the simplest of experimental probability models, the Bernoulli trial. Here, denote X as the outcome of an experiment which can have only two outcomes. Let the first outcome be $X = 1$ which occurs with probability p, and the second outcome $X = 0$, which occurs with probability $q = 1 - p$. Although a sequence of Bernoulli trials are a sequence of these experiments, each outcome is independent of the outcome of any other experiment in the sequence, and the concern is with the occurrence of only one trial. Observe at once that the probability generating function is

$$G(s) = \sum_{k=0}^{\infty} p_k s^k = \sum_{k=0}^{1} p_k s^k = qs^0 + ps^1 = q + ps \tag{2.131}$$

Note that in this computation we can collapse the range over which the sum is expressed from the initial range of 0 to ∞ to the range of 0 to 1 since $p_k=0$ for $k=2, 3, 4, \ldots$

2.9.2 Binomial distribution

The binomial model represents an experiment that is the sum of independent Bernoulli trials. Here, for $k = 0, 1, 2, \ldots, n$

$$P[X = k] = \binom{n}{k} p^k (1-p)^{n-k} \tag{2.132}$$

The generating function for this model is

$$G(s) = \sum_{k=0}^{\infty} p_k s^k = \sum_{k=0}^{n} P[X = k] s^k = \sum_{k=0}^{n} \binom{n}{k} p^k (1-p)^{n-k} s^k$$
$$= \sum_{k=0}^{n} \binom{n}{k} (ps)^k (1-p)^{n-k} \tag{2.133}$$

now using the binomial theorem for the last step.

$$\sum_{k=0}^{n} \binom{n}{k} (ps)^k (1-p)^{n-k} = (q + ps)^n \tag{2.134}$$

2.9.3 The geometric distribution

The geometric distribution continues to use the Bernoulli trial paradigm, but addresses a different event. The question that arises in the geometric distribution is "what is the probability that the first success occurs after k failures, k = 0, 1, 2, ..., ?" If X is the trial on which the first success occurs, then

$$P[X = k] = q^k p \tag{2.135}$$

We need to convince ourselves that this really does sum to one over the entire range of events (i.e., all non-negative integers). Work in this chapter has prepared us for this simple calculation.

$$\sum_{k=0}^{\infty} q^k p = p \sum_{k=0}^{\infty} q^k = p \left[\frac{1}{1-q} \right] = 1 \tag{2.136}$$

and computing the generating function as follows

$$G(s) = \sum_{k=0}^{\infty} s^k q^k p = p \sum_{k=0}^{\infty} (qs)^k = p \left[\frac{1}{1-qs} \right] = \frac{p}{1-qs} \tag{2.137}$$

2.9.4 Negative binomial distribution

As a generalization of the geometric distribution it is of interest to compute the probability that the rth success occurs on the kth trial. It is useful to think of this event as the event of there being r - 1 successes in k - 1 trials, followed by one additional trial for which there is a success. Then, if X is the probability that the kth success occurs on the rth trial, then

$$P[X=k]=\left[\binom{k-1}{r-1}p^{r-1}q^{k-r}\right]p=\binom{k-1}{r-1}p^{r}q^{k-r} \tag{2.138}$$

The probability generating function can be derived as follows

$$\begin{aligned} G(s) &= \sum_{k=r}^{\infty} s^{k}\binom{k-1}{r-1}p^{r}q^{k-r} = \sum_{k=r}^{\infty}\binom{k-1}{r-1}s^{k-r}s^{r}p^{r}q^{k-r} \\ &= \sum_{k=r}^{\infty}\binom{k-1}{r-1}(ps)^{r}(qs)^{k-r} = (ps)^{r}\sum_{k=r}^{\infty}\binom{k-1}{r-1}(qs)^{k-r} \end{aligned} \tag{2.139}$$

Continuing

$$\begin{aligned} &\sum_{k=r}^{\infty}\binom{k-1}{r-1}(qs)^{k-r} = \sum_{k=r}^{\infty}\binom{k-1}{k-r}(qs)^{k-r} = \sum_{n=0}^{\infty}\binom{r+n-1}{n}(qs)^{n} \\ &= \frac{1}{(1-qs)^{r}} \end{aligned} \tag{2.140}$$

The last equality in equation (2.140) comes from equation (2.99). Thus

$$G(s)=\left[\frac{ps}{1-qs}\right]^{r} \tag{2.141}$$

This development is described in Woodroofe [2].

2.9.5 Poisson distribution

The Poisson distribution is used to describe the occurrence of independent events that are rare. A classic example is the arrival of particles to a Geiger counter. Suppose the average arrival rate in the time interval $(t, t + \Delta t)$ is $\lambda\Delta t$. If X is the number of particles which arrive in time interval $(t, t + \Delta t)$ then

$$P[X = k] = \frac{(\lambda\Delta t)^k}{k!} e^{-\lambda\Delta t} \tag{2.142}$$

for all non-negative integers k. One key for this being a proper probability distribution is the relationship

$$\sum_{k=0}^{\infty} \frac{x^k}{k!} = e^x \tag{2.143}$$

With this result it is easy to show that the sum of Poisson probabilities is one.

$$\sum_{k=0}^{\infty} P[X = k] = \sum_{k=0}^{\infty} \frac{(\lambda\Delta t)^k}{k!} e^{-\lambda\Delta t} = e^{-\lambda\Delta t} \sum_{k=0}^{\infty} \frac{(\lambda\Delta t)^k}{k!} = e^{-\lambda\Delta t} \cdot e^{\lambda\Delta t} = 1 \tag{2.144}$$

and the generating function follows analogously

$$\begin{aligned} G(s) = \sum_{k=0}^{\infty} s^k P[X = k] &= \sum_{k=0}^{\infty} s^k \frac{(\lambda\Delta t)^k}{k!} e^{-\lambda\Delta t} \\ &= e^{-\lambda\Delta t} \sum_{k=0}^{\infty} \frac{(\lambda s\Delta t)^k}{k!} = e^{-\lambda\Delta t} \cdot e^{\lambda\Delta t s} = e^{\lambda\Delta t(s-1)} \end{aligned} \tag{2.145}$$

These are the probability generating functions for the most commonly used discrete probability distributions. They will be important in the use of difference equations to identify the probability distribution of the number of patients with a disease in a population. Although the development of these equations in Chapters 11 and 12 of this book will appear complex, the recognition of the resultant generating functions as the probability generating functions of the probability distributions developed in this section will add to our understanding of the spread of a the disease process.

Problems

Using the scaling tool and the sliding tool, invert the following nine generating functions. The solution should provide all of the information necessary to identify each member of the infinite sequence $\{y_k\}$.

1. $G(s) = \frac{1}{3s - 7}$
2. $G(s) = \frac{1}{2 + 6s}$
3. $G(s) = \frac{1}{5s + 9}$
4. $G(s) = \frac{1}{3s - 7}$
5. $G(s) = \frac{1}{s - 9}$
6. $G(s) = \frac{3}{3 + 2s}$
7. $G(s) = \frac{12}{2s - 7}$
8. $G(s) = \frac{7}{3s + 5}$
9. $G(s) = \frac{1}{1 + 3s}$

Invert the following generating functions:

10. $G(s) = \dfrac{1}{(s+1)(s-1)}$

11. $G(s) = \dfrac{3}{(s-7)(s-8)}$

12. $G(s) = \dfrac{2}{(3s-1)(4s+2)}$

13. $G(s) = \dfrac{16}{(3s+4)(s-2)}$

14. $G(s) = \dfrac{8}{(s+1)(2s-1)(3s-1)}$

15. $G(s) = \dfrac{4}{(2s-6)(s+1)(s+6)}$

16. $G(s) = \dfrac{6}{s(s+4)(s+1)}$

17. $G(s) = \dfrac{11}{(s-1)(2+s)(s-6)}$

18. $G(s) = \dfrac{9}{(s-1)(s-3)(s-4)(s-5)}$

19. $G(s) = \dfrac{3}{(3s-2)(s+6)(s-6)(2s+1)}$

20. $G(s) = \dfrac{12}{(s-2)(s-3)(s+1)(3s+7)}$

21. $G(s) = \dfrac{8}{(2s-1)(3s-2)(4s+4)(s-1)}$

Invert the following:

22. $G(s) = \dfrac{3s}{(s-2)^2(s+4)^2}$

23. $G(s) = \dfrac{4s^2}{(2s-2)^3}$

24. $G(s) = \dfrac{6}{(s-1)^3}$

25. $G(s) = \dfrac{13}{(2s+7)^3}$

26. $G(s) = \dfrac{4}{(1-6s)^3}$

27. $G(s) = \dfrac{3}{(s+4)^4}$

28. $G(s) = \dfrac{7}{(2s-1)^2}$

29. $G(s) = \dfrac{13}{(s+1)^3}$

30. $G(s) = \dfrac{1}{(4s-3)^2}$

31. $G(s) = \dfrac{7}{(3+s)^4}$

32. $G(s) = \dfrac{9}{(9+6s)^3}$

33. $G(s) = \dfrac{2}{(3+7s)^2}$

34. $G(s) = \dfrac{3}{(12s+1)^4}$

35. $G(s) = \dfrac{5}{(3s+4)^2}$

36. $G(s) = \dfrac{6}{(9s+3)^3}$

37. $G(s) = \dfrac{7}{(5+2s)^2}$

Invert the following generating functions using the convolution principle and the sliding tool:

38. $G(s) = \dfrac{s}{(s+2)(s+3)}$

39. $G(s) = \dfrac{s^2}{(s-1)(2s-6)}$

40. $G(s) = \dfrac{2s^5}{(2s+3)(s-8)}$

41. $G(s) = \dfrac{6s^3}{(s-3)(s+3)}$

42. $G(s) = \dfrac{3s}{(s-1)(s-1)(s-1)}$

43. $G(s) = \dfrac{2}{s^2(s+1)(s-1)(s+3)}$

44. $G(s) = \dfrac{7}{s^4(3s+6)(s-1)(s+5)}$

45. Verify by using both the derivative approach and the product principle that if

$$\frac{1}{(a+bs)^3} > \{f_1(k)\} \text{ and } \left(\frac{1}{a+bs}\right)\left(\frac{1}{a+bs}\right)\left(\frac{1}{a+bs}\right) > \{f_2(k)\}$$

then $f_1(k) = f_2(k)$.

46. In section 2.8 it was pointed out that if x is a random variable which follows a negative binomial distribution then its probability generating function G(s) is

$$G(s) = \left[\frac{ps}{1-qs}\right]^r$$

Applying successive derivatives to G(s) and the scaling tool, compute the probability P[X=k] for r = 4, 5, and 6.

47. Consider this lemma to the convolution principle. Consider two generating functions $G_1(s) = \sum_{j=0}^{\infty} a_j s^j$ and $G_2(s) = \sum_{j=0}^{J} b_j s^j$. Then prove

$$G_1(s)G_2(s) > \left\{\sum_{j=0}^{\min(k,J)} a_{k-j} b_j\right\}$$

48. It was shown is section 2.95 that

$$e^{\lambda t(1-s)} > \left\{\frac{(-\lambda t)^k}{k!} e^{-\lambda t}\right\}$$

Show that

a.

$$e^{\lambda t(s-1)} > \left\{\frac{(\lambda t)^k}{k!} e^{-\lambda t}\right\}$$

b.

$$e^{\lambda t(s^{-1}-1)} > \left\{\frac{(\lambda t)^{-k}}{(-k)!} e^{-\lambda t} I_{k \le 0}\right\}$$

References

1. Goldberg S. Introduction to Difference Equations. New York. John Wiley and Sons. 1958.

3

Generating Functions II: Coefficient Collection

3.1 Introduction

The procedures developed in Chapter 2 have provided a collection of important tools for inverting generating functions. Already, there are many generating functions that we can invert using a combination of the scaling and sliding tools, in combination with the addition and convolution principles. These tools include the ability to identify and manipulate the generating functions of many commonly used discrete probability distributions. However, there is an important class of generating functions that we have not attempted to invert, but that occur very frequently in the solution of difference equations. In fact, this class is among the most commonly occurring difference equations that one encounters. In addition, they have historically provided important complications in the solution of difference equations. This class of generating functions are those that contain general polynomials in the denominator, and will be the subject of this entire chapter. At the conclusion of the chapter, the reader should be able to invert a generating function whose denominator is a polynomial in s.

3.2 Polynomial Generators

Defining a polynomial generator as a generating function that has a polynomial in s in its denominator, written as

$$G(s) = \frac{f(s)}{a_0 + a_1 s + a_2 s^2 + a_3 s^3 + \ldots + a_p s^p} \tag{3.1}$$

The denominator is a polynomial of order p (p an integer > 0). The following represent examples of these polynomial generators

$$G(s) = \frac{s}{1 - \frac{1}{3}s} \tag{3.2}$$

$$G(s) = \frac{4s - 1}{-3s^2 + 4s - 1} \tag{3.3}$$

$$G(s) = \frac{s^2 - 5s}{4s^5 - 2s^4 + s^3 - s^2 + 17s - 12} \tag{3.4}$$

Each of equations (3.2), (3.3), and (3.4) contains a polynomial denominator and each represents an infinite sequence $\{y_k\}$ that must be identified.

Chapter 2 provides some background in the inversion of polynomial generators. We know from Chapter 2 how to invert the generating function in equation (3.2). For equation (3.3), since $-3s^2 + 4s - 1 = (3s - 1)(1 - s)$, will enable the use of the principle of generating function convolutions to invert G(s). However, no clue is available as to how to invert the generating function in equation (3.4) unless the denominator can be factored. If the denominator can be factored into a number of simpler polynomials of the form $(a + bs)^m$, the convolution principle of generating function inversion discussed in Chapter 2 could be invoked. If this cannot be done, there will be a need to create some additional tools to invert these more complicated polynomial denominators.

We will begin the study of polynomial generators, beginning with the simplest of polynomials, building up to more complicated polynomials.

3.3 Second-Order Polynomials

3.3.1 The role of factorization

A clear example of the generating function decomposition is in the factoring of a complicated polynomial into the product of at least two terms. Consider as a first example:

$$G(s) = \frac{1}{6s^2 - 5s + 1} \tag{3.5}$$

From what has been seen and accomplished thus far, little is known about how to proceed with the inversion of G(s) as written. However, the ability to factor the denominator opens up some additional options that would allow completion of the inversion procedure. Rewrite equation (3.5) as

$$G(s) = \frac{1}{6s^2 - 5s + 1} = \frac{1}{(1-3s)(1-2s)} = \left(\frac{1}{1-3s}\right)\left(\frac{1}{1-2s}\right) \tag{3.6}$$

Once the factoring is complete, there are two major procedures that can be followed to complete the inversion. The initial factorization is the important first step.

3.3.2 The use of partial fractions

An important traditional tool in the inversion of polynomials is the use of partial fractions [1]. This procedure allows us to write a ratio with a polynomial in the denominator as sum of several ratios of polynomials, each resultant polynomial being of lower order than the original. From equation (3.7)

$$\frac{1}{6s^2 - 5s + 1} = \frac{1}{(1-3s)(1-2s)} = \frac{A}{1-3s} + \frac{B}{1-2s} \tag{3.7}$$

Note that the inversion may be completed if the constants A and B could be determined. Begin by multiplying both sides by (1 - 3s)(1 - 2s) to find

$$1 \equiv A(1-2s) + B(1-3s) \tag{3.8}$$

Since this must be true for all s, choose convenient values for s to solve for the quantities A and B. For example, letting s =1/2

$$\begin{aligned} 1 &= A\left(1-2\left(\frac{1}{2}\right)\right) + B\left(1-3\left(\frac{1}{2}\right)\right) \\ 1 &= \frac{-B}{2} \text{ or } B = -2 \end{aligned} \tag{3.9}$$

Similarly, allowing s = 1/3 in equation (3.8) removes the term involving B, and gives A = 3. Thus rewriting equation (3.7) as

$$G(s) = \frac{1}{(1-3s)(1-2s)} = \frac{3}{1-3s} - \frac{2}{1-2s} \tag{3.10}$$

and proceeding with the inversion

$$G(s) = \frac{1}{6s^2-5s+1} = \frac{1}{(1-3s)(1-2s)} = \frac{3}{1-3s} - \frac{2}{1-2s} \tag{3.11}$$
$$\triangleright \left\{3^{k+1} - 2^{k+1}\right\}$$

3.3.3 The use of convolutions

The fact that G(s) from equation (3.5) can be factored allows applying the convolution argument directly.

$$G(s) = \frac{1}{6s^2-5s+1} = \frac{1}{(1-3s)(1-2s)} = \left(\frac{1}{1-3s}\right)\left(\frac{1}{1-2s}\right) \tag{3.12}$$
$$\triangleright \left\{\sum_{j=0}^{k} 3^j 2^{k-j}\right\}$$

Factorization makes the process much easier.

3.3.4 Comparison of partial fractions vs. convolutions

Of course, either of these two processes is applicable, provided the answers are equivalent. As a check, let k = 3. From equation (3.11) $y_3 = 3^4 - 2^4 = 65$. From equation (3.12)

$$y_3 = 3^0 2^3 + 3^1 2^2 + 3^2 2^1 + 3^3 2^0 = 8 + 12 + 18 + 27 = 65 \tag{3.13}$$

Equations (3.11) and equation (3.12) are equivalent.* Of the two, however, the form of the answer provided by partial fractions is the easiest to understand.

---

* $\frac{a^{k+1} - b^{k+1}}{a-b} = \frac{(a-b)\sum_{j=0}^{k} a^j b^{k-j}}{a-b} = \sum_{j=0}^{k} a^j b^{k-j}, a \neq b$

3.3.5 Coefficient collection

Yet another procedure is the method of coefficient collection. This procedure is a very general approach using the binomial and/or multinomial theorems. Return to the example provided in equation (3.12). Rather than factor the denominator, rewrite it as

$$\frac{1}{6s^2-5s+1}=\frac{1}{1-(5-6s)s}=\frac{1}{1-w_s s} \tag{3.14}$$

where w_s = 5 - 6s. If we invert $\frac{1}{1-w_s s}$, treating w_s as a constant, $\frac{1}{1-w_s s} \triangleright_s \{w_s^k\}$ permitting the observation that the kth term of the sequence is $w_s^k s^k = (5-6s)^k s^k$. However, although this finding is correct, it is not helpful, since powers of s are produced from the (5 - 6s) term in the denominator. We use the notation $\triangleright_s$ to denote that the inversion is incomplete since w_s is a function of s. The difficulty with this incomplete inversion becomes apparent at once. As an example, examine the coefficient for the k = 3 term. Note that

$$\begin{aligned}(5-6s)^3 s^3 &= \left(125-450s+540s^2+216s^3\right)s^3 \\ &= 125s^3-450s^4+540s^5+216s^6\end{aligned} \tag{3.15}$$

So an examination of the k = 3 term does not yield only coefficients of s^3, but coefficients of other powers of s as well. In addition, just as the k = 3 term yields powers other than those of s^3, it will also be true that powers of s^3 will arise from the examination of other terms in this expression other than k = 3. For example, expansion of the k = 2 term reveals $25s^2 - 60s^3 + 36s^4$. We are going to have to find a way to collect all of the coefficients from every place in this sequence where s^3 occurs. This will be a two-step process, and in general is very straightforward.

To begin this process, first find a more succinct way to identify all of the powers of s from the term $(5 - 6s)^k s^k$. For this, return to the binomial theorem, which states that, for any constants a and b and any positive integer k

$$(a+b)^k=\sum_{j=0}^{k}\binom{k}{j}a^j b^{k-j} \tag{3.16}$$

where $\binom{k}{j} = \frac{k!}{j!(k-j)!}$. Applying the binomial theorem to $(5 - 6s)^k$

$$(5-6s)^k = \sum_{j=0}^{k} \binom{k}{j} (-6s)^j 5^{k-j} = \sum_{j=0}^{k} \binom{k}{j} (-6)^j 5^{k-j} s^j \qquad (3.17)$$

and

$$(5-6s)^k s^k = \sum_{j=0}^{k} \binom{k}{j} (-6)^j 5^{k-j} s^{k+j} \qquad (3.18)$$

This transformation gives exactly the coefficients of s^k that are produced by the kth term of the sequence $(5 - 6s)^k s^k$. For k = 0, the only exponent of s is for k = 0 and j = 0, and therefore the only coefficient of s^0 produced is one. When k = 1, j can be either 0 or 1. For j = 0, $s^{k+j} = s^1$ and for j = 1, $s^{k+j} = s^2$. Continuing with these observations allows the generation of a table (Table 3.1) from which a pattern of the appearances of exponents of s may be noted.

Thus, the coefficients of s^3 will be

$$\binom{2}{1} (-6)^1 5^1 + \binom{3}{0} (-6)^0 5^3 = -60 + 375 = 315 \qquad (3.19)$$

The process followed here for s^3 is the process that should be followed in general to collect the terms s^k. The pattern by which powers of s^k are generated is therefore identified. As defined in Chapter 2, consider the indicator function $I_{x=A}$ as

$$I_{x=A} = \begin{cases} 1 \text{ if } x = A \\ 0 \text{ otherwise} \end{cases} \qquad (3.20)$$

Table 3.1 Relationship Between Values of k in the Series and the Power of s Produced

k	j	(k+j) Power of s
0	0	0
1	0	1
1	1	2
2	0	2
2	1	3
2	2	4
3	0	3
3	1	4
3	2	5
3	3	6

In general, all of the coefficients of s^k are represented by the expression

$$\sum_{m=0}^{k}\sum_{j=0}^{m}\binom{m}{j}(-6)^j 5^{m-j} I_{m+j=k} \tag{3.21}$$

The indicator variable accumulates all of the appropriate (m, j) combinations. The above equation indicates that terms of the form $\binom{m}{j}(-6)^j 5^{m-j}$ for which m + j = k and $j \leq m$ are the only ones that can be collected. For example, to identify the coefficients of s^4 from equation (3.21), the (m, j) combinations of interest are (4, 0), (3, 1), and (2, 2). Collecting coefficients

$$\begin{aligned} y_4 &= \binom{4}{0}(-6)^0 5^4 + \binom{3}{1}(-6)^1 5^2 + \binom{2}{2}(-6)^2 5^2 \\ &= (1)(1)(625) - (3)(6)(25) + (1)(36)(125) \\ &= 4675 \end{aligned} \tag{3.22}$$

Thus, the inversion may be written as

$$G(s) = \frac{1}{6s^2 - 5s + 1} \triangleright \left\{ \sum_{m=0}^{k} \sum_{j=0}^{m} \binom{m}{j} (-6)^j 5^{m-j} I_{m+j=k} \right\} \tag{3.23}$$

and the indicator function indicates precisely which of the summands in the double summation of equation (3.23) need to be evaluated to bring together all of the terms involving s^k.

In general, using this process of collecting coefficients, any second-order polynomial generator G(s) of the form $\frac{1}{a_0 + a_1 s + a_2 s^2}$ may be inverted. First rewrite G(s) as follows.

$$\begin{aligned} G(s) &= \frac{1}{a_0 + a_1 s + a_2 s^2} = \frac{\frac{1}{a_0}}{\frac{1}{a_0}\left(a_0 + a_1 s + a_2 s^2\right)} = \frac{\frac{1}{a_0}}{1 + \frac{a_1}{a_0} s + \frac{a_2}{a_0} s^2} \\ &= \frac{\frac{1}{a_0}}{1 - \left(\frac{-a_1}{a_0} + \frac{-a_2}{a_0} s\right) s} = \frac{\frac{1}{a_0}}{1 - \left(\frac{a_1}{a_0} + \frac{a_2}{a_0} s\right)(-1)s} \end{aligned} \tag{3.24}$$

Rewriting G(s) as in equation (3.24) shows that the inversion of G(s) requires that collection of terms for the sequence whose k^{th} term is $\left(\frac{a_1}{a_0} + \frac{a_2}{a_0} s\right)^k (-1)^k s^k$, i.e.

$$G(s) \triangleright_s \left\{ \left[\frac{1}{a_0}\right] \left(\frac{a_1}{a_0} + \frac{a_2}{a_0} s\right)^k (-1)^k \right\} \tag{3.25}$$

The job now is to collect all of the coefficients of s^k in an orderly manner. Begin by proceeding as before, invoking the binomial theorem for $\left(\frac{a_1}{a_0} + \frac{a_2}{a_0} s\right)^k$ as follows.

$$\left(\frac{a_1}{a_0} + \frac{a_2}{a_0} s\right)^k = \sum_{j=0}^{k} \binom{k}{j} \left(\frac{a_2}{a_0}\right)^j \left(\frac{a_1}{a_0}\right)^{k-j} s^j \tag{3.26}$$

Holding aside for a moment the constants $\frac{1}{a_0}$ and $(-1)^k$ from equation (3.25), note that G(s) generates the sequence

$$\left(\frac{a_1}{a_0}+\frac{a_2}{a_0}s\right)^k s^k = \sum_{j=0}^{k}\binom{k}{j}\left(\frac{a_2}{a_0}\right)^j\left(\frac{a_1}{a_0}\right)^{k-j} s^j s^k = \frac{1}{a_0^k}\sum_{j=0}^{k}\binom{k}{j}a_2^j a_1^{k-j}s^{k+j} \quad (3.27)$$

Focus on the last expression in equation (3.27). An examination of this term reveals that powers of s are generated not just from an examination of k but from an evaluation of k+j. Thus, there are different combinations of (k, j) pairs that will generate the same power of s. For example, (1, 1) and (2, 0) each generate the power of 2. Thus, for every (k, j) combination, the term $\frac{1}{a_0^k}\binom{k}{j}a_2^j a_1^{k-j}$ must be included as a coefficient of that power of s. These observation lead to the conclusion that the task of gathering powers of s (termed here collecting coefficients of the powers of s), is one of organization. The last expression in equation (3.27) displays the method for gathering the coefficients of s^k. Define m and j such that $0 \leq m \leq k$, and $0 \leq j \leq m$, and note that the term $\frac{1}{a_0^m}\binom{m}{j}a_2^j a_1^{m-j}$ will contribute to the coefficient of s^k whenever $m + j = k$. The sum of these coefficient terms can be expressed as

$$\sum_{m=0}^{k}\sum_{j=0}^{m}\frac{1}{a_0^m}\binom{m}{j}a_2^j a_1^{m-j} I_{m+j=k} = \sum_{m=0}^{k}\frac{1}{a_0^m}\sum_{j=0}^{m}\binom{m}{j}a_2^j a_1^{m-j} I_{m+j=k} \quad (3.28)$$

Now return to the original G(s), recalling that

$$G(s) \rhd_s \left\{\left[\frac{1}{a_0}\right]\left(\frac{a_1}{a_0}+\frac{a_2}{a_0}s\right)^k (-1)^k\right\} \quad (3.29)$$

it is only necessary to incorporate the constants $\frac{1}{a_0}$ and $(-1)^k$ into equation (3.28) to write

$$G(s) = \frac{1}{a_0 + a_1 s + a_2 s^2} \quad \triangleright \quad \left\{ \frac{1}{a_0} \sum_{m=0}^{k} \frac{(-1)^m}{a_0^m} \frac{1}{} \sum_{j=0}^{m} \binom{m}{j} a_2^j a_1^{m-j} I_{m+j=k} \right\}$$
$$= \left\{ \sum_{m=0}^{k} \frac{(-1)^m}{a_0^{m+1}} \sum_{j=0}^{m} \binom{m}{j} a_2^j a_1^{m-j} I_{m+j=k} \right\} \tag{3.30}$$

3.3.6 Second-order polynomials that cannot be factored

Most second-order polynomials cannot be factored. When this type of commonly occurring polynomial appears in the denominator of the generating function, neither the method of partial fractions nor the convolution principle applies, as each of them are based on the ability to factor the polynomial portion of the polynomial generator G(s). However, the method of coefficient collection is not dependent on factoring the quadratic. The only restriction is that $a_0 \neq 0$.

Thus, if $G(s) = \frac{1}{s^2 - 6s + 1}$, both the partial fraction procedure and the direct application of the convolution principle will fail, while the process of coefficient collection can be applied directly. Begin by writing

$$G(s) = \frac{1}{s^2 - 6s + 1} = \frac{1}{1 - (6 - s)s} \triangleright_s \left\{ (6 - s)^k \right\} \tag{3.31}$$

This notation means that each term of the sequence $G(s) = \sum_{k=0}^{\infty} y_k s^k = \sum_{k=0}^{\infty} (6 - s)^k s^k$. Now use the binomial theorem to write

$$(6 - s)^k = \sum_{j=0}^{k} \binom{k}{j} (-1)^j 6^{k-j} s^j \tag{3.32}$$

Thus, G(s) may be written as

$$G(s) = \sum_{k=0}^{\infty} y_k s^k = \sum_{k=0}^{\infty} (6 - s)^k s^k = \sum_{k=0}^{\infty} \sum_{j=0}^{k} \binom{k}{j} (-1)^j 6^{k-j} s^j s^k$$
$$= \sum_{k=0}^{\infty} \sum_{j=0}^{k} \binom{k}{j} (-1)^j 6^{k-j} s^{k+j} \tag{3.33}$$

What remains is to collect the coefficients of each unique power of s. Begin by creating the integer valued variables m and j such that $0 \leq m \leq k$, $0 \leq j \leq m$, observing that the $\binom{m}{j}(-1)^j 6^{m-j}$ will be a coefficient for s^k whenever $m + j = k$. Thus the coefficient of s^k from the rightmost expression in equation (3.33) is $\sum_{m=0}^{k}\sum_{j=0}^{m}\binom{m}{j}(-1)^j 6^{m-j} I_{m+j=k}$. The inversion of G(s) is now complete.

$$G(s) = \frac{1}{s^2 - 6s + 1} \triangleright \left\{\sum_{m=0}^{k}\sum_{j=0}^{m}\binom{m}{j}(-1)^j 6^{m-j} I_{m+j=k}\right\} \tag{3.34}$$

When $a_0 = 0$, the process equation (3.30) breaks down because $1/a_0$ is not defined. However, when $a_0 = 0$, the process of collecting coefficients for a second-order polynomial need not be invoked, and the inversion of G(s) simplifies. In this case, we observe

$$G(s) = \frac{1}{a_2 s^2 + a_1 s} = \frac{1}{s(a_2 s + a_1)} = \frac{s^{-1}}{(a_2 s + a_1)} \tag{3.35}$$

The denominator becomes

$$\frac{1}{(a_2 s + a_1)} = \frac{1/a_1}{1/a_1 (a_2 s + a_1)} = \frac{1/a_1}{1 - \left(-a_2/a_1\right)s} \triangleright \left\{\left[\frac{1}{a_1}\right]\left[\frac{-a_2}{a_1}\right]^k\right\} \tag{3.36}$$

Use the sliding rule to see that, when $a_1 \neq 0$,

$$G(s) = \frac{1}{a_2 s^2 + a_1 s} \triangleright \left\{\left[\frac{1}{a_1}\right]\left[\frac{-a_2}{a_1}\right]^{k+1}\right\} \tag{3.37}$$

3.4 Third-Order Polynomials

In this section we consider polynomial generators of the form

$$G(s) = \frac{1}{a_3 s^3 + a_2 s^2 + a_1 s + a_0} \tag{3.38}$$

It can be seen that these equations will be handled in the same fashion as the second-order equations. There are, in general, three methods of inverting this generating function. Two of them (method of partial fractions and the principle of convolutions) lead to easily articulated solutions as long as the function can be factored. Since most cubic equations cannot be factored, however, the utility of these approaches is, in general, limited. After providing examples of each of the partial fraction and convolution approach, the balance of the discussion in this section will be on the method of coefficient collection.

3.4.1 Partial fractions

Consider the third-order polynomial generating function

$$G(s) = \frac{1}{-s^3 + s^2 + 14s - 24} = \frac{1}{(s-3)(2-s)(s+4)} \tag{3.39}$$

Since G(s) can be factored, apply the method of partial fractions and proceed as in the previous section, writing

$$\frac{1}{(s-3)(2-s)(s+4)} = \frac{A}{s-3} + \frac{B}{2-s} + \frac{C}{s+4} \tag{3.40}$$

Multiply throughout by the three terms in the denominator,

$$1 \equiv A(2-s)(s+4) + B(s-3)(s+4) + C(s-3)(2-s) \tag{3.41}$$

The task is somewhat more complicated since each of the constants multiplies not one but two terms. Letting s = 2 removes the a and c terms, and from the resulting equation find B = -1/6. Similarly, s = 3 leads to the conclusion that A = -1/7, and finally s = -4 leads to C = -1/42. Equation (3.41) can now be written as

$$\frac{1}{(s-3)(2-s)(s+4)} = -\frac{\frac{1}{7}}{(s-3)} - \frac{\frac{1}{6}}{(2-s)} - \frac{\frac{1}{42}}{(s+4)} \tag{3.42}$$

Invoking the addition principle reveals at once that

$$\begin{aligned} G(s) &= \frac{1}{(s-3)(2-s)(s+4)} \\ &> \left\{ \frac{1}{7}\left(\frac{1}{3}\right)^{k+1} - \frac{1}{6}\left(\frac{1}{2}\right)^{k+1} - \frac{1}{42}\left(\frac{1}{4}\right)\left(\frac{-1}{4}\right)^{k} \right\} \end{aligned} \tag{3.43}$$

When the polynomial is factorable, the partial fractions solution is quite simple.

3.4.2 Principles of convolutions

The use of convolutions for G(s) when the generating function has a third-order coefficient in the denominator is a straightforward process, although the final solution for y_k can be complicated. To begin, write

$$G(s) = \frac{1}{-24+14s+s^2-s^3} = \left[\frac{1}{s-3}\right]\left[\frac{1}{2-s}\right]\left[\frac{1}{s+4}\right] \tag{3.44}$$

Begin by inverting $\left[\frac{1}{s-3}\right]\left[\frac{1}{2-s}\right]$ in a first application of the convolution principle. Since

$$\frac{1}{s-3} = \frac{1}{-3+s} = \frac{-\frac{1}{3}}{1-\left(\frac{1}{3}\right)s} > \left\{-\left[\frac{1}{3}\right]^{k+1}\right\} \tag{3.45}$$

and

$$\frac{1}{2-s} = \frac{\frac{1}{2}}{1-\frac{1}{2}s} > \left[\frac{1}{2}\right]^{k+1} \tag{3.46}$$

write

$$\left[\frac{1}{s-3}\right]\left[\frac{1}{2-s}\right] \triangleright \left\{\sum_{j=0}^{k}(-1)\left[\frac{1}{3}\right]^{j+1}\left[\frac{1}{2}\right]^{k+1-j}\right\} \tag{3.47}$$

Also see that the last factor in equation (3.44) can be inverted as

$$\frac{1}{s+4} = \frac{\frac{1}{4}}{1-\left(-\frac{1}{4}\right)s} \triangleright \left\{\frac{(-1)^k}{4^{k+1}}\right\} \tag{3.48}$$

Now let $w_k = \sum_{j=0}^{k}(-1)\left[\frac{1}{3}\right]^{j+1}\left[\frac{1}{2}\right]^{k+1-j}$ and observe that G(s) represents a final convolution of the sequence $\sum_{j=0}^{\infty} w_k s^k$ and the sequence whose coefficient of s^k is $\left\{\frac{(-1)^k}{4^{k+1}}\right\}$. Performing the final convolution

$$\left[\sum_{k=0}^{\infty} w_k s^k\right]\left[\sum_{k=0}^{\infty}\frac{(-1)^k}{4^{k+1}} s^k\right] \triangleright \left\{\sum_{j=0}^{k} w_j \frac{(-1)^{k-j}}{4^{k-j+1}}\right\} \tag{3.49}$$

and substituting for w_k

$$G(s) = \frac{1}{-24+14s+s^2-s^3} \triangleright \left\{\sum_{j=0}^{k}\sum_{h=0}^{j}(-1)\left[\frac{1}{3}\right]^{h+1}\left[\frac{1}{2}\right]^{j+1-h}\frac{(-1)^{k-j}}{4^{k-j+1}}\right\} \tag{3.50}$$

3.4.3 Coefficient collection

Based on the developments thus far, the plan for the inversion of polynomial generator $G(s) = \frac{1}{-s^3+s^2+14s-24}$ is to apply the method of coefficient collection. Begin by writing

$$G(s) = \frac{1}{-s^3 + s^2 + 14s - 24} = \frac{\frac{-1}{24}}{1 - \frac{14}{24}s - \frac{1}{24}s^2 + \frac{1}{24}s^3}$$

$$= \frac{\frac{-1}{24}}{1 - \left(\frac{14}{24} + \frac{1}{24}s - \frac{1}{24}s^2\right)s} = \frac{\frac{-1}{24}}{1 - w_s s} \tag{3.51}$$

Proceeding as before, observe that by letting $w_s = \frac{14}{24} + \frac{1}{24}s - \frac{1}{24}s^2$, then $G(s) \triangleright_s \{w_s^k\}$. In order to be sure to collect each of the coefficients of s^k, it is necessary to expand the w_s^k term. However, the binomial theorem will not permit the full expansion. With three summands in w_s, we turn instead to the multinomial theorem. The multinomial theorem states that, given three quantities a, b, and c, and an integer k,

$$(a + b + c)^k = \sum_{i=0}^{k}\sum_{j=0}^{k-i}\binom{k}{i\ \ j} a^i b^j c^{k-i-j} \tag{3.52}$$

where

$$\binom{k}{i\ \ j} = \frac{k!}{i!j!(k-i-j)!} \tag{3.53}$$

Applying the multinomial theorem to w_s, find that

$$\left(\frac{14}{24} + \frac{1}{24}s - \frac{1}{24}s^2\right)^k$$
$$= \sum_{i=0}^{k}\sum_{j=0}^{k-i}\binom{k}{i\ \ j}\left[-\frac{1}{24}s^2\right]^i\left[\frac{1}{24}s\right]^j\left[\frac{14}{24}\right]^{k-i-j} \tag{3.54}$$
$$= \sum_{i=0}^{k}\sum_{j=0}^{k-i}\binom{k}{i\ \ j}\left[-\frac{1}{24}\right]^i\left[\frac{1}{24}\right]^j\left[\frac{14}{24}\right]^{k-i-j} s^{2i+j}$$

and the kth term of the inverted sequence $w_s^k s^k$ can be written as

$$\left[\frac{1}{24}\right]^k \sum_{i=0}^{k}\sum_{j=0}^{k-i}\binom{k}{i\ \ j}[-1]^i[14]^{k-i-j}s^{k+2i+j} \tag{3.55}$$

The task before us now is to collect the powers of s carefully, identifying which terms in this sequence will contribute a term to the coefficient of a power of s. Following the procedure developed in the previous section for expanding powers of s from quadratic polynomials, define the integer variables m, i, and j such that $0 \le m \le k$, $0 \le i \le m$, and $0 \le j \le m-i$. This allows the expression $\left[\frac{1}{24}\right]^m \binom{m}{i\ \ j}[-1]^i[14]^{m-i-j}$ as a coefficient of s^k whenever $m + 2i + j = k$.

Therefore write the inversion of G(s) as

$$G(s) = \frac{1}{-s^3 + s^2 + 14s - 24}$$
$$\triangleright \left\{\sum_{m=0}^{k}\left[\frac{1}{24}\right]^m \sum_{i=0}^{m}\sum_{j=0}^{m-i}\binom{m}{i\ \ j}[-1]^i[14]^{m-i-j} I_{m+2i+j=k}\right\} \tag{3.56}$$

This process did not consider whether the denominator of G(s) could be factored, and therefore provided a solution that is nontrivially expressed. However, the advantage of the coefficient collection mechanism is that the denominator of the generating function need not be factorable. The solution is complicated, but the method of collecting coefficient has the advantage of always working.

3.4.4 Coefficient collection and the general cubic equation

In the previous section, a specific cubic polynomial generator (i.e., a generating function whose denominator contains a cubic polynomial) in s was inverted. We can now use the coefficient collection procedure to invert the general cubic polynomial generator

$$G(s) = \frac{1}{a_0 + a_1 s + a_2 s^2 + a_3 s^3} \qquad a_0 \ne 0 \tag{3.57}$$

The plan will be to invert $G(s) \triangleright_s \{w_s\}$ where w_s is a coefficient containing s. Utilizing the multinomial theorem to rewrite w_s as a product, and finally gathering the coefficients of s^k using an indicator function as in the previous section. Write

$$G(s) = \frac{1}{a_0 + a_1 s + a_2 s^2 + a_3 s^3} = \frac{\frac{1}{a_0}}{\frac{1}{a_0}\left(a_0 + a_1 s + a_2 s^2 + a_3 s^3\right)}$$

$$= \frac{\frac{1}{a_0}}{1 + \frac{a_1}{a_0}s + \frac{a_2}{a_0}s^2 + \frac{a_3}{a_0}s^3} = \frac{\frac{1}{a_0}}{1 - \left(\frac{-a_1}{a_0} + \frac{-a_2}{a_0}s + \frac{-a_3}{a_0}s^2\right)s} \tag{3.58}$$

leading to

$$G(s) = \frac{\frac{1}{a_0}}{1 - \left(\frac{a_1}{a_0} + \frac{a_2}{a_0}s + \frac{a_3}{a_0}s^2\right)(-1)s} \rhd_s \left\{\frac{1}{a_0} w_s^k (-1)^k\right\} \tag{3.59}$$

Now use the multinomial theorem to write

$$w_s^k = \left(\frac{a_1}{a_0} + \frac{a_2}{a_0}s + \frac{a_3}{a_0}s^2\right)^k = \sum_{i=0}^{k}\sum_{j=0}^{k-i}\binom{k}{i\ \ j}\left(\frac{a_3}{a_0}\right)^i\left(\frac{a_2}{a_0}\right)^j\left(\frac{a_1}{a_0}\right)^{k-i-j} s^{2i+j} \tag{3.60}$$

Carry out the inversion one additional step by incorporating equation (3.60) into equation (3.59)

$$\frac{1}{a_0}[w(s)]^k (-1)^k s^k$$

$$= \frac{1}{a_0}(-1)^k \sum_{i=0}^{k}\sum_{j=0}^{k-i}\binom{k}{i\ \ j}\left(\frac{a_3}{a_0}\right)^i\left(\frac{a_2}{a_0}\right)^j\left(\frac{a_1}{a_0}\right)^{k-i-j} s^{k+2i+j} \tag{3.61}$$

$$= \frac{(-1)^k}{a_0^{k+1}} \sum_{i=0}^{k}\sum_{j=0}^{k-i}\binom{k}{i\ \ j} a_3^{\,i} a_2^{\,j} a_1^{\,k-i-j} s^{k+2i+j}$$

The only task that remains is to accrue the correct coefficients using the indicator function.

$$G(s) \triangleright \left\{ \sum_{m=0}^{k} \frac{(-1)^m}{a_0^{m+1}} \sum_{i=0}^{m} \sum_{j=0}^{m-i} \binom{m}{i\ \ j} a_3^i a_2^j a_1^{m-i-j} I_{m+2i+j=k} \right\} \tag{3.62}$$

Equation (3.62) represents the solution for the cubic polynomial generator. When the denominator of this generating function can be factored, then the method of partial fractions provides the most recognizable solution. However, as in most circumstances, the cubic equation cannot be factored. When this is the case, the method of coefficient collection will work as long as $a_0 \neq 0$. If $a_0 = 0$, then G(s) reduces

$$G(s) = \frac{1}{a_1 s + a_2 s^2 + a_3 s^3} = \frac{s^{-1}}{a_1 + a_2 s + a_3 s^2} \tag{3.63}$$

which is a quadratic generator, and the methods of the previous section may be applied invoking the sliding principle when considering the numerator of s^{-1}.

3.4.5 Partial factorization of a cubic generator

We have presented the issue of factorization for a cubic polynomial generator as though the process is "all or none," i.e., either one can identify all of the roots of the polynomial or the reader only knows some of them. However, knowledge of any root of the polynomial residing in the denominator of G(s) can provide an important simplification in the inversion. For example, consider a cubic equation that has one real root and two complex roots, and we know the identify of the real root. The cubic generator G(s) is such a case where

$$G(s) = \frac{1}{s^3 - 3s^2 + s + 2} \tag{3.64}$$

It can be verified that one of the roots of the denominator is s = 2. However, this finding is not sufficient to apply the partial fraction argument alone. Also the direct appeal to the convolution argument will fail, since one of the factors is a quadratic that cannot be easily factored. Since all of the roots are not known, neither the procedure using partial fractions nor the solution using the convolution principle alone will work. However, proceeding by factoring the denominator as $s^3 - 3s^2 + s + 2 = (s-2)(s^2 - s - 1)$ and writing G(s) as

$$G(s) = \frac{1}{s^3 - 3s^2 + s + 2} = \left(\frac{1}{s-2}\right)\left(\frac{1}{s^2 - s - 1}\right) \tag{3.65}$$

use a convolution argument to complete this inversion. Taking each factor a piece at a time

$$\frac{1}{s-2} \triangleright \left\{\frac{-1}{2^{k+1}}\right\} \tag{3.66}$$

Now use the coefficient collection method to invert the second-order polynomial generator $\frac{1}{s^2 - s - 1}$

$$\frac{1}{s^2 - s - 1} \triangleright \left\{-\sum_{m=0}^{k}\sum_{j=0}^{m}\binom{m}{j}(-1)^{m-j} I_{m+j=k}\right\} \tag{3.67}$$

Next, complete the inversion through a convolution argument

$$\begin{aligned} G(s) &= \frac{1}{s^3 - 3s^2 + s + 2} = \left(\frac{1}{s-2}\right)\left(\frac{1}{s^2 - s - 1}\right) \\ &\triangleright \left\{\sum_{i=0}^{k}\frac{(-1)}{2^{i+1}}\sum_{m=0}^{k-i}\sum_{j=0}^{m}\binom{m}{j}(-1)^{m-j} I_{m+j=k-i}\right\} \end{aligned} \tag{3.68}$$

3.4.6 Fourth-order and higher polynomial generators

The inversion of fourth-order, fifth-order, and higher order polynomial generators follows the procedures that were outlined for second- and third-order polynomials. If all of the roots in the polynomial residing in the denominator of G(s) can be identified, then the easiest method to apply for its inversion is that of partial fractions. The application is straightforward, and the inversion is most easily interpreted. However, when all of the roots cannot be found, the coefficient collection method works very well. It can be applied nicely in situations where the simpler procedures of partial fractions and convolution arguments cannot complete the inversion. An example of an interesting case is

$$G(s) = \frac{1}{3s^4 - 2s^3 - 29s^2 - 16s - 55} \tag{3.69}$$

The polynomial denominator has no easily identified roots. Nevertheless it can be factored into the product of two quadratics. Performing this factorization leads to

$$G(s) = \left(\frac{1}{3s^2 + s + 5}\right)\left(\frac{1}{s^2 - s - 11}\right) \tag{3.70}$$

Consider the application of a convolution argument once each of the factors is inverted. Using equation (3.30) proceed with the factor inversion for the first term.

$$G_1(s) = \frac{1}{3s^2 + s + 5} = \frac{\frac{1}{5}}{1 - \frac{1}{5}(1+3s)(-1)s} \triangleright_s \left\{\frac{1}{5}\left[\frac{1}{5}\right]^k (-1)^k (1+3s)^k\right\}$$

$$(1+3s)^k = \sum_{j=0}^{k}\binom{k}{j} 3^j s^j \tag{3.71}$$

$$G_1(s) \triangleright \left\{\sum_{m=0}^{k}\left[\frac{1}{5}\right]^{m+1}(-1)^m \sum_{j=0}^{m}\binom{m}{j} 3^j I_{m+j=k}\right\} = \{\alpha_k\}$$

The inverse of the second term proceeds analogously Begin by writing

$$G_2(s) = \frac{1}{s^2 - s - 11} = \frac{\frac{-1}{11}}{1 - \left(\frac{1}{11}\right)(s-1)s} \triangleright_s \left\{\frac{-1}{11^{k+1}}(s-1)^k\right\} \tag{3.72}$$

Proceeding,

$$(s-1)^k = \sum_{j=0}^{k}\binom{k}{j}(-1)^{k-j} s^j \tag{3.73}$$

$$G_2(s) \triangleright \left\{-\sum_{m=0}^{k}\frac{1}{11^{m+1}}\sum_{j=0}^{m}\binom{m}{j}(-1)^{m-j} I_{m+j=k}\right\} = \{\beta_k\}$$

and a convolution completes the inversion. Begin by writing

$$G_1(s)G_2(s) \triangleright \sum_{h=0}^{k} \alpha_h \beta_{k-h} \tag{3.74}$$

and

$$G(s) \triangleright \left\{ \sum_{h=0}^{k} \left[\left(\sum_{m_1=0}^{h} (-1)^{m_1} \left[\frac{1}{5}\right]^{m_1+1} \sum_{j_1=0}^{m_1} \binom{m_1}{j_1} 3^{j_1} I_{m_1+j_1=h} \right) \times \left(-1 \sum_{m_2=0}^{k-h} \left(\frac{1}{11^{m_2-h+1}} \right) \sum_{j_2=0}^{m_2} \binom{m_2}{j_2} (-1)^{m_2-j_2} I_{m_2+j_2=k-h} \right) \right] \right\} \tag{3.75}$$

These finite sums are easily evaluated.

3.5 The Polynomial Generator Inversion Theorem

The method of collecting coefficients is a useful tool in the inversion of generating functions G(s) when G(s) contains a polynomial in the denominator. Its use can be formalized in the form of a theorem, whose result can from this point forward be invoked when the inversion of a polynomial generator is required.

Theorem: Polynomial Generator Inversion.

If G(s) be a generating function of the form

$$G(s) = \frac{1}{1 + a_1 s + a_2 s^2 + a_3 s^3 + \ldots + a_L s^L} \tag{3.76}$$

where $a_i \neq 0$ for i = 1 to L, then the inversion formula for G(s) may be written as

$$G(s) \triangleright \left\{ \sum_{m=0}^{k} (-1)^{m} \sum_{j_1=0}^{m} \sum_{j_2=0}^{m-j_1} \sum_{j_3}^{m-j_1-j_2} \cdots \sum_{j_{L-1}}^{m-\sum_{i=1}^{m-2} j_i} CJ \right\} \tag{3.77}$$

where

$$C = \begin{pmatrix} & & m & & \\ j_1 & j_2 & j_3 & \cdots & j_{L-1} \end{pmatrix} \tag{3.78}$$

and

$$J = \left[\prod_{i=2}^{L} a_i^{j_{i-1}} \right] a_1^{m-\sum_{i=1}^{L-1} j_i} I_{m+\sum_{i=1}^{L-1} ij_i=k} \tag{3.79}$$

Proof:

The proof will mirror the development of the coefficient collection method that has been demonstrated in this chapter. Begin by writing G(s) as

$$\begin{aligned} G(s) &= \frac{1}{1+a_1s+a_2s^2+a_3s^3+\ldots+a_Ls^L} \\ &= \frac{1}{1-(-1)\left(a_1+a_2s+a_3s^2+\ldots+a_Ls^{L-1}\right)s} = \frac{1}{1-\left[(-1)\sum_{i=1}^{L} a_i s^{i-1}\right]s} \end{aligned} \tag{3.80}$$

We begin the inversion process by carrying out a partial inversion, writing

$$\frac{1}{1-\left[(-1)\sum_{i=1}^{L} a_i s^{i-1}\right]s} \triangleright_s \left\{ (-1)^k \left[\sum_{i=1}^{L} a_i s^{i-1} \right]^k \right\} \tag{3.81}$$

where the "coefficient" of s^k is itself a function of s. The next task is to expand the expression on the right side of equation (3.81). Invoking the multinomial theorem write

$$\left[\sum_{i=1}^{L} a_i s^{i-1}\right]^k$$
$$= \sum_{j_1=0}^{k}\sum_{j_2=0}^{k-j_1}\sum_{j_3}^{k-j_1-j_2}\cdots\sum_{j_{L-1}}^{k-\sum_{i=1}^{L-2} j_i}\binom{k}{j_1\quad j_2\quad j_3\quad \cdots\quad j_{L-1}}\left[\prod_{i=2}^{L} a_i^{j_{i-1}}\right] a_1^{k-\sum_{i=1}^{L-1} j_i}\, s^{\sum_{i=1}^{L-1} i j_i} \tag{3.82}$$

Since the kth term in the inversion sequence is $\left[\sum_{i=1}^{L} a_i s^{i-1}\right]^k s^k$, multiply each side of equation (3.82) by s^k to obtain

$$\left[\sum_{i=1}^{L} a_i s^{i-1}\right]^k s^k = \sum_{j_1=0}^{k}\sum_{j_2=0}^{k-j_1}\sum_{j_3}^{k-j_1-j_2}\cdots\sum_{j_{L-1}}^{k-\sum_{i=1}^{L-2} j_i}\binom{k}{j_1\quad j_2\quad j_3\quad \cdots\quad j_{L-1}}\left[\prod_{i=2}^{L} a_i^{j_{i-1}}\right] a_1^{k-\sum_{i=1}^{L-1} j_i}\, s^{k+\sum_{i=1}^{L-1} i j_i} \tag{3.83}$$

The only task remaining is the coefficient collection process. Define m such that $0 \le m \le k$ and write

$$G(s) \triangleright \left\{\sum_{m=0}^{k}(-1)^m \sum_{j_1=0}^{m}\sum_{j_2=0}^{m-j_1}\sum_{j_3}^{m-j_1-j_2}\cdots\sum_{j_{L-1}}^{m-\sum_{i=1}^{m-2} j_i}\binom{m}{j_1\quad j_2\quad j_3\quad \cdots\quad j_{L-1}}\left[\prod_{i=2}^{m} a_i^{j_{i-1}}\right] a_1^{m-\sum_{i=1}^{L-1} j_i}\, I_{m+\sum_{i=1}^{L-1} i j_i = k}\right\} \tag{3.84}$$

Problems

Invert the following generating functions by both factorization and convolution and by coefficient collection:

1. $G(s)=\frac{1}{s^2-3s+2}$

2. $G(s)=\frac{1}{3s^2+s-2}$

3. $G(s)=\frac{1}{-16s^2+8s-1}$

4. $G(s)=\frac{1}{s^2+3s+2}$

5. $G(s)=\frac{1}{2s^2+7s-4}$

6. $G(s)=\frac{1}{s^2+s-2}$

7. $G(s)=\frac{1}{6s^2-10s-4}$

8. $G(s)=\frac{1}{4s^2+11s+7}$

9. $G(s)=\frac{1}{-s^2+6s-9}$

10. $G(s)=\frac{1}{s^2-4s-12}$

11. $G(s)=\frac{1}{4s^2+27s-7}$

12. $G(s)=\frac{1}{2s^2+s-21}$

13. $G(s)=\frac{1}{15s^2-35s+15}$

14. $G(s)=\frac{1}{18s^2+17s+4}$

Invert the following generating functions, G(s) given the supplied root for its denominator:

15. $G(s) = \frac{1}{s^3 - 2s^2 + 4s - 3} : s = 1$

16. $G(s) = \frac{1}{4s^3 + 3s^2 - 7s - 276} : s = 4$

17. $G(s) = \frac{1}{7s^3 - s^2 + s + 195} : s = -3$

18. $G(s) = \frac{1}{2s^3 + 5s^2 + 2s - 40} : s = 2$

19. $G(s) = \frac{1}{s^3 - 3s^2 - 6s - 1224} : s = 12$

20. $G(s) = \frac{1}{4s^3 - 7s^2 + 3s + 690} : s = -5$

21. $G(s) = \frac{1}{s^3 + s^2 + s + 301} : s = -7$

22. $G(s) = \frac{1}{3s^3 - 2s^2 - 2s - 1392} : s = 8$

23. $G(s) = \frac{1}{11s^3 - s^2 + 4s + 14564} : s = 11$

24. $G(s) = \frac{1}{6s^3 - 4s^2 + s - 129} : s = 3$

25. $G(s) = \frac{1}{6s^3 - 4s^2 + s - 129} : s = 3$

24. $G(s) = \frac{1}{s^3 - 3s^2 + 2s + 24} : s = -2$

25. $G(s) = \frac{1}{2s^3 + 2s^2 + s + 595} : s = -7$

26. $G(s) = \frac{1}{s^3 - 3s^2 - 6s - 432} : s = 9$

27. Prove:

$$\sum_{m=0}^{k}\sum_{j=0}^{m}\binom{m}{j}c^{j}b^{m-j}I_{m+j=k} = \sum_{j=0}^{\left[\frac{k}{2}\right]}\binom{\left[\frac{k}{2}\right]+1+j}{\left[\frac{k}{2}\right]-j}c^{\left[\frac{k}{2}\right]-j}b^{2j+1} \quad \text{for k odd}$$

28. Prove:

$$\sum_{m=0}^{k}\sum_{j=0}^{m}\binom{m}{j}c^{j}b^{m-j}I_{m+j=k} = b^{k}\sum_{j=0}^{\frac{k}{2}}\binom{\frac{k}{2}+j}{\frac{k}{2}-j}\left(\frac{c}{b}\right)^{j} \quad \text{for k even}$$

References

1. Goldberg S. Introduction to Difference Equations. New York. John Wiley and Sons. 1958.

4

Difference Equations: Invoking the Generating Functions

4.1 Introduction

Chapter 2 was focused on generating functions. Beginning with the simplest of infinite series, we developed familiarity with the features of generating functions and became comfortable with the idea of shuttling back and forth between generating functions and infinite series whose sum they represent. While rewarding in itself, the long range goal all along has been to apply this new knowledge to the solution of difference equations. This chapter will concentrate on applying these new found skills to the classes of difference equations introduced in Chapter 1. At the conclusion of this chapter, the reader's ability to solve generating functions of all orders, both homogeneous and non-homogeneous difference equations should be quite advanced. As has been our habit, we will start with the application of generating functions to the simplest of difference equations, gradually increasing our ability as we move to more complex ones. Throughout our work, we will remain focused on the three steps of the application of generating functions to difference equations: conversion, condensation, and inversion. The reader should feel free to return to Chapters 2 and 3 to review necessary sections of the development of generating functions as our requirements of them increase.

4.2 The Simplest Case

At the beginning of Chapter 2, a development for the solution of the following family of difference equations was initiated.

$$y_{k+1} = 3y_k \tag{4.1}$$

$k = 0,1,2,\ldots,\infty$; y_0 is a known constant. Recalling the development from Chapter 2, a generating function argument was invoked and condensed this collection of an infinite number of equations down to one equation in terms of the generating function

$$\begin{aligned} G(s) - y_0 &= 3sG(s) \\ G(s) &= \frac{y_0}{1-3s} \end{aligned} \tag{4.2}$$

This is where the beginning of Chapter 2 ended. However, with the generating function skills acquired in Chapters 2 and 3, this simple inversion can be carried out:

$$G(s) = \frac{y_0}{1-3s} \triangleright \{y_0 3^k\} \tag{4.3}$$

Note that the solution to the difference equation $y_{k+1} = 3y_k$, $k = 0,1,2,\ldots,\infty$; y_0 is a known constant is $y_k = y_0 3^k$ for $k \geq 0$. This approach is now used to solve the general family of first-order nonhomogenous difference equations. Express any member of this family as

$$ay_{k+1} = by_k + c \tag{4.4}$$

for k equal zero to infinity, and y_0 is a known constant. As in the previous example, multiply each side of equation (4.4) by s^k.

$$as^k y_{k+1} = bs^k y_k + cs^k \tag{4.5}$$

Since there are an infinite number of equations, k = 0 to ∞, summing over all of these equations

$$\sum_{k=0}^{\infty} as^k y_{k+1} = \sum_{k=0}^{\infty} bs^k y_k + \sum_{k=0}^{\infty} cs^k \tag{4.6}$$

We can rewrite each term in equation (4.6) in terms of $G(s) = \sum_{k=0}^{\infty} s^k y_k$

$$\begin{aligned}\sum_{k=0}^{\infty} as^k y_{k+1} &= a\sum_{k=0}^{\infty} s^k y_{k+1} = as^{-1}\sum_{k=0}^{\infty} s^{k+1} y_{k+1} \\ &= as^{-1}\left[\sum_{k=0}^{\infty} s^k y_k - y_0\right] = as^{-1}\left[G(s) - y_0\right]\end{aligned} \tag{4.7}$$

$$\sum_{k=0}^{\infty} bs^k y_k = b\sum_{k=0}^{\infty} s^k y_k = bG(s) \tag{4.8}$$

$$\sum_{k=0}^{\infty} cs^k = c\sum_{k=0}^{\infty} s^k = \frac{c}{1-s} \tag{4.9}$$

Thus, equation (4.6) may be written as

$$as^{-1}\left[G(s) - y_0\right] = bG(s) + \frac{c}{1-s} \tag{4.10}$$

This equation can be solved for G(s) very easily

$$\begin{aligned} a\left[G(s) - y_0\right] &= bsG(s) + \frac{cs}{1-s} \\ aG(s) - ay_0 &= bsG(s) + \frac{cs}{1-s} \\ G(s)(a - bs) &= \frac{cs}{1-s} + ay_0 \end{aligned} \tag{4.11}$$

leading to

$$G(s) = \frac{cs}{(1-s)(a-bs)} + \frac{ay_0}{(a-bs)} \tag{4.12}$$

The inversion is straightforward. After inverting each term, use the summation principle to complete the inversion of G(s). Each of these two terms requires focus on the expression $\frac{1}{a-bs}$. Its inversion is simply seen as

$$\frac{1}{a-bs}=\frac{1/a}{1-\frac{b}{a}s} \triangleright \left\{\left[\frac{1}{a}\right]\left[\frac{b}{a}\right]^k\right\} \tag{4.13}$$

Attending to the second term on the right of equation (4.12)

$$\frac{ay_0}{(a-bs)} \triangleright \left\{y_0\left[\frac{b}{a}\right]^k\right\} \tag{4.14}$$

The inversion of the first term requires the use of the convolution principle and the sliding tool.

$$\frac{cs}{(1-s)(a-bs)} \triangleright \left\{\left[\frac{c}{a}\right]\sum_{j=0}^{k-1}\left[\frac{b}{a}\right]^j\right\} \tag{4.15}$$

The final solution can now be written as

$$G(s)=\frac{cs}{(1-s)(a-bs)}+\frac{ay_0}{(a-bs)} \triangleright \left\{\left[\frac{c}{a}\right]\sum_{j=0}^{k-1}\left[\frac{b}{a}\right]^j+y_0\left[\frac{b}{a}\right]^k\right\} \tag{4.16}$$

which is the solution to the general, first-order, nonhomogenous difference equation.

4.3 Second-Order Homogeneous Difference Equations–Example 1

This generating function procedure can be applied to a very simple family of second-order homogeneous equations that are very easy to solve. Consider the family of difference equations given by

$$y_{k+2} = y_{k+1} + y_k \tag{4.17}$$

for k = 0, 1, 2, 3, ...,∞: where y_0 and y_1 are known constants. Following the approach that worked for the simple first-order homogeneous equation of the previous section, begin the conversion process by multiplying each term in this equation by s

$$s^k y_{k+2} = s^k y_{k+1} + s^k y_k \tag{4.18}$$

Using the definition, $G(s) = \sum_{k=0}^{\infty} s^k y_k$, the task is to condense the infinite number of equations involving the y_k's to one equation involving G(s). As before, sum each term in equation (4.18) over the range of k (i.e., from zero to infinity) to obtain

$$\sum_{k=0}^{\infty} s^k y_{k+2} = \sum_{k=0}^{\infty} s^k y_{k+1} + \sum_{k=0}^{\infty} s^k y_k \tag{4.19}$$

Recognizing that each of these terms contains G(s), work with the three infinite sums one at a time.

$$\begin{aligned}
\sum_{k=0}^{\infty} s^k y_{k+2} &= \sum_{k=0}^{\infty} s^{-2} s^2 s^k y_{k+2} = s^{-2} \sum_{k=0}^{\infty} s^{k+2} y_{k+2} \\
&= s^{-2} \sum_{k=0}^{\infty} s^{k+2} y_{k+2} + s^0 y_0 + s y_1 - s^0 y_0 - s y_1 \\
&= s^{-2} \left[\sum_{k=0}^{\infty} s^k y_k - s^0 y_0 - s y_1 \right] \\
&= s^{-2} \left[G(s) - y_0 - s y_1 \right]
\end{aligned} \tag{4.20}$$

Continue the process of condensation

$$\begin{aligned}
\sum_{k=0}^{\infty} s^k y_{k+2} &= s^{-2} \left[G(s) - y_0 - s y_1 \right] \\
\sum_{k=0}^{\infty} s^k y_{k+1} &= s^{-1} \left[G(s) - y_0 \right]
\end{aligned} \tag{4.21}$$

Equation (4.19) can be written as

$$
\begin{aligned}
s^{-2}\left[G(s) - y_0 - sy_1\right] &= s^{-1}\left[G(s) - y_0\right] + G(s) \\
G(s)\left[1 - s - s^2\right] &= y_0 + s\left(y_1 - y_0\right)
\end{aligned}
\tag{4.22}
$$

and it is seen that

$$
G(s) = \frac{y_0}{1 - s - s^2} + \frac{s\left(y_1 - y_0\right)}{1 - s - s^2} \tag{4.23}
$$

4.3.1 Inversion of the denominator of G(s)

In order to invert G(s) in equation (4.23), we must focus on its denominator. Once the denominator is inverted, the scaling, translation, and addition principles of generating function manipulation can be used to complete the inversion of G(s). Using the binomial theorem, recognize

$$
\begin{aligned}
&\frac{1}{1 - s - s^2} = \frac{1}{1 - (1+s)s} \;\triangleright_s\; \left\{(1+s)^k\right\} \\
&(1+s)^k s^k = \sum_{j=0}^{k} \binom{k}{j} s^{k+j}
\end{aligned}
\tag{4.24}
$$

In order to proceed further, the coefficients of s^k must be collected. If we introduce another index m, then note that we need to collect all terms of the form

$\binom{m}{j}$ when $0 \le m \le k, 0 \le j \le m, m + j = k$. Thus,

$$
\frac{1}{1 - s - s^2} \;\triangleright\; \left\{\sum_{m=0}^{k}\sum_{j=0}^{m}\binom{m}{j} I_{m+j=k}\right\} \tag{4.25}
$$

This is the key to the inversion of the generating function G(s) in equation (4.23). Completing the inversion process requires only the use of the addition and translation principles of generating function manipulation.

$$G(s) = \frac{y_0}{1-s-s^2} + \frac{s(y_1-y_0)}{1-s-s^2}$$

$$\triangleright \left\{ y_0 \sum_{m=0}^{k}\sum_{j=0}^{m}\binom{m}{j} I_{m+j=k} + (y_1-y_0)\sum_{m=0}^{k-1}\sum_{j=0}^{m}\binom{m}{j} I_{m+j=k-1} \right\} \qquad (4.26)$$

4.3.2 Checking the solution

This is the solution to the difference equation $y_{k+2} = y_{k+1} + y_k$ for k = 0, 1, 2, 3, ...,∞: y_0, y_1 are known constants. The solution appears complicated for this relatively straightforward second-order homogeneous equation, raising the question, "Does the solution derived through this inversion procedure actually work?" For this equation, of course, y_0, y_1 are known. We can easily compute $y_2 = y_1 + y_0$, and $y_3 = 2y_1 + y_0$. Does the kth term in the infinite series of G(s) as denoted in equation (4.26) reproduce these solutions? For k = 2, the solution from equation (4.26) is

$$y_0 \sum_{m=0}^{2}\sum_{j=0}^{m}\binom{m}{j} I_{m+j=2} + (y_1-y_0)\sum_{m=0}^{1}\sum_{j=0}^{m}\binom{m}{j} I_{m+j=1} \qquad (4.27)$$

For each double summation, identify each of the admissible (m, j) combinations and include them as part of the solution when the (m, j) combination meets the criteria provided by the indicator function. For example, when considering the first double summation, note that the entire set of (m, j) combinations are (0, 0), (1, 0), (1, 1), (2, 0), (2, 1), and (2, 2). Of these terms, only the two terms (1,1) and (2,0) meet the criteria that m + j = 2. Thus from the first term

$$y_0\left[\binom{1}{1} + \binom{2}{0}\right] = 2y_0 \qquad (4.28)$$

The second double summation term can be similarly simplified. Note that the generating function translation principle has changed the focus in this term from k = 2 to k = 1. Thus, the only candidate (m, j) combinations are (0, 0), (1, 0), and (1, 1), and of these three terms, only (1, 0) meets the criteria that m + j=1. Thus, the second term reduces to $(y_1 - y_0)\binom{1}{0} = y_1 - y_0$. The solution for y_2 is $y_2 = 2y_0 + y_1 - y_0 = y_1 + y_0$, verifying the solution found directly. Note that, of the six possible combinatoric terms that arise from $0 \leq j \leq m \leq 2$, only two terms meet the constraint that m + j = 2. Thus, the computation only has to be carried through for two of the six terms, substantially simplifying the computing effort. This manner of simplification will occur repeatedly as we use the coefficient collection approach to solve generating functions. Using this same approach, it can be verified that the solution for $y_3 = 2y_1 + y_0$ as

$$\begin{aligned} y_3 &= y_0 \sum_{m=0}^{3}\sum_{j=0}^{m}\binom{3}{j} I_{m+j=3} + (y_1 - y_0)\sum_{m=0}^{2}\sum_{j=0}^{m}\binom{2}{j} I_{m+j=2} \\ &= y_0\left[\binom{2}{1} + \binom{3}{0}\right] + (y_1 - y_0)\left[\binom{1}{1} + \binom{2}{0}\right] \\ &= 3y_0 + 2y_1 - 2y_0 = 2y_1 + y_0 \end{aligned} \tag{4.29}$$

4.4 Second-Order Homogeneous Difference Equation – Example 2

Now consider a second-order equation that was used to help define the order of a difference equation from Chapter 1. The family of difference equations is

$$y_{k+2} = 6y_{k+1} - 3y_k \tag{4.30}$$

for k = 1,2,...,∞: y_0, y_1 are known. The generating function approach will be used to find a general solution. As before,define the generating function as $G(s) = \sum_{k=0}^{\infty} y_k s^k$. Begin by multiplying each side of the equation by s^k.

$$s^k y_{k+2} = 6s^k y_{k+1} - 3s^k y_k \tag{4.31}$$

For $k = 0, 1, 2, \ldots, \infty$. Now consolidate by summing over the index k for $k = 0$ to infinity, the range over which the intrasequence relationships operate.

$$\sum_{k=0}^{\infty} s^k y_{k+2} = 6\sum_{k=0}^{\infty} s^k y_{k+1} - 3\sum_{k=0}^{\infty} s^k y_k \tag{4.32}$$

Take each of these terms separately, moving from left to right, and perform what algebra is required to rewrite these in terms of the generating function G(s). Begin with the first term.

$$\begin{aligned}\sum_{k=0}^{\infty} s^k y_{k+2} &= s^{-2}\sum_{k=0}^{\infty} s^{k+2} y_{k+2} = s^{-2}\left[\sum_{k=0}^{\infty} s^k y_k - s^0 y_0 - s y_1\right] \\ &= s^{-2}\left[G(s) - s^0 y_0 - s y_1\right]\end{aligned} \tag{4.33}$$

This sequence of steps is identical to those taken in the preceding example in this chapter. First, adjust the power of s so that its power aligns with the subscript for y in the summation. Then add in the terms needed to convert the summation to G(s), and subtract these terms to maintain the equality.

Proceed similarly for the next term in equation (4.33).

$$6\sum_{k=0}^{\infty} s^k y_{k+1} = 6s^{-1}\sum_{k=0}^{\infty} s^{k+1} y_{k+1} = 6s^{-1}\left[\sum_{k=0}^{\infty} s^k y_k - s^0 y_0\right] = 6s^{-1}\left[G(s) - y_0\right] \tag{4.34}$$

The last term is $3\sum_{k=0}^{\infty} s^k y_k = 3G(s)$, and the conclusion of the consolidation phase reveals

$$s^{-2}\left[G(s) - s^0 y_0 - s y_1\right] = 6s^{-1}\left(G(s) - y_0\right) - 3G(s) \tag{4.35}$$

Solve for G(s) by rearranging terms:

$$
\begin{aligned}
s^{-2}\left[G(s) - s^0 y_0 - s y_1\right] &= 6s^{-1}\left(G(s) - y_0\right) - 3G(s) \\
G(s) - s^0 y_0 - s y_1 &= 6s\left(G(s) - y_0\right) - 3s^2 G(s) \\
G(s)\left[3s^2 - 6s + 1\right] &= y_0 + s\left(y_1 - 6y_0\right) \\
G(s) &= \frac{y_0 + s\left(y_1 - 6y_0\right)}{3s^2 - 6s + 1}
\end{aligned}
\tag{4.36}
$$

This is all of the preparation required for the inversion process. Using our experience from Chapter 2 for guidance, rewrite G(s)

$$
G(s) = \frac{y_0 + s\left(y_1 - 6y_0\right)}{3s^2 - 6s + 1} = \frac{y_0}{3s^2 - 6s + 1} + \frac{s\left(y_1 - 6y_0\right)}{3s^2 - 6s + 1} \tag{4.37}
$$

and write

$$
\frac{y_0}{3s^2 - 6s + 1} = \frac{y_0}{1 - s\left(6 - 3s\right)} \triangleright_s \left\{y_0\left(6 - 3s\right)^k\right\} \tag{4.38}
$$

To evaluate the kth term in this sequence, use the binomial theorem to convert the power of a sum into the sum of products

$$
\left(6 - 3s\right)^k s^k = \sum_{j=0}^{k} \binom{k}{j} \left(-3s\right)^j 6^{k-j} s^k = \sum_{j=0}^{k} \binom{k}{j} \left(-3\right)^j 6^{k-j} s^{k+j} \tag{4.39}
$$

and define integer j and m such that $0 \le j \le m \le k$ and $m + j = k$, collect the appropriate coefficients $\binom{m}{j}\left(-3\right)^j 6^{m-j}$. Thus,

$$
\begin{aligned}
\frac{y_0}{1 - s\left(6 - 3s\right)} &\triangleright_s \left\{y_0\left(6 - 3s\right)^k\right\} \\
&= \left\{y_0 \sum_{m=0}^{k} \sum_{j=0}^{m} \binom{m}{j} \left(-3\right)^j 6^{m-j} I_{m+j=k}\right\}
\end{aligned}
\tag{4.40}
$$

For the second term of G(s), use the principle of addition and translation of generating functions to see that

$$\frac{s(y_1 - 6y_0)}{3s^2 - 6s + 1} \triangleright \left\{(y_1 - 6y_0)\sum_{m=0}^{k-1}\sum_{j=0}^{m}\binom{m}{j}(-3)^j 6^{m-j} I_{m+j=k-1}\right\} \quad (4.41)$$

and

$$\frac{y_0}{3s^2 - 6s + 1} \triangleright \left\{y_0\sum_{m=0}^{k}\sum_{j=0}^{m}\binom{m}{j}(-3)^j 6^{m-j} I_{m+j=k}\right\} \quad (4.42)$$

Since $G(s) = \frac{y_0}{3s^2 - 6s + 1} + \frac{s(y_1 - 6y_0)}{3s^2 - 6s + 1}$ complete the problem by computing

$$\begin{aligned} y_k &= y_0\sum_{m=0}^{k}\sum_{j=0}^{m}\binom{m}{j}(-3)^j 6^{m-j} I_{m+j=k} \\ &+ (y_1 - 6y_0)\sum_{m=0}^{k-1}\sum_{j=0}^{m}\binom{m}{j}(-3)^j 6^{m-j} I_{m+j=k-1} \end{aligned} \quad (4.43)$$

for $k = 2, 3, 4, \ldots, \infty$

4.4.1 Second-order nonhomogeneous equations and generating functions

Having worked through specific examples of second-order difference equations, we are now ready to provide a general solution for these systems. A family of second-order, nonhomogeneous difference equation can be written as

$$ay_{k+2} + by_{k+1} + cy_k = d \quad (4.44)$$

for $k = 0, 1, 2, 3, \ldots, \infty$: y_0, y_1, y_2 are known constants. Assume that a, b, c, or d, are known, fixed, and not equal to zero. During this development the impact of the nonhomogeneous part of this family of difference equations will be noted. As before, multiply each term in the equation by s^k, and sum over the range $0 \le k \le \infty$.

$$as^{-2}\sum_{k=0}^{\infty}s^{k+2}y_{k+2} + bs^{-1}\sum_{k=0}^{\infty}s^{k+1}y_{k+1} + c\sum_{k=0}^{\infty}s^{k}y_{k} = d\sum_{k=0}^{\infty}s^{k} \qquad (4.45)$$

The first three terms contain no difficulties as we convert to expressions involving G(s). The last term represents the nonhomogeneous component $\frac{d}{1-s}$. Converting each summation in equation (4.45) to a term involving G(s), equation (4.45) becomes

$$\begin{aligned} &as^{-2}\left[G(s) - y_0 - y_1 s - y_2 s^2\right] + bs^{-1}\left[G(s) - y_0 - y_1 s\right] \\ &+ cG(s) = d\sum_{k=0}^{\infty}s^{k} \end{aligned} \qquad (4.46)$$

Rearranging terms reveals

$$\begin{aligned} G(s)\left[cs^2 + bs + a\right] &= ay_0 + s\left(ay_1 + by_0\right) \\ &+ s^2\left(ay_2 + by_1\right) + \frac{ds^2}{1-s} \end{aligned} \qquad (4.47)$$

and the solution for G(s) is

$$\begin{aligned} G(s) &= \frac{ay_0}{cs^2 + bs + a} + \frac{s\left(ay_1 + by_0\right)}{cs^2 + bs + a} \\ &+ \frac{s^2\left(ay_2 + by_1\right)}{cs^2 + bs + a} + \frac{ds^2}{\left(cs^2 + bs + a\right)\left(1-s\right)} \end{aligned} \qquad (4.48)$$

The nonhomogeneous component of this family of difference equations has introduced the rightmost term on the right side of equation (4.48). In order to invert it, invoke the scaling, translation, and product principles of generating function inversion. However, for each term in G(s), we need to invert $\frac{1}{a + bs + cs^2}$. Proceed with the inversion

$$\frac{1}{a+bs+cs^2} = \frac{\frac{1}{a}}{1-(-1)\left(\frac{b}{a}+\frac{c}{a}s\right)s} \tag{4.49}$$

recognizing that the expression on the right side of equation (4.49) is the sum of the series whose k^{th} term is $\frac{1}{a}(-1)^k\left(\frac{b}{a}+\frac{c}{a}s\right)^k s^k$. Using the binomial theorem to reevaluate this expression reveals

$$\begin{aligned}\frac{(-1)^k}{a}\left(\frac{b}{a}+\frac{c}{a}s\right)^k s^k &= \frac{(-1)^k}{a^{k+1}}(b+cs)^k s^k \\ &= \frac{(-1)^k}{a^{k+1}}\sum_{j=0}^{k}\binom{k}{j}c^j b^{k-j} s^{k+j}\end{aligned} \tag{4.50}$$

The next task is to pull together the coefficients of s^k. Introducing the new index variables j and m such that $0 \le j \le m \le k$, observe that we must accumulate the coefficient $\binom{m}{j}c^j b^{m-j}$ whenever $m + j = k$. Writing

$$\frac{1}{cs^2+bs+a} \triangleright \left\{\sum_{m=0}^{k}\frac{(-1)^m}{a^{m+1}}\sum_{j=0}^{m}\binom{m}{j}c^j b^{m-j} I_{m+j=k}\right\} \tag{4.51}$$

we are now in a position to completely invert G(s), with repeated use of the scaling, addition, translation, and multiplication principles of generating functions.

$$G(s) = \frac{ay_0}{cs^2+bs+a} + \frac{s(ay_1+by_0)}{cs^2+bs+a} + \frac{s^2(ay_2+by_1)}{cs^2+bs+a} + \frac{ds^2}{(cs^2+bs+a)(1-s)}$$

$$\triangleright \left\{ \begin{aligned} & ay_0 \sum_{m=0}^{k} \frac{(-1)^m}{a^{m+1}} \sum_{j=0}^{m} \binom{m}{j} c^j b^{m-j} I_{m+j=k} \\ & + (ay_1+by_0) \sum_{m=0}^{k-1} \frac{(-1)^m}{a^{m+1}} \sum_{j=0}^{m} \binom{m}{j} c^j b^{m-j} I_{m+j=k-1} \\ & + (ay_2+by_1) \sum_{m=0}^{k-2} \frac{(-1)^m}{a^{m+1}} \sum_{j=0}^{m} \binom{m}{j} c^j b^{m-j} I_{m+j=k-2} \\ & + d \sum_{h=0}^{k-2} \sum_{m=0}^{h} \frac{(-1)^m}{a^{m+1}} \sum_{j=0}^{m} \binom{m}{j} c^j b^{m-j} I_{m+j=h} \end{aligned} \right\}$$

(4.52)

This is the solution for the second-order nonhomogeneous, difference equation introduced earlier with y_0, y_1, and y_2 known. Some simplification is afforded by noting that the common term in this solution is $\sum_{m=0}^{k} \sum_{j=0}^{m} \binom{m}{j} c^j b^{m-j} I_{m+j=k}$. For k an even integer it is easy to show

$$\sum_{m=0}^{k} \sum_{j=0}^{m} \binom{m}{j} c^j b^{m-j} I_{m+j=k} = b^k \sum_{j=0}^{\frac{k}{2}} \binom{k/2+j}{k/2-j} \left[\frac{c}{b}\right]^j \tag{4.53}$$

and for k an odd integer

$$\sum_{m=0}^{k} \sum_{j=0}^{m} \binom{m}{j} c^j b^{m-j} I_{m+j=k} = b^k \sum_{j=0}^{\left[\frac{k}{2}+1\right]} \binom{\left[k/2+1\right]+j}{\left[k/2+1\right]-j} \left[\frac{c}{b}\right]^j \tag{4.54}$$

where $\left[k/2+1\right]$ is the greatest integer in k/2 +1.

4.5 Example: Unusual Heart Rhythm

We will pause here for a moment to apply our ability to work with difference equations to an unusual problem in cardiology involving heart rhythms. This is a preamble to a more complicated discussion of modeling dangerous heart rhythms that will occur in Chapter 9. Typically, the heart follows a regular rhythm, but occasionally, in normal hearts, that regular rhythm is interrupted by different kind of beat (called an ectopic beat). If the probability of a normal beat is p and the probability of an abnormal beat is q where $p + q = 1$, we need to calculate the probability that the heart rhythm is an alternating one, with each regular beat followed by one irregular (ectopic) beat, which itself is followed by a regular beat.

Further assume that the first heart beat is a regular one. Given this framework, we want to find the probability, A_k that, a sequence of k consecutive heartbeats is composed of all alternating beats, assuming that the first beat is normal. Begin with the observation that $A_0 = 0$ and $A_1 = p$, and $A_2 = pq$. In general,

$$A_{k+2} = pqA_k \tag{4.55}$$

for k = 0 to infinity. This is a second-order, homogeneous difference equation. Define G(s) as

$$G(s) = \sum_{k=0}^{\infty} s^k A_k \tag{4.56}$$

and develop the conversion and consolidation of this second-order, homogeneous difference equation.

$$\begin{aligned}
A_{k+2} &= pqA_k \\
s^k A_{k+2} &= pqs^k A_k \\
\sum_{k=0}^{\infty} s^k A_{k+2} &= pq\sum_{k=0}^{\infty} s^k A_k \\
s^{-2}\sum_{k=0}^{\infty} s^{k+2} A_{k+2} &= pq\sum_{k=0}^{\infty} s^k A_k \\
s^{-2}\left[G(s) - A_0 - sA_1\right] &= pqG(s)
\end{aligned} \tag{4.57}$$

Since $A_0 = 0$ and $A_1 = p$ rewrite the preceding equation as

$$s^{-2}\left[G(s)-ps\right]=pqG(s)$$
$$G(s)-ps=pqG(s)s^2 \tag{4.58}$$

and solving for G(s),

$$G(s)[1-pqs^2]=ps$$
$$G(s)=\frac{ps}{1-pqs^2} \tag{4.59}$$

Begin by seeing that the denominator can quickly be inverted

$$\frac{1}{1-pqs^2}=1+(pqs)s+(pqs)^2s^2+(pqs)^3s^3+\ldots$$
$$=1+pqs^2+(pq)^2s^4+(pq)^3s^6 \tag{4.60}$$
$$\triangleright\left\{(pq)^{\frac{k}{2}}I_{k\bmod 2=0}\right\}$$

Thus,

$$\frac{ps}{1-pqs^2}\triangleright\left\{p(qp)^{\frac{k-1}{2}}I_{(k-1)\bmod 2=0}\right\} \tag{4.61}$$

Thus,

$$A_k=p(qp)^{\frac{k-1}{2}}I_{(k-1)\bmod 2=0} \tag{4.62}$$

Note that, even for large q which is the probability of an ectopic beat, the probability of this rare arrhythmia becomes exceedingly small.

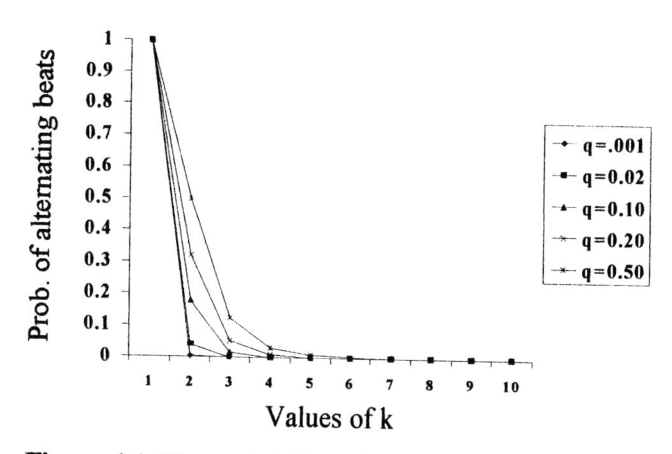

Figure 4.1 The probability of alternating heartbeats increases as a function of q, the probability of an irregular beat.

4.6 Third-Order Difference Equations and Their Generating Functions

Now expand our solutions to consider families of third-order difference equations. Consider the family of difference equations given by

$$4y_{k+3} = 3y_{k+2} - 5y_{k+1} + 7y_k - 12 \tag{4.63}$$

for k = 0, 1, 2, 3, ...,∞: y_0, y_1, y_2 are known constants. As before, proceed sequentially through conversion, consolidation, and inversion. Begin by multiplying each side by s^k to observe that

$$4s^k y_{k+3} = 3s^k y_{k+2} - 5s^k y_{k+1} + 7s^k y_k - 12s^k \tag{4.64}$$

for k = 0, 1, 2, 3, ..., ∞, and proceed to the consolidation step

$$4\sum_{k=0}^{\infty} s^k y_{k+3} = 3\sum_{k=0}^{\infty} s^k y_{k+2} - 5\sum_{k=0}^{\infty} s^k y_{k+1} + 7\sum_{k=0}^{\infty} s^k y_k - 12\sum_{k=0}^{\infty} s^k \quad (4.65)$$

The next step is to extract $G(s) = \sum_{k=0}^{\infty} s^k y_k$ from each term of (4.56). Proceeding from left to right,

$$4\sum_{k=0}^{\infty} s^k y_{k+3} = 4s^{-3}\sum_{k=0}^{\infty} s^{k+3} y_{k+3} = 4s^{-3}\left[\sum_{k=0}^{\infty} s^k y_k - y_0 - sy_1 - s^2 y_2\right]$$
$$= 4s^{-3}\left[G(s) - y_0 - sy_1 - s^2 y_2\right] \quad (4.66)$$

$$3\sum_{k=0}^{\infty} s^k y_{k+2} = 3s^{-2}\left[G(s) - y_0 - sy_1\right] \quad (4.67)$$

$$5\sum_{k=0}^{\infty} s^k y_{k+1} = 5s^{-1}\left[G(s) - y_0\right] \quad (4.68)$$

$$7\sum_{k=0}^{\infty} s^k y_k = 7G(s) \quad (4.69)$$

The last term will not involve G(s), only s

$$12\sum_{k=0}^{\infty} s^k = \frac{12}{1-s} \quad (4.70)$$

As seen for the second-order nonhomogeneous difference equation, this last term reflects the contribution of the nonhomogeneous part of the equation to the generating function solution. At this point, the family of an infinite number of difference equations reduces to one equation involving G(s).

$$4s^{-3}\left[G(s) - y_0 - sy_1 - s^2 y_2\right]$$
$$= 3s^{-2}\left[G(s) - y_0 - sy_1\right] - 5s^{-1}\left[G(s) - y_0\right] + 7G(s) - \frac{12}{1-s} \quad (4.71)$$

Multiplying through by s^3 shows

$$4\left[G(s) - y_0 - sy_1 - s^2y_2\right]$$
$$= 3s\left[G(s) - y_0 - sy_1\right] - 5s^2\left[G(s) - y_0\right] + 7s^3G(s) - \frac{12s^3}{1-s} \tag{4.72}$$

Continuing

$$G(s)\left[-7s^3 + 5s^2 - 3s + 4\right]$$
$$= 4y_0 + s\left(4y_1 - 3y_0\right) + s^2\left(4y_2 - 3y_1 + 5y_0\right) - \frac{12s^3}{1-s} \tag{4.73}$$

Solving for G(s,)

$$\begin{aligned} G(s) = & \frac{4y_0}{-7s^3 + 5s^2 - 3s + 4} + \frac{s\left(4y_1 - 3y_0\right)}{-7s^3 + 5s^2 - 3s + 4} \\ & + \frac{s^2\left(4y_2 - 3y_1 + 5y_0\right)}{-7s^3 + 5s^2 - 3s + 4} - \frac{12s^3}{\left(-7s^3 + 5s^2 - 3s + 4\right)\left(1-s\right)} \end{aligned} \tag{4.74}$$

It is useful to note the contribution of the nonhomogeneous component of the family of difference equations to the solution of G(s). The inversion will involve a translation of the generating function using s^3, and also will require invocation of the product rule.

The key to this inversion process focuses on the denominator, namely

$$\frac{1}{-7s^3 + 5s^2 - 3s + 4} \triangleright ? \tag{4.75}$$

We can expect some additional terms to appear when an attempt is made to invert a function whose denominator contains a third-degree polynomial. The tools of Chapter 2 will serve well here. Begin by writing

$$\frac{1}{-7s^3+5s^2-3s+4} = \frac{1}{4-3s+5s^2-7s^3} = \frac{\frac{1}{4}}{1-\frac{3}{4}s+\frac{5}{4}s^2-\frac{7}{4}s^3} = \frac{\frac{1}{4}}{1-s\left(\frac{3}{4}-\frac{5}{4}s+\frac{7}{4}s^2\right)} \tag{4.76}$$

and the kth term from the series whose sum is $\frac{1}{-7s^3+5s^2-3s+4}$ is $\left[\frac{1}{4}\right]\left(\frac{3}{4}-\frac{5}{4}s+\frac{7}{4}s^2\right)^k s^k$. Invoking the multinomial theorem and some algebra reveals

$$\begin{aligned}
&\left(\tfrac{3}{4}-\tfrac{5}{4}s+\tfrac{7}{4}s^2\right)^k s^k \\
&= \sum_{j=0}^{k}\sum_{h=0}^{k-j}\binom{k}{j\ \ h}\left(\tfrac{7}{4}s^2\right)^j\left(-\tfrac{5}{4}s\right)^h\left(\tfrac{3}{4}\right)^{k-j-h} s^k \\
&= \sum_{j=0}^{k}\sum_{h=0}^{k-j}\binom{k}{j\ \ h}\left(\tfrac{7}{4}\right)^j\left(-\tfrac{5}{4}\right)^h\left(\tfrac{3}{4}\right)^{k-j-h} s^{k+2j+h}
\end{aligned} \tag{4.77}$$

and the collection of the coefficients for each power of s can now begin. Introducing three new index variables, h, j, and m such that $0 \le m \le k$, $0 \le j \le m$, and $0 \le h \le m-j$, note that it is required to collect a coefficient for the kth power of s when $m + 2j + h = k$. Thus

$$\frac{1}{-7s^3+5s^2-3s+4} \triangleright \left\{\left[\frac{1}{4}\right]\sum_{m=0}^{k}\sum_{j=0}^{m}\sum_{h=0}^{m-j}\binom{m}{j\ \ h}\left(\tfrac{7}{4}\right)^j\left(-\tfrac{5}{4}\right)^h\left(\tfrac{3}{4}\right)^{m-j-h} I_{m+2j+h=k}\right\} \tag{4.78}$$

and

$$\frac{4y_0}{-7s^3+5s^2-3s+4} \triangleright \left\{ y_0 \sum_{m=0}^{k}\sum_{j=0}^{m}\sum_{h=0}^{m-j}\binom{m}{j\ h}\left(\tfrac{7}{4}\right)^j\left(-\tfrac{5}{4}\right)^h\left(\tfrac{3}{4}\right)^{m-j-h} I_{m+2j+h=k} \right\} \tag{4.79}$$

The inversion process is now almost completed. Moving through the remaining terms of G(s) inverting term by term using the translation principle of generating functions

$$\frac{s(4y_1-3y_0)}{-7s^3+5s^2-3s+4} \triangleright \left\{ (4y_1-3y_0)\left[\frac{1}{4}\right] \sum_{m=0}^{k-1}\sum_{j=0}^{m}\sum_{h=0}^{m-j}\binom{m}{j\ h}\left(\tfrac{7}{4}\right)^j\left(-\tfrac{5}{4}\right)^h\left(\tfrac{3}{4}\right)^{m-j-h} I_{m+2j+h=k-1} \right\} \tag{4.80}$$

For the next term,

$$\frac{s^2(4y_2-3y_1+5y_0)}{-7s^3+5s^2-3s+4} \triangleright \left\{ (4y_2-3y_1+5y_0)\left[\frac{1}{4}\right] \sum_{m=0}^{k-2}\sum_{j=0}^{m}\sum_{h=0}^{m-j}\binom{m}{j\ h}\left(\tfrac{7}{4}\right)^j\left(-\tfrac{5}{4}\right)^h\left(\tfrac{3}{4}\right)^{m-j-h} I_{m+2j+h=k-2} \right\} \tag{4.81}$$

For the last term, $\frac{12s^3}{(-7s^3+5s^2-3s+4)(1-s)}$, application of first the multiplication principle followed by the translation principle and the sliding tool gives

$$\frac{12s^3}{(-7s^3+5s^2-3s+4)(1-s)} \triangleright \left\{ 3\sum_{L=0}^{k-3}\sum_{m=0}^{L}\sum_{j=0}^{m}\sum_{h=0}^{k-j}\binom{m}{j\ h}\left(\tfrac{7}{4}\right)^j\left(-\tfrac{5}{4}\right)^h\left(\tfrac{3}{4}\right)^{m-j-h} I_{m+2j+h=L} \right\} \tag{4.82}$$

and the solution of y_k of the difference equation $4y_{k+4} = 3y_{k+2} - 5y_{k+1} + 7y_k - 12$ for $k = 0, 1, 2, 3, \ldots, \infty$: y_0, y_1, y_2 are known constants is

$$
\begin{aligned}
y_k = {} & y_0 \sum_{m=0}^{k}\sum_{j=0}^{m}\sum_{h=0}^{m-j}\binom{m}{j\ h}\left(\tfrac{7}{4}\right)^j\left(-\tfrac{5}{4}\right)^h\left(\tfrac{3}{4}\right)^{m-j-h} I_{m+2j+h=k} \\
& + (4y_1 - 3y_0)\left[\frac{1}{4}\right]\sum_{m=0}^{k-1}\sum_{j=0}^{m}\sum_{h=0}^{m-j}\binom{m}{j\ h}\left(\tfrac{7}{4}\right)^j\left(-\tfrac{5}{4}\right)^h\left(\tfrac{3}{4}\right)^{m-1-j-h} I_{m+2j+h=k-1} \\
& + (4y_2 - 3y_1 + 5y_0)\left[\frac{1}{4}\right]\sum_{m=0}^{k-2}\sum_{j=0}^{m}\sum_{h=0}^{m-j}\binom{m}{j\ h}\left(\tfrac{7}{4}\right)^j\left(-\tfrac{5}{4}\right)^h\left(\tfrac{3}{4}\right)^{m-2-j-h} I_{m+2j+h=k-2} \\
& - 3\sum_{L=0}^{k-3}\sum_{m=0}^{L}\sum_{j=0}^{m}\sum_{h=0}^{m-j}\binom{m}{j\ h}\left(\tfrac{7}{4}\right)^j\left(-\tfrac{5}{4}\right)^h\left(\tfrac{3}{4}\right)^{m-j-h} I_{m+2j+h=L}
\end{aligned}
\tag{4.83}
$$

This is the complete solution for y_k. However, the solution itself is complicated, and its calculation appears to be a daunting task. The hidden simplification in this solution is the indicator variable. For example, the restriction $m + 2j + h = k$ profoundly reduces the number of terms that must be considered in these computations. For example, consider the case for $k = 5$. In this circumstance, there are 34 combinations of (m, j, h) such that $0 \le j \le m \le 5$ and $0 \le h \le m - j$. However, the only (m, j, h) combinations one need consider are (2, 1, 1), (3, 0, 2), (3, 1, 0), and (4, 0, 1). For $k = 7$, there are many more possible combinations of (m, j, h) such that $0 \le j \le m \le 7$ and $0 \le h \le m - j$. However, there are only four relevant contributions ([3, 1, 2], [3, 2, 0], [4, 0, 3], and [4, 1, 1]). In carrying out the computations involved in equation (4.83) one should first identify the combinations of (m, j, h) that satisfy the indicator variables. This leads to a profound simplification in the calculation of difference equation solutions that are expressed in terms of the coefficient collection method.

As an additional example of the ability to solve difference equations that may be difficult to solve when other methods are used, consider the family of third-order, nonhomogeneous difference equations

$$4y_k - 6y_{k+1} - 13y_{k+2} - 15y_{k+3} = 9 \tag{4.84}$$

for k equal zero to infinity and y_0, y_1, and y_2 as known constants. Proceeding as before, multiplying each term in equation (4.84) by s^k.

$$-15y_{k+3} - 13y_{k+2} - 6y_{k+1} + 4y_k = 9$$
$$-15s^k y_{k+3} - 13s^k y_{k+2} - 6s^k y_{k+1} + 4s^k y_k = 9s^k \quad (4.85)$$

and summing this equation from k = 0 to ∞,

$$-15\sum_{k=0}^{\infty} s^k y_{k+3} - 13\sum_{k=0}^{\infty} s^k y_{k+2} - 6\sum_{k=0}^{\infty} s^k y_{k+1} + 4\sum_{k=0}^{\infty} s^k y_k = 9\sum_{k=0}^{\infty} s^k \quad (4.86)$$

Proceeding with the definition of the generating function developed in Chapter 2, $G(s) = \sum_{k=0}^{\infty} s^k y_k$ and our consolidation progresses.

$$-15\sum_{k=0}^{\infty} s^k y_{k+3} = -15s^{-3}\left[G(s) - y_0 - y_1 s - y_2 s^2\right]$$
$$-13\sum_{k=0}^{\infty} s^k y_{k+2} = -13s^{-2}\left[G(s) - y_0 - y_1 s\right]$$
$$-6\sum_{k=0}^{\infty} s^k y_{k+1} = -6s^{-1}\left[G(s) - y_0\right] \quad (4.87)$$
$$4\sum_{k=0}^{\infty} s^k y_k = 4G(s)$$
$$= 9\sum_{k=0}^{\infty} s^k = \frac{9}{1-s}$$

Now rewrite equation (4.86) as

$$-15s^{-3}\left[G(s) - y_0 - y_1 s - y_2 s^2\right] - 13s^{-2}\left[G(s) - y_0 - y_1 s\right]$$
$$-6s^{-1}\left[G(s) - y_0\right] + 4G(s) = \frac{9}{1-s} \quad (4.88)$$

With some algebraic manipulation

$$G(s)\left[4s^3 - 6s^2 - 13s - 15\right]$$
$$= -\left[15y_1 + 13y_0\right]s - \left[15y_2 + 13y_1 + 6y_0\right]s^2 + \frac{9s^3}{1-s} - 15y_0 \quad (4.89)$$

and proceed to solve for G(s)

$$G(s) = \frac{-[15y_1 + 13y_0]s}{4s^3 - 6s^2 - 13s - 15} - \frac{[15y_2 + 13y_1 + 6y_0]s^2}{4s^3 - 6s^2 - 13s - 15} + \frac{9s^3}{[1-s][4s^3 - 6s^2 - 13s - 15]} - \frac{15y_0}{4s^3 - 6s^2 - 13s - 15} \tag{4.90}$$

As we plan for the inversion of G(s) the key to this procedure will be the cubic polynomial $4s^3 - 6s^2 - 13s - 15$. From our experience with the inversion of these generating functions, it is known that using the method of coefficient collection will lead to a solution, but one that is complicated. The solution can be simplified by observing that one root of $4s^3 - 6s^2 - 13s - 15$ is s = 3. This leads to the factorization $4s^3 - 6s^2 - 13s - 15 = (s-3)(4s^2 + 6s + 5)$. Since this cubic polynomial will yield no more real root, begin the coefficient collection for the quadratic $4s^2 + 6s + 5$. Begin with

$$\frac{1}{4s^2 + 6s + 5} = \frac{1/5}{1 - \left[-\frac{6}{5} - \frac{4}{5}s\right]s} \triangleright_s \left\{\frac{1}{5}\left[\frac{-2}{5}\right]^k (3+2s)^k\right\} \tag{4.91}$$

Now apply the binomial theorem to the term $(3+2s)^k$ to find

$$(3+2s)^k = \sum_{j=0}^{k} \binom{k}{j} 2^j 3^{k-j} s^j \tag{4.92}$$

and $(3+2s)^k s^k = \sum_{j=0}^{k} \binom{k}{j} 2^j 3^{k-j} s^{k+j}$. What is now required is to collect coefficients of s^k from this latter term. By defining the two integers m and j such that $0 \le j \le m \le k$, note that we will need to collect the coefficient $\binom{m}{j} 2^j 3^{m-j}$ whenever $m + j = k$. Now proceed with the inversion by writing

$$\frac{1}{4s^2+6s+5} \triangleright \left\{ \frac{1}{5}\sum_{m=0}^{k}\left[\frac{-2}{5}\right]^m \sum_{j=0}^{m}\binom{m}{j}2^j 3^{m-j} I_{m+j=k} \right\} \tag{4.93}$$

Using the convolution principle to see that

$$\frac{1}{4s^3-6s^2-13s-15} = \frac{1}{(s-3)(4s^2+6s+5)}$$
$$\triangleright \left\{ \frac{1}{5}\sum_{h=0}^{k}(-1)\left(\frac{1}{3}\right)^{h+1} \sum_{m=0}^{k-h}\left[\frac{-2}{5}\right]^m \sum_{j=0}^{m}\binom{m}{j}2^j 3^{m-j} I_{m+j=k-h} \right\} \tag{4.94}$$

Apply this term to each of the component parts of equation (4.90)

$$\frac{-[15y_1+13y_0]s}{4s^3-6s^2-13s-15}$$
$$\triangleright \left\{ -[15y_1+13y_0]\frac{1}{5}\sum_{h=0}^{k-1}(-1)\left(\frac{1}{3}\right)^{h+1} \sum_{m=0}^{k-h-1}\left[\frac{-2}{5}\right]^m \sum_{j=0}^{m}\binom{m}{j}2^j 3^{m-j} I_{m+j=k-h-1} \right\} \tag{4.95}$$

and

$$-\frac{[15y_2+13y_1+6y_0]s^2}{4s^3-6s^2-13s-15}$$
$$\triangleright \left\{ \frac{-[15y_2+13y_1+6y_0]}{5}\sum_{h=0}^{k-2}(-1)\left(\frac{1}{3}\right)^{h+1} \sum_{m=0}^{k-2-h}\left[\frac{-2}{5}\right]^m \sum_{j=0}^{m}\binom{m}{j}2^j 3^{m-j} I_{m+j=k-h-2} \right\}$$
$$\tag{4.96}$$

These two computations are two components that are required to complete the solution of the family of difference equations as given in equation (4.84) whose generating function is described by equation (4.89). Further evaluation of the generating function requires the recognition that

$$\frac{9s^3}{[1-s][4s^3-6s^2-13s-15]} \triangleright \left\{\frac{9}{5}\sum_{L=0}^{k-3}\sum_{h=0}^{L}(-1)\left(\frac{1}{3}\right)^{h+1}\sum_{m=0}^{L-h}\left[\frac{-2}{5}\right]^m\sum_{j=0}^{m}\binom{m}{j}2^j3^{m-j}I_{m+j=L-h}\right\} \tag{4.97}$$

This last expression requiring a simple convolution. The final term is a simple use of the scaling tool.

$$-\frac{15y_0}{4s^3-6s^2-13s-15} \triangleright \left\{-3y_0\sum_{h=0}^{k}(-1)\left(\frac{1}{3}\right)^{h+1}\sum_{m=0}^{k-h}\left[\frac{-2}{5}\right]^m\sum_{j=0}^{m}\binom{m}{j}2^j3^{m-j}I_{m+j=k-h}\right\} \tag{4.98}$$

Now using the summation principle to invert the generating function from equation (4.90) to see that the solution to $4y_k - 6y_{k+1} - 13y_{k+2} - 15y_{k+3} = 9$ is

$$\begin{aligned} y_k = &-[15y_1+13y_0]\frac{1}{5}\sum_{h=0}^{k-1}(-1)\left(\frac{1}{3}\right)^{h+1}\sum_{m=0}^{k-h-1}\left[\frac{-2}{5}\right]^m\sum_{j=0}^{m}\binom{m}{j}2^j3^{m-j}I_{m+j=k-h-1} \\ &\frac{-[15y_2+13y_1+6y_0]}{5}\sum_{h=0}^{k-2}(-1)\left(\frac{1}{3}\right)^{h+1}\sum_{m=0}^{k-h-2}\left[\frac{-2}{5}\right]^m\sum_{j=0}^{m}\binom{m}{j}2^j3^{m-j}I_{m+j=k-h-2} \\ &+\frac{9}{5}\sum_{L=0}^{k-3}\sum_{h=0}^{L}(-1)\left(\frac{1}{3}\right)^{h+1}\sum_{m=0}^{L-h}\left[\frac{-2}{5}\right]^m\sum_{j=0}^{m}\binom{m}{j}2^j3^{m-j}I_{m+j=L-h} \\ &-3y_0\sum_{h=0}^{k}(-1)\left(\frac{1}{3}\right)^{h+1}\sum_{m=0}^{k-h}\left[\frac{-2}{5}\right]^m\sum_{j=0}^{m}\binom{m}{j}2^j3^{m-j}I_{m+j=k-h} \end{aligned} \tag{4.99}$$

4.7 Factoring Generating Functions

Although the tools covered in Chapter 2 are very useful in solving generating functions, the solutions identified are sometimes difficult to decipher easily. The third-order nonhomogeneous equation we just solved does have a solution that is in terms of indices, and the boundary conditions y_0, y_1, and y_2, but the overall expression is somewhat complicated, even though its component pieces are easy to compute. There are some higher order difference equations that are somewhat easier to solve. For example, consider the family of difference equations

$$-7y_{k+2} + 27y_{k+1} + 4y_k = 0 \tag{4.100}$$

for k = 0, 1, 2, 3, ... ,∞; y_0, y_1, y_2 are known constants.

4.7.1 Coefficient collection: the binomial theorem approach

We proceed as before to reduce this infinite number of equations into one equation involving the generating function $G(s) = \sum_{k=0}^{\infty} s^k y_k$, following our procedures of conversion and consolidation as follows:

$$\begin{aligned}
-7s^k y_{k+2} + 27s^k y_{k+1} + 4s^k y_k &= 0 \\
-7\sum_{k=0}^{\infty} s^k y_{k+2} + 27\sum_{k=0}^{\infty} s^k y_{k+1} + 4\sum_{k=0}^{\infty} s^k y_k &= 0 \\
-7s^{-2}\left[G(s) - y_0 - sy_1\right] + 27s^{-1}\left[G(s) - y_0\right] + 4G(s) &= 0 \\
-7\left[G(s) - y_0 - sy_1\right] + 27s\left[G(s) - y_0\right] + 4s^2 G(s) &= 0
\end{aligned} \tag{4.101}$$

Further simplification reveals

$$G(s)\left[4s^2 + 27s - 7\right] = -7y_0 + s\left(27y_0 - 7y_1\right) \tag{4.102}$$

And solving for G(s), find

$$\begin{aligned}
G(s) &= \frac{-7y_0 + s\left(27y_0 - 7y_1\right)}{4s^2 + 27s - 7} \\
&= \frac{-7y_0}{4s^2 + 27s - 7} + \frac{s\left(27y_0 - 7y_1\right)}{4s^2 + 27s - 7}
\end{aligned} \tag{4.103}$$

and observe that for the inversion to proceed, we must identify the coefficients of the infinite series whose sum is $\frac{1}{4s^2+27s-7}$. Of course one approach is to proceed as we have before, noting that

$$\frac{1}{4s^2+27s-7} = \frac{-1/7}{1-s\left(27/7+4/7\,s\right)} \cdot \tag{4.104}$$

The kth term from this series whose sum this represents is $\frac{-1}{7}\left(\frac{27}{7}+\frac{4}{7}s\right)^k s^k$.

Using the binomial theorem, write this as

$$\frac{-1}{7}\left(\frac{27}{7}+\frac{4}{7}s\right)^k s^k = \frac{-1}{7}\sum_{j=0}^{k}\binom{k}{j}\left(\frac{4}{7}s\right)^j\left(\frac{27}{7}\right)^{k-j} s^k = \frac{-1}{7^{k+1}}\sum_{j=0}^{k}\binom{k}{j}4^j 27^{k-j} s^{k+j} \tag{4.105}$$

and, gathering the coefficients for each power of s, we see that by introducing another index m, we accrue a coefficient when $0 \le m \le k$, $0 \le j \le m$, $k+j=m$. Thus,

$$\frac{1}{4s^2+27s-7} \triangleright \left\{\sum_{m=0}^{k}\frac{-1}{7^{m+1}}\sum_{j=0}^{m}\binom{m}{j}4^j 27^{m-j} I_{m+j=k}\right\} \tag{4.106}$$

4.7.2 Simplification afforded by factoring

A simpler approach comes from the recognition that $4s^2 + 27s - 7 = (4s-1)(7+s)$. Thus, we can also write

$$\begin{aligned}\frac{1}{4s^2+27s-7} &= \frac{1}{(4s-1)(7+s)} = \left(\frac{1}{4s-1}\right)\left(\frac{1}{7+s}\right) \\ &= \left(\frac{-1}{1-4s}\right)\left(\frac{1/7}{1-\left(-1/7\right)s}\right)\end{aligned} \tag{4.107}$$

and proceed with the inversion invoking not the binomial theorem, but the simple product rule.

$$\frac{1}{4s^2+27s-7} = \left(\frac{-1}{1-4s}\right)\left(\frac{\frac{1}{7}}{1-\left(-\frac{1}{7}\right)s}\right) \tag{4.108}$$
$$\triangleright \left\{\frac{1}{7}\sum_{j=0}^{k}(-1)4^j\left(-\frac{1}{7}\right)^{k-j}\right\}$$

Now invert G(s) quite simply finding the solution y_k for $k = 2, 3, 4, \ldots, \infty$ as

$$G(s) = \frac{-7y_0 + s(27y_0 - 7y_1)}{4s^2+27s-7}$$
$$\triangleright \left\{-y_0\sum_{j=0}^{k}(-1)4^j\left(-\frac{1}{7}\right)^{k-j} + \frac{27y_0-7y_1}{7}\sum_{j=0}^{k-1}(-1)4^j\left(-\frac{1}{7}\right)^{k-1-j}\right\} \tag{4.109}$$

This is a much more direct, comprehensible solution than when we approach the inversion without any attempt to factor the denominator G(s).

4.7.3 Partial factorization

As we have seen, the solution of difference equations (homogeneous and nonhomogeneous) involves the ability to manipulate polynomials. The more roots that can be identified of the polynomial, the crisper the solution. In some circumstances, just being able to find one of the roots of the denominator of G(s) can lead to a major simplification of the inversion process. As an example of this utility of partial factorization, consider, for example, the family of fourth-order, nonhomogeneous difference equations given by

$$270y_{k+4} - 129y_{k+3} - 95y_{k+2} + 7y_{k+1} + 3y_k = 11 \tag{4.110}$$

for $k = 0, 1, 2, 3, \ldots, \infty$: y_0, y_1, y_2, y_3 known constants.

4.7.4 The multinomial approach

We begin to apply the generating function approach by multiplying this fourth-order, nonhomogeneous equation by s^k to find that

$$270s^k y_{k+4} - 129s^k y_{k+3} - 95s^k y_{k+2} + 7s^k y_{k+1} + 3s^k y_k = 11s^k \quad (4.111)$$

Defining $G(s) = \sum_{k=0}^{\infty} s^k y_k$, apply the sum to each of the terms in the above equation to find

$$270\sum_{k=0}^{\infty} s^k y_{k+4} - 129\sum_{k=0}^{\infty} s^k y_{k+3} - 95\sum_{k=0}^{\infty} s^k y_{k+2} + 7\sum_{k=0}^{\infty} s^k y_{k+1} + 3\sum_{k=0}^{\infty} s^k y_k = 11\sum_{k=0}^{\infty} s^k \quad (4.112)$$

and follow the standard procedure of consolidation for solving for $G(s)$ to show that

$$\begin{aligned} G(s)\left[3s^4 + 7s^3 - 95s^2 - 129s + 270\right] &= 270y_0 \\ &+ s\left[270y_1 - 129y_0\right] + s^2\left[270y_2 - 129y_1 - 95y_0\right] \\ &+ s^3\left[270y_3 - 129y_2 - 95y_1 + 7y_0\right] + \frac{11}{1-s} \end{aligned} \quad (4.113)$$

or $G(s) = A(s) + B(s)$ where

$$\begin{aligned} A(s) &= \frac{270y_0}{3s^4 + 7s^3 - 95s^2 - 129s + 270} \\ &+ \frac{s\left[270y_1 - 129y_0\right]}{3s^4 + 7s^3 - 95s^2 - 129s + 270} \end{aligned} \quad (4.114)$$

and

$$B(s) = \frac{s^2[270y_2 - 129y_1 - 95y_0]}{3s^4 + 7s^3 - 95s^2 - 129s + 270} + \frac{s^3[270y_3 - 129y_2 - 95y_1 + 7y_0]}{3s^4 + 7s^3 - 95s^2 - 129s + 270} + \frac{11}{(3s^4 + 7s^3 - 95s^2 - 129s + 270)(1-s)} \quad (4.115)$$

The next task is the inversion of $\frac{1}{3s^4 + 7s^3 - 95s^2 - 129s + 270}$. Once this is accomplished, G(s) can be inverted through the sequential application of the addition, translation, and product principles from Chapter 2 as follows:

$$\frac{1}{3s^4 + 7s^3 - 95s^2 - 129s + 270} = \frac{1/270}{1-\left(129/270 + 95/270\, s - 7/270\, s^2 - 3/270\, s^3\right)s} \quad (4.116)$$

that, when inverted, has as a kth term

$\left(129/270 + 95/270\, s - 7/270\, s^2 - 3/270\, s^3\right)^k s^k$. As before, applying the

multinomial theorem to write this as

$$\left(129/270 + 95/270\, s - 7/270\, s^2 - 3/270\, s^3\right)^k s^k$$

$$= \sum_{i=0}^{k}\sum_{j=0}^{k-i}\sum_{h=0}^{k-i-j}\binom{k}{i\ j\ h}\left(-3/270\right)^i\left(-7/270\right)^j\left(95/270\right)^h\left(129/270\right)^{k-i-j-h} s^{k+3i+2j+h}$$

(4.117)

The job now remains to identify how these coefficients occur. This can be accomplished by first defining indices m, i, j, h when $0 \le m \le k$, $0 \le i \le m$, $0 \le j$

$\leq m - i$, and $0 \leq h \leq m - j - i$. Collecting the coefficient from equation (4.117) when $m + 3i + 2j + h = k$ gives

$$\frac{1}{3s^4 + 7s^3 - 95s^2 - 129s + 270}$$

$$\triangleright \left\{ \sum_{m=0}^{k} \sum_{i=0}^{m} \sum_{j=0}^{m-i} \sum_{h=0}^{m-i-j} \binom{m}{i\ j\ h} \left(-\tfrac{3}{270}\right)^i \left(-\tfrac{7}{270}\right)^j \left(\tfrac{95}{270}\right)^h \left(\tfrac{129}{270}\right)^{m-i-j-h} I_{m+3i+2j+h=k} \right\}$$

$$\triangleright \quad \{w_k\} \tag{4.118}$$

denoting this collection of coefficients w_k, then, through the application of the sliding, convolution, and addition principles of generating functions find

$$\begin{aligned} G(s) = {} & \frac{270y_0}{3s^4 + 7s^3 - 95s^2 - 129s + 270} + \frac{s[270y_1 - 129y_0]}{3s^4 + 7s^3 - 95s^2 - 129s + 270} \\ & + \frac{s^2[270y_2 - 129y_1 - 95y_0]}{3s^4 + 7s^3 - 95s^2 - 129s + 270} + \frac{s^3[270y_3 - 129y_2 - 95y_1 + 7y_0]}{3s^4 + 7s^3 - 95s^2 - 129s + 270} \\ & + \frac{11}{(3s^4 + 7s^3 - 95s^2 - 129s + 270)(1-s)} \end{aligned}$$

$$\triangleright \left\{ \begin{aligned} & 270y_0 w_k + [270y_1 - 129y_0]w_{k-1} + [270y_2 - 129y_1 - 95y_0]w_{k-2} \\ & + [270y_3 - 129y_2 - 95y_1 + 7y_0]w_{k-3} + 11\sum_{j=0}^{k} w_j \end{aligned} \right\} \tag{4.119}$$

where

$$w_k = \sum_{m=0}^{k} \sum_{i=0}^{m} \sum_{j=0}^{m-i} \sum_{h=0}^{m-i-j} \binom{m}{i\ j\ h} \left(-\tfrac{3}{270}\right)^i \left(-\tfrac{7}{270}\right)^j \left(\tfrac{95}{270}\right)^h \left(\tfrac{129}{270}\right)^{m-i-j-h} I_{m+3i+2j+h=k} \tag{4.120}$$

4.7.5 The use of partial factorization

This solution to (4.111) equaion using the method of coefficient collection is very complicated. It is eased if we find that two of the roots of $3s^4 - 7s^3 - 95s^2 - 129s + 270 = 0$ are s = 5 and s = -6. Applying some algebra reveals that this expression can be factored as follows

$$3s^4 + 7s^3 - 95s^2 - 129s + 270 = (s-5)(s+6)(3s^2+4s-9) \qquad (4.121)$$

and use the product principle for the inversion. $\frac{1}{s-5} \triangleright \frac{-1}{5^{k+1}}$, and $\frac{1}{s+6} \triangleright \frac{(-1)^k}{6^{k+1}}$.

Thus,

$$\left(\frac{1}{s-5}\right)\left(\frac{1}{s+6}\right) \triangleright \left\{\sum_{j=0}^{k}\left(\frac{-1}{5^{j+1}}\right)\left(\frac{(-1)^{k-j}}{6^{k-j+1}}\right)\right\} \qquad (4.122)$$

Only the final term in the product needs interpretation. Proceeding easily

$$\frac{1}{3s^2+4s-9} = \frac{-1/9}{1-\left[4/9 + 1/3\,s\right]s} \qquad \text{note that}$$

$\frac{1}{3s^2+4s-9}$ has as a kth term $\left(-1/9\right)\left[4/9 + 1/3\,s\right]^k s^k$. Using the binomial theorem, write this term as

$$\left(-1/9\right)\left[4/9 + 1/3\,s\right]^k s^k = \left(-1/9\right)\sum_{j=0}^{k}\binom{k}{j}\left(\frac{1}{3}s\right)^j\left(\frac{4}{9}\right)^{k-j} s^k$$

$$= \sum_{j=0}^{k}\binom{k}{j}\left(\frac{-1}{9}\right)\left(\frac{1}{3}\right)^j\left(\frac{4}{9}\right)^{k-j} s^{k+j} \qquad (4.123)$$

and it can be seen at once that these coefficients can be harvested when $0 \le m \le k$, $0 \le j \le m$ and $m + j = k$. Thus,

$$\frac{1}{3s^2+4s-9} \triangleright \left\{\sum_{m=0}^{k}\sum_{j=0}^{m}\binom{m}{j}\left(\frac{-1}{9}\right)\left(\frac{1}{3}\right)^j\left(\frac{4}{9}\right)^{m-j} I_{m+j=k}\right\} \qquad (4.124)$$

therefore write

$$\frac{1}{3s^4+7s^3-95s^2-129s+270} = \left(\frac{1}{s-5}\right)\left(\frac{1}{s+6}\right)\left(\frac{1}{3s^2+4s-9}\right)$$

$$\triangleright \left\{\sum_{i=0}^{k}\left[\sum_{h=0}^{i}\left(\frac{-1}{5^{h+1}}\right)\left(\frac{-1^{i-h}}{6^{i-h+1}}\right)\right]\left[\sum_{m=0}^{k-i}\sum_{j=0}^{m}\binom{m}{j}\left(\frac{-1}{9}\right)\left(\frac{1}{3}\right)^{j}\left(\frac{4}{9}\right)^{m-j} I_{m+j=k-i}\right]\right\} \quad (4.125)$$

$$\triangleright \{\alpha_k\}$$

This is a much simpler solution than equation (4.120), whose development was facilitated by the identification of two of the four roots of the quartic equation. If α_k is substituted for w_k in equation (4.119) , the solution is much simpler.

4.8 Expanding the Types of Generating Functions

One of the advantages of the application of generating functions is in avoiding the complications of other approaches to their inversion. As an example, consider the family of difference equations $y_{k+2} - y_k = 0$ for k= 0, 1, 2, 3, …, ∞: y_0 and y_1 are known, constants. Proceeding as before

$$\begin{aligned} s^k y_{k+2} &= s^k y_k \\ \sum_{k=0}^{\infty} s^k y_{k+2} &= \sum_{k=0}^{\infty} s^k y_k \\ s^{-2}\left[G(s) - y_0 - sy_1\right] &= G(s) \\ G(s)\left[1-s^2\right] &= y_0 + sy_1 \end{aligned} \quad (4.126)$$

and

$$G(s) = \frac{y_0}{1-s^2} + \frac{sy_1}{1-s^2} \quad (4.127)$$

There are two convenient ways to invert $\frac{1}{1-s^2}$. The first is to proceed by using the product principle of generating functions.

$$\frac{1}{1-s^2} = \left(\frac{1}{1-s}\right)\left(\frac{1}{1+s}\right) = \left(\frac{1}{1-s}\right)\left(\frac{1}{1-(-1)s}\right) \triangleright \{I_{k\bmod 2}\} \tag{4.128}$$

which is either 0 or 1 depending on whether k is odd or even. A more direct approach involves simply observing that $\frac{1}{1-s^2} = \frac{1}{1-(s)s}$ which has terms of s^{2k}. This sequence has terms only for the even powers of k. Thus the inversion of $\frac{1}{1-s^2}$ results in a sequence which has coefficients equal to zero for the odd powers of k and one for the even powers of k. We can denote this using modulus, as introduced in Chapter 2.* Applying modulus 2 to the solution here. The sequence whose sum is $\frac{1}{1-s^2}$ has coefficients which are 1 if $k \bmod 2$ is 0, otherwise, the coefficient is zero. Therefore rewriting using the indicator function.

$$\frac{1}{1-s^2} \triangleright \{I_{k\bmod 2=0}\} \tag{4.129}$$

and

$$G(s) = \frac{y_0}{1-s^2} + \frac{sy_1}{1-s^2} \triangleright \{y_0 I_{k\bmod 2=0} + y_1 I_{k\bmod 2=1}\} \tag{4.130}$$

For this simple family of difference equations, $y_{k+2} - y_k = 0$ for k= 0, 1, 2, 3, ..., ∞: y_0 and y_1 are known, it is easy to verify that this is the correct solution.

---

* The mod M function reduces a number x to another number r between 0 and M - 1 where x mod M is the remainder of x after division by M. Thus 7 mod 4 = 3 since the remainder of 7/4 is 3. Here we are concerned with modulus 2. For any number x, x mod 2 can be only 0 if x is even, or 1 if x is odd.

4.9 Generating Function Derivatives and Difference Equations

A sequence of examples will be provided here which demonstrate the use of generating function derivatives in the solution of difference equations. This will further expand ability to solve difference equations. In particular, recognition of generating function derivatives will permit the solution of some difference equations whose coefficients are not constants, but functions of k.

4.9.1 Example 1: Derivatives in inverting generating functions of difference equations

Another example in this genre is the family of second-order, nonhomogeneous difference equations given by

$$2y_{k+2} + 2y_{k+1} + y_k = k \tag{4.131}$$

for $k = 0, 1, 2, 3, \ldots, \infty$ and y_0, y_1 are known constants. Up until this point in the discussions, the right side of the difference equation has been a constant. Examine the complications introduced by this more general term and proceed through the generating function argument, first multiplying by k

$$2s^k y_{k+2} + 2s^k y_{k+1} + s^k y_k = ks^k \tag{4.132}$$

then summing from k equal zero to infinity,

$$2s^{-2}\left[G(s) - y_0 - sy_1\right] + 2s^{-1}\left[G(s) - y_0\right] + G(s) = \sum_{k=0}^{\infty} ks^k \tag{4.133}$$

The term on the right side of equation (4.133) is the new complication. In order to proceed, we must pause to remember from Chapter 2 that $\frac{s}{(1-s)^2} \triangleright \{k\}$, so

$$\sum_{k=0}^{\infty} ks^k = \frac{s}{(1-s)^2}.$$

With this relationship the solution for G(s) proceeds

$$G(s)\left[s^2 + 2s + 2\right] = 2y_0 + s\left[2(y_1 + y_0)\right] + \frac{s^3}{(1-s)^2} \tag{4.134}$$

revealing

$$G(s) = \frac{2y_0}{s^2+2s+2} + \frac{s\left[2\left(y_1+y_0\right)\right]}{s^2+2s+2} + \frac{s^3}{\left(s^2+2s+2\right)\left(1-s\right)^2} \quad (4.135)$$

and the job is reduced to inversion of G(s), with effort concentrated on s^2+2s+2. Previous algorithms used to solve generating functions involved themselves in the analysis of complex numbers [1]. Such a method requires identification of the roots of this quadratic equation using the quadratic formula.

$$\frac{-b \pm \sqrt{b^2-4ac}}{2a} = \frac{-2 \pm \sqrt{4-8}}{2} = -1 \pm i \quad (4.136)$$

Each root is complex. This complication does not arise using the method of coefficient collection. Write without difficulty

$\dfrac{1}{s^2+2s+2} = \dfrac{1/2}{1-\left(-1-\frac{1}{2}s\right)s}$ and see that the kth term of the infinite series that must be evaluated is $(-1)^k\left(1+\frac{1}{2}s\right)^k s^k$. Evaluation of this expression presents no problem using the binomial theorem.

$$(-1)^k\left(1+\tfrac{1}{2}s\right)^k s^k = (-1)^k \sum_{j=0}^{k} \binom{k}{j}\left(\frac{1}{2}\right)^j s^{k+j} \quad (4.137)$$

Collect the coefficients for s^k and observe that a coefficient will be required when for a new index m, $0 \le m \le k$, $0 \le j \le m$, and $m+j=k$. Thus,

$$\frac{1}{s^2+2s+2} \triangleright \left\{\frac{1}{2}\sum_{m=0}^{k}(-1)^m \sum_{j=0}^{m}\binom{m}{j}\left(\frac{1}{2}\right)^j I_{m+j=k}\right\} \quad (4.138)$$

The inversion of the first two terms on the right side of equation (4.135) may now proceed with only the additional use of the scaling tool and the sliding tool. However, before the inversion can be completed, more attention is needed on the last term in G(s). Writing

$$\frac{s^3}{\left(s^2+2s+2\right)\left(1-s\right)^2} = \left(\frac{1}{s^2+2s+2}\right)\left(\frac{s^3}{\left(1-s\right)^2}\right) \tag{4.139}$$

it can be seen at once the ease with which the convolution principle handles this situation

$$\begin{aligned} \frac{s^3}{\left(s^2+2s+2\right)\left(1-s\right)^2} &= s^2\left(\frac{1}{s^2+2s+2}\right)\left(\frac{s}{\left(1-s\right)^2}\right) \\ \triangleright &\left\{\sum_{i=0}^{k-2}\left[\frac{1}{2}\sum_{m=0}^{i}(-1)^m\sum_{j=0}^{m}\binom{m}{j}\left(\frac{1}{2}\right)^j I_{m+j=i}\right][k-i-2]\right\} \end{aligned} \tag{4.140}$$

Thus

$$\begin{aligned} G(s) &= \frac{2y_0}{s^2+2s+2}+\frac{s\left[2\left(y_1+y_0\right)\right]}{s^2+2s+2}+\frac{s^3}{\left(s^2+2s+2\right)\left(1-s\right)^2} \\ \triangleright &\left\{\begin{aligned} &2y_0\frac{1}{2}\sum_{m=0}^{k}(-1)^m\sum_{j=0}^{m}\binom{m}{j}\left(\frac{1}{2}\right)^j I_{m+j=k} \\ &+\left[2\left(y_1+y_0\right)\right]\frac{1}{2}\sum_{m=0}^{k-1}(-1)^m\sum_{j=0}^{m}\binom{m}{j}\left(\frac{1}{2}\right)^j I_{m+j=k-1} \\ &+\sum_{i=0}^{k-2}\left[\frac{1}{2}\sum_{m=0}^{i}(-1)^m\sum_{j=0}^{m}\binom{m}{j}\left(\frac{1}{2}\right)^j I_{m+j=i}\right][k-i-2] \end{aligned}\right\} \end{aligned} \tag{4.141}$$

The factorization approach would have provided a solution, but in terms of complex numbers. The method of coefficient collection permits a solution in terms of finite sums of real numbers.

4.9.2 Additional example of derivative use

As an additional example of this approach, consider the family of difference equations given by

$$y_{k+3} + 3y_{k+2} = (k+2)(k+1) \tag{4.142}$$

for k = 0, 1, 2, 3, ..., ∞; y_0, y_1, y_2 are known constants. As is the custom, multiply throughout by s^k, and condense:

$$\begin{aligned} y_{k+3} + 3y_{k+2} &= (k+2)(k+1) \\ s^k y_{k+3} + 3s^k y_{k+2} &= s^k(k+2)(k+1) \end{aligned} \tag{4.143}$$

Then take the sum of each of the equations in the previous expression from k = 0 to ∞ to find

$$\sum_{k=0}^{\infty} s^k y_{k+3} + 3\sum_{k=0}^{\infty} s^k y_{k+2} = \sum_{k=0}^{\infty} s^k (k+2)(k+1) \tag{4.144}$$

Pause here to define $G(s) = \sum_{k=0}^{\infty} s^k y_k$ and substitute this expression into each term of equation (4.144)

$$s^{-3}\left[G(s) - y_0 - sy_1 - s^2 y_2\right] + 3s^{-2}\left[G(s) - y_0 - sy_1\right] = \sum_{k=0}^{\infty} s^k (k+2)(k+1) \tag{4.145}$$

Pause here to evaluate $\sum_{k=0}^{\infty} s^k (k+2)(k+1)$. Recall from Chapter 2 that if the coefficient of s^k is a product involving k, then the sum of the series is likely to be related to be a derivative of $\frac{1}{1-s}$. Also from Chapter 2, recall that $\frac{2}{(1-s)^3} \triangleright \{(k+2)(k+1)\}$. With this information, proceed with the condensation

$$s^{-3}\left[G(s)-y_0-sy_1-s^2y_2\right] + 3s^{-2}\left[G(s)-y_0-sy_1\right]=\frac{2}{(1-s)^3}$$

$$G(s)[1+3s] = y_0+s(y_1+3y_0)+s^2(y_2+3y_1)+\frac{2s^3}{(1-s)^3} \tag{4.146}$$

$$G(s) = \frac{y_0}{1+3s}+\frac{s(y_1+3y_0)}{1+3s}+\frac{s^2(y_2+3y_1)}{1+3s}+\frac{2s^3}{(1+3s)(1-s)^3}$$

The only term that will present a new challenge is the last term $\frac{2s^3}{(1+3s)(1-s)^3}$, which will be evaluated using the product principle of generating functions, i.e.,

$$\frac{2s^3}{(1+3s)(1-s)^3}=\left(\frac{1}{1+3s}\right)\left(\frac{2s^3}{(1-s)^3}\right) \triangleright \left\{\sum_{j=0}^{k-3}(j+2)(j+1)(-3)^{k-3-j}\right\} \tag{4.147}$$

The required solution therefore is:

$$G(s) = \frac{y_0}{1+3s}+\frac{s(y_1+3y_0)}{1+3s}+\frac{s^2(y_2+3y_1)}{1+3s}+\frac{2s^3}{(1+3s)(1-s)^3}$$

$$\triangleright \left\{\begin{matrix} y_0(-3)^k+(y_1+3y_0)(-3)^{k-1} \\ +(y_2+3y_1)(-3)^{k-2}+\sum_{j=0}^{k-3}(j+2)(j+1)(-3)^{k-3-j}\end{matrix}\right\} \tag{4.148}$$

which can be simplified to

$$\triangleright \left\{(-3)^{k-2}y_2+\sum_{j=0}^{k-3}(j+2)(j+1)(-3)^{k-3-j}\right\} \tag{4.149}$$

Finally, consider the following difference equation

$$6y_{k+5} + 2y_k = 7(k+1)^3 \tag{4.150}$$

for k = 0, 1, 2, 3, ..., ∞ and y_0, y_1, y_2, y_3, and y_4 are known constants. This is a nonhomogeneous, fifth-order difference equation, however, several intermediate values of y are missing. This observation will translate into some simplification in the condensation process, and a few of these terms will involve G(s). Proceed by multiplying each term in equation (4.150) and summing over all values of k from zero to infinity.

$$\begin{aligned} &6y_{k+5} + 2y_k = 7(k+1)^3 \\ &6s^k y_{k+5} + 2s^k y_k = 7s^k(k+1)^3 \\ &6\sum_{k=0}^{\infty} s^k y_{k+5} + 2\sum_{k=0}^{\infty} s^k y_k = 7\sum_{k=0}^{\infty} s^k(k+1)^3 \end{aligned} \tag{4.151}$$

The first two terms may be recognized as being functions of G(s). The last term suggests that its sum will be a function of a derivative. In fact, from Chapter 2, remember that

$$\sum_{k=0}^{\infty} s^k(k+1)^3 = \frac{s^2+4s+1}{(1-s)^4} \tag{4.152}$$

Proceed with the consolidation.

$$\begin{aligned} 6\sum_{k=0}^{\infty} s^k y_{k+5} + 2\sum_{k=0}^{\infty} s^k y_k &= 7\frac{s^2+4s+1}{(1-s)^4} \\ 6s^{-5}\left[G(s) - y_0 - sy_1 - s^2y_2 - s^3y_3 - s^4y_4\right] + 2G(s) &= 7\frac{s^2+4s+1}{(1-s)^4} \\ \left[G(s) - y_0 - sy_1 - s^2y_2 - s^3y_3 - s^4y_4\right] + \frac{1}{3}s^5G(s) &= \frac{7}{6}s^5\left[\frac{s^2+4s+1}{(1-s)^4}\right] \end{aligned} \tag{4.153}$$

Solving for G(s) reveals

$$G(s) + \frac{1}{3}s^5G(s)$$

$$= y_0 + sy_1 + s^2y_2 + s^3y_3 + s^4y_4 + \frac{7}{6}s^5\left[\frac{s^2+4s+1}{(1-s)^4}\right] \tag{4.154}$$

and

$$G(s) = \frac{y_0 + sy_1 + s^2y_2 + s^3y_3 + s^4y_4}{1+\frac{1}{3}s^5} + \frac{7}{6}s^5\left[\frac{1}{1+\frac{1}{3}s^5}\right]\left[\frac{s^2+4s+1}{(1-s)^4}\right] \tag{4.155}$$

The term on which the inversion focuses is $\frac{1}{1+\frac{1}{3}s^5}$. The first summand involves the translation and summation principles. Also, notice the last summand in equation (4.151). By factoring it, we are portraying our intension to approach the inversion by using the product principle, which will be straightforward after inversion of $\frac{1}{1+\frac{1}{3}s^5}$. In fact, the term $\frac{s^2+4s+1}{(1-s)^4}$ which came from the term $(k+1)^3$ will be inverted back to $(k+1)^3$, finding itself embedded in either the application of product principle of generating functions, or the fourth derivative of the geometric series inversion. Begin with

$$\frac{1}{1+\frac{1}{3}s^5} = \frac{1}{1-\left(-\frac{1}{3}s^4\right)s} = 1 + \left(-\frac{1}{3}s^4\right)s + \left(-\frac{1}{3}s^4\right)^2 s^2 + \left(-\frac{1}{3}s^4\right)^3 s^3 + \ldots +$$

$$= 1 + \left(-\frac{1}{3}\right)s^5 + \left(-\frac{1}{3}\right)^2 s^{10} + \left(-\frac{1}{3}\right)^3 s^{15} + \ldots +$$

$$\triangleright \left\{\left(-\frac{1}{3}\right)^{\frac{k}{5}} I_{k \bmod 5=0}\right\} \tag{4.156}$$

We are now prepared to complete the inversion.

$$G(s) = \frac{y_0 + sy_1 + s^2y_2 + s^3y_3 + s^4y_4}{1+\frac{1}{3}s^5} + \frac{7}{6}s^5\left[\frac{1}{1+\frac{1}{3}s^5}\right]\left[\frac{s^2+4s+1}{(1-s)^4}\right]$$

$$\triangleright \left\{ \begin{array}{l} y_0\left(-\frac{1}{3}\right)^{\frac{k}{5}} I_{k \bmod 5=0} + y_1\left(-\frac{1}{3}\right)^{\frac{k-1}{5}} I_{(k-1) \bmod 5=0} \\ +y_2\left(-\frac{1}{3}\right)^{\frac{k-2}{5}} I_{(k-2) \bmod 5=0} + y_3\left(-\frac{1}{3}\right)^{\frac{k-3}{5}} I_{(k-3) \bmod 5=0} \\ +y_4\left(-\frac{1}{3}\right)^{\frac{k-4}{5}} I_{(k-4) \bmod 5=0} \\ +\frac{7}{6}\sum_{j=0}^{k-5}\left[\left(-\frac{1}{3}\right)^{\frac{j}{5}} I_{j \bmod 5=0}\right][k-4-j]^3 \end{array} \right\} \quad (4.157)$$

4.10 Monthly Payments Revisited

Consider the mortgage example of Chapter 1. Recall that we were working through a debt repayment, in which the clinic owners arranged for a business loan of \$500,000 for equipment and operations costs to be paid back in 30 years for the construction of a screening clinic for women at risk of breast and cervical cancer. The clinic has arranged for an annual interest rate of r percent, and we had to find the monthly annual payments for the next thirty years. The notation was that m_k, the total monthly due, had two components. The first is the interest payment for that month, i_k. The second component is the principle that is due for the kth month*, p_k. Then $m_k = i_k + p_k$. for k = 1, 2, 3, ..., 360.

It was shown that

$$p_{k+1} = \frac{1200+r}{1200} p_k \quad (4.158)$$

* For simplicity, we will ignore premium mortgage insurance, property insurance, and taxes, that would increase the monthly payment even more.

for k = 1,2,...360. So, the sequence of 360 principle payments has embedded in them a relationship reflected by a difference equation. Recognize that this equation is a first-order, homogeneous equation with constant coefficients and is similar to the equation $y_{k+1} = ay_k$, for which the solution is known. Applying the solution $(y_k = a^k y_1)$ here, find that

$$p_k = \left(\frac{1200+r}{1200}\right)^{k-1} p_1 \tag{4.159}$$

Now p_1 can be determined. Note that the sum of the principle payments over the 360 payments must be the total amount borrowed. Thus

$$\begin{aligned}
\$500{,}000 &= \sum_{k=1}^{360} p_k = \sum_{k=1}^{360}\left(\frac{1200+r}{1200}\right)^{k-1} p_1 = p_1\sum_{k=1}^{360}\left(\frac{1200+r}{1200}\right)^{k-1} = \\
&= p_1\sum_{k=0}^{359}\left(\frac{1200+r}{1200}\right)^{k} = p_1\frac{1-\left(\frac{1200+r}{1200}\right)^{360}}{1-\left(\frac{1200+r}{1200}\right)} = \\
&= p_1\frac{1200\left[\left(\frac{1200+r}{1200}\right)^{360}-1\right]}{r}
\end{aligned} \tag{4.160}$$

and

$$p_1 = \frac{500{,}000r}{1200\left[1-\left(\frac{1200+r}{1200}\right)^{360}\right]} \tag{4.161}$$

So, with $i_1 = 500{,}000\left(\frac{r}{1200}\right)$ one can compute the monthly payment $m_1 = i_1 + p_1$. If r = 12%, the first month's interest will be \$5000, the first month's principle will be \$143, and the monthly payments for the clinic will be \$5143. If the interest rate is 5%, the interest payment for the first month will be \$2083, p_1 = \$597, and the monthly payment will be reduced from \$5143 to \$2680.

4.11 The Random Walk Problem

One of the most informative exercises in understanding stochastic processes is the random walk problem. Stated simply, consider the one dimensional movement of a particle. Its movement is determined by a random process. The process moves forward one unit with probability p and backward with probability q, where $p + q = 1$. The task is to compute the probability that the particle is in location k. Formulating this model in terms of a difference equation, let y_k be the probability that the particle is in location k. Then it can only be in location k if one of two events occurs. The first is that the particle was in location $k - 1$ and with probability p it moved to location k. Similarly, it could have been in location $k + 1$ and moved to location k with probability q. Assume y_0 is known and $y_k = 0$ for $k < 0$. Write this as

$$y_k = py_{k-1} + qy_{k+1} \tag{4.162}$$

for $k = 0$ to ∞. The solution of this equation is straightforward at this point in the development. Begin with

$$\begin{aligned}
y_{k+1} &= py_{k-1} + qy_{k+1} \\
s^k y_{k+1} &= ps^k y_{k-1} + qs^k y_{k+1} \\
\sum_{k=0}^{\infty} s^k y_{k+1} &= p\sum_{k=0}^{\infty} s^k y_{k-1} + q\sum_{k=0}^{\infty} s^k y_{k+1}
\end{aligned} \tag{4.163}$$

and writing $G(s) = \sum_{k=0}^{\infty} s^k y_k$

$$\begin{aligned}
\sum_{k=0}^{\infty} s^k y_k &= p\sum_{k=0}^{\infty} s^k y_{k-1} + q\sum_{k=0}^{\infty} s^k y_{k+1} \\
G(s) &= ps\sum_{k=0}^{\infty} s^{k-1} y_{k-1} + qs^{-1}\sum_{k=0}^{\infty} s^{k+1} y_{k+1} \\
G(s) &= psG(s) + qs^{-1}\left[G(s) - y_0\right] \\
sG(s) &= ps^2 G(s) + q\left[G(s) - y_0\right] \\
G(s)[-q + s - ps^2] &= -qy_0
\end{aligned} \tag{4.164}$$

or

$$G(s) = \frac{qy_0}{q - s + ps^2} \tag{4.165}$$

Inverting this by the method of coefficient collection, write

$$G(s) = \frac{qy_0}{q - s + ps^2} = \frac{y_0}{1 - \frac{1}{q}s + \frac{p}{q}s^2} = \frac{y_0}{1 - \frac{1}{q}(1+ps)s} \tag{4.166}$$

$$\triangleright_s \left\{ y_0 \left(\frac{1}{q}\right)^k (1+ps)^k \right\}$$

Using the binomial theorem to expand $(1+ps)^k$

$$(1+ps)^k = \sum_{j=0}^{k} \binom{k}{j} p^j s^j \tag{4.167}$$

and write

$$\left\{ y_0 \left(\frac{1}{q}\right)^k (1+ps)^k s^k \right\} = y_0 \left(\frac{1}{q}\right)^k \sum_{j=0}^{k} \binom{k}{j} p^j s^{k+j} \tag{4.168}$$

Collecting coefficients we have

$$G(s) = \frac{qy_0}{q - s + ps^2} \triangleright y_0 \sum_{m=0}^{k} \left(\frac{1}{q}\right)^m \sum_{j=0}^{m} \binom{m}{j} p^j I_{m+j=k} \tag{4.169}$$

4.12 Clinic Visits

Consider several physicians who are interested in opening an urgent care clinic in a community. The philosophy of an urgent care clinic is to see patients with health problems that are not emergencies (and therefore do not require visits to the emergency room), but require immediate care. The nature of these injuries include lacerations, minor eye injuries (e.g., easily removable foreign bodies and mild corneal abrasions), traumatic injuries, fractures that do not require surgery, and sprains. In addition, these clinics will see patients with essential hypertension, diabetes mellitus, arthritis, and other conditions that can be managed on an outpatient basis. As an introduction, consider the following simple model.

The clinic physicians would like to predict the growth in their patient census, or the number of patients they will be seeing each day. The clinic sees essentially two types of patients: The first type of patients are those seen for the first time on a particular day. The second type of patients are those who were seen earlier, but have returned for a scheduled revisit. These return visits are primarily for follow-up examinations of injuries seen previously. (For example, lower back strains, eye injuries, and lacerations must be seen the next day). Lower back injuries will sometimes be seen on two consecutive days. Fractures can be seen the day following the injuries. Sprains will be seen in two days. Patients who have bad upper respiratory tract infections may be seen in two days. Patients who have started a new medication may be seen in two days.

Assume that 50% of patients seen on a day return for a clinic visit the next day, 30% of the patients return for a visit two days later, and 20% of the patients return for a visit three days later. Let the number of new patients seen on a given day be denoted by d. The clinic physicians understand that by increasing the new patients, their clinic will be seeing growth. However, many of these new patients will be seen on several days, and the impact of this carryover effect is uncertain.

Let y_k be the number of patient visits seen on the kth day. Representing this scenario as a difference equation

$$y_{k+3} = 0.50y_{k+2} + 0.30y_{k+1} + 0.20y_k + d \tag{4.170}$$

k=0 to ∞; y_0, y_1, y_2 are known constants. The goal is to solve this equation in terms of y_0, y_1, y_2, and d. Begin as before:

$$\begin{aligned}
&y_{k+3} = 0.50y_{k+2} + 0.30y_{k+1} + 0.20y_k + d \\
&G(s)\left[1 - 0.50s - 0.30s^2 - 0.20s^3\right] \\
&= y_0 + s\left[y_1 - 0.50y_0\right] + s^2\left[y_2 - 0.50y_1 - 0.30y_0\right] + \frac{ds^3}{1-s} \qquad (4.171)\\
&G(s) = \frac{y_0 + s\left[y_1 - 0.50y_0\right] + s^2\left[y_2 - 0.50y_1 - 0.30y_0\right]}{1 - 0.50s - 0.30s^2 - 0.20s^3} \\
&\quad + \frac{ds^3}{(1-s)\left(1 - 0.50s - 0.30s^2 - 0.20s^3\right)}
\end{aligned}$$

The inversion hinges on the evaluation of the denominator

$$1 - 0.50s - 0.30s^2 - 0.20s^3 \tag{4.172}$$

We are aided in the evaluation by the observation that s=1 is a root of this cubic polynomial. Use this observation to factor the cubic and proceed with the inversion.

$$1-0.50s-0.30s^2-0.20s^3=(s-1)\left(-0.20s^2-0.50s-1\right)$$
$$\frac{1}{1-0.50s-0.30s^2-0.20s^3}=\left[\frac{1}{s-1}\right]\left[\frac{1}{-0.20s^2-0.50s-1}\right] \tag{4.173}$$

The inversion of this last factor may be conducted as follows:

$$\begin{aligned}\frac{1}{-0.20s^2-0.50s-1}&=\frac{-1}{0.20s^2+0.50s+1}\\&=\frac{-1}{1-(0.50+0.20s)(-1)s}\\&\triangleright_s\left\{-\sum_{j=0}^{k}\binom{k}{j}(0.20)^j(0.50)^{k-j}s^j\right\}\\&\triangleright\left\{\sum_{m=0}^{k}(-1)^{m+1}\sum_{j=0}^{m}\binom{m}{j}(0.20)^j(0.50)^{m-j}I_{m+j=k}\right\}\end{aligned} \tag{4.174}$$

Thus,

$$\begin{aligned}&\frac{1}{1-0.50s-0.30s^2-0.20s^3}=\left[\frac{1}{s-1}\right]\left[\frac{1}{-0.20s^2-0.50s-1}\right]\\&\triangleright\left\{\sum_{i=0}^{k}\sum_{m=0}^{i}(-1)^{m+1}\sum_{j=0}^{m}\binom{m}{j}(0.20)^j(0.50)^{m-j}I_{m+j=i}\right\}\end{aligned} \tag{4.175}$$

The inversion can now be completed as

$$G(s) = \frac{y_0 + s[y_1 - 0.50y_0] + s^2[y_2 - 0.50y_1 - 0.30y_0]}{1 - 0.50s - 0.30s^2 - 0.20s^3} + \frac{ds^3}{(1-s)(1 - 0.50s - 0.30s^2 - 0.20s^3)}$$

$$\triangleright \left\{ \begin{array}{l} y_0 \sum_{i=0}^{k} \sum_{m=0}^{i} (-1)^{m+1} \sum_{j=0}^{m} \binom{m}{j} (0.20)^j (0.50)^{m-j} I_{m+j=i} \\ + [y_1 - 0.50y_0] \sum_{i=0}^{k-1} \sum_{m=0}^{i} (-1)^{m+1} \sum_{j=0}^{m} \binom{m}{j} (0.20)^j (0.50)^{m-j} I_{m+j=i} \\ + [y_2 - 0.50y_1 - 0.30y_0] \sum_{i=0}^{k-2} \sum_{m=0}^{i} (-1)^{m+1} \sum_{j=0}^{m} \binom{m}{j} (0.20)^j (0.50)^{m-j} I_{m+j=i} \\ + d \sum_{h=0}^{k-3} \sum_{i=0}^{h} \sum_{m=0}^{i} (-1)^{m+1} \sum_{j=0}^{m} \binom{m}{j} (0.20)^j (0.50)^{m-j} I_{m+j=i} \end{array} \right\} \tag{4.176}$$

We are interested in examining the behavior of this solution as a function of d. As the beginning of the existence of the clinic, it is not unreasonable to assume that the initial daily clinic census is low, since it takes time for the news about the availability of this new clinic to spread. Assume $y_0 = 1$, $y_1 = 2$, $y_2 = 3$, meaning one patient on the first day, two patients visit on the second day, and 3 patients visit on the third day. The finite sums here are evaluated fairly easily, leading to an examination of the relationship between daily patient clinic visits and the number of days the clinic has been open. The difference equations provide the anticipated result, i.e. as d (the number of new daily clinic visits) increased from four to ten, the daily clinic visits increased (Figure 4.2).

This model, however, may be used to examine an additional complexity. This description of the number of daily clinic visits is a function not just of the new visits that day, but of patients who have been seen on preceding days. If for example, the day under consideration is Thursday, the model assumes that 50% of Wednesday's patients will be seen, as well as 30% of Tuesday's patients, and 20% of Monday's patients. We can examine the degree to which these assumptions changes the clinic census. This is a valuable examination, since the ability to see patients evaluated on previous days helps to encourage relationships with the community members and builds good will. We can adjust this mix by changing the initial parameter values from 0.50, 0.30, and 0.20. However, the solution to the difference equation requires that the sum of these parameter estimates be one, since they represent the probabilities of previous visits for patients seen on a given day. Changing the parameter estimates to

0.70, 0.20, and 0.10, increases the likelihood that patients will be seen from the prior day, but decreases the likelihood that they will be seen from two days and three days ago.

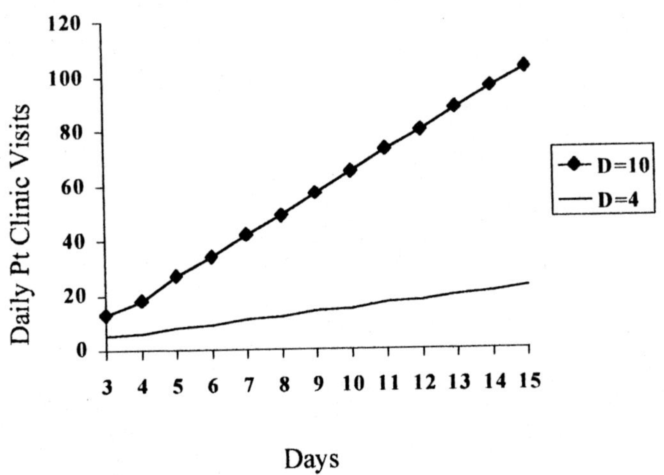

Figure 4.2 The daily clinic census is a linear function of the number of new patients seen each day at the clinic.

Problems

Solve these first-order nonhomogeneous difference equations:

1. $92y_{k+1} + -25y_k = -60$
2. $-17y_{k+1} + 64y_k = 88$
3. $4y_{k+1} + -22y_k = 48$
4. $-65y_{k+1} + 11y_k = -23$
5. $-100y_{k+1} + 6y_k = -73$
6. $68y_{k+1} + 45y_k = -63$
7. $39y_{k+1} + 27y_k = 24$
8. $92y_{k+1} + -75y_k = -31$
9. $55y_{k+1} + -86y_k = 44$
10. $-4y_{k+1} + -28y_k = -65$
11. $89y_{k+1} + -24y_k = 87$
12. $-63y_{k+1} + -63y_k = 68$
13. $-60y_{k+1} + 71y_k = 31$

14. $-61y_{k+1} + -55y_k = -71$
15. $-14y_{k+1} + 81y_k = -70$
16. $36y_{k+1} + 3y_k = 56$
17. $-26y_{k+1} + 29y_k = 71$
18. $28y_{k+1} + -92y_k = 24$
19. $67y_{k+1} + -76y_k = -93$
20. $-48y_{k+1} + -52y_k = 74$
21. $-27y_{k+1} + -28y_k = 31$
22. $39y_{k+1} + -45y_k = 81$
23. $-30y_{k+1} + -70y_k = -10$
24. $-51y_{k+1} + 88y_k = 46$
25. $-84y_{k+1} + -49y_k = -72$

Solve each of the second-order difference equations, assuming y_0, y_1, and y_2 are known constants:

26. $72y_{k+2} + -93y_{k+1} -72y_k = -51$
27. $-50y_{k+2} + -72y_{k+1} -15y_k = 34$
28. $9y_{k+2} + -76y_{k+1}\ 58y_k = 38$
29. $-13y_{k+2} + -26y_{k+1}\ 22y_k = 9$
30. $-27y_{k+2} + -94y_{k+1} -15y_k = -9$
31. $60y_{k+2} + 77y_{k+1} -54y_k = 23$
32. $44y_{k+2} + -76y_{k+1} -22y_k = 94$
33. $78y_{k+2} + -79y_{k+1}\ 80y_k = 69$
34. $-70y_{k+2} + -20y_{k+1} -15y_k = 76$
35. $82y_{k+2} + -78y_{k+1} -49y_k = -8$
36. $-93y_{k+2} + -21y_{k+1} -76y_k = -29$
37. $-92y_{k+2} + 67y_{k+1}\ 6y_k = -2$
38. $53y_{k+2} + -46y_{k+1} -25y_k = 24$
39. $-83y_{k+2} + -72y_{k+1}\ 42y_k = -60$
40. $9y_{k+2} + 66y_{k+1}\ 74y_k = 23$
41. $-66y_{k+2} + 33y_{k+1} -98y_k = -87$
42. $45y_{k+2} + -19y_{k+1} -85y_k = -95$
43. $25y_{k+2} + 77y_{k+1} -23y_k = -92$
44. $66y_{k+2} + -19y_{k+1} -69y_k = 8$
45. $-15y_{k+2} + -53y_{k+1} -83y_k = -87$
46. $574y_{k+2} + -83y_{k+1}\ 9y_k = -55$
47. $-97y_{k+2} + 26y_{k+1}\ 57y_k = 92$
48. $-61y_{k+2} + 66y_{k+1} -26y_k = -53$
49. $-56y_{k+2} + -39y_{k+1}\ 33y_k = -84$
50. $-42y_{k+2} + 75y_{k+1}\ 67y_k = 38$

Solve each of the third-order difference equations, assuming y_0, y_1, y_2, and y_3 are known constants:

51. $-87y_{k+3} + 56y_{k+2} + 42y_{k+1} + 9y_k = 40$
52. $-87y_{k+3} + -38y_{k+2} + 58y_{k+1} + -75y_k = -12$
53. $94y_{k+3} + 18y_{k+2} + 27y_{k+1} + 38y_k = -21$
54. $22y_{k+3} + -76y_{k+2} + -41y_{k+1} + -96y_k = -49$
55. $44.54y_{k+3} + 21y_{k+2} + -97y_{k+1} + -89y_k = 65$
56. $91y_{k+3} + -11y_{k+2} + 59y_{k+1} + 52y_k = -53$
57. $-70y_{k+3} + 69y_{k+2} + -28y_{k+1} + -27y_k = 88$
58. $-12y_{k+3} + -49y_{k+2} + -98y_{k+1} + 42y_k = 24$
59. $-67y_{k+3} + 81y_{k+2} + -1y_{k+1} + 62y_k = 13$
60. $63y_{k+3} + 14y_{k+2} + -9y_{k+1} + 40y_k = 72$
61. $-20y_{k+3} + 38y_{k+2} + 21y_{k+1} + -31y_k = -73$
62. $36y_{k+3} + -14y_{k+2} + -44y_{k+1} + -76y_k = 4$
63. $35y_{k+3} + 5y_{k+2} + -99y_{k+1} + 21y_k = -21$
64. $-99y_{k+3} + 43y_{k+2} + 4y_{k+1} + -64y_k = -35$
65. $72y_{k+3} + 27y_{k+2} + -94y_{k+1} + -92y_k = 16$
66. $-99y_{k+3} + 95y_{k+2} + -97y_{k+1} + 75y_k = -100$
67. $-18y_{k+3} + -25y_{k+2} + 25y_{k+1} + 67y_k = -51$
68. $-66y_{k+3} + -74y_{k+2} + 77y_{k+1} + 66y_k = 90$
69. $85y_{k+3} + 57y_{k+2} + 68y_{k+1} + -11y_k = -38$
70. $-88y_{k+3} + -30y_{k+2} + -69y_{k+1} + 5y_k = 16$
71. $27y_{k+3} + 63y_{k+2} + 65y_{k+1} + -6y_k = 64$
72. $85y_{k+3} + 6y_{k+2} + -78y_{k+1} + 15y_k = 17$
73. $64y_{k+3} + 93y_{k+2} + -37y_{k+1} + -87y_k = -32$
74. $87y_{k+3} + 55y_{k+2} + 70y_{k+1} + -94y_k = 32$
75. $67y_{k+3} + -25y_{k+2} + -96y_{k+1} + 99y_k = 5$

Solve each of the fourth-order difference equations, assuming y_0, y_1, y_2, y_3, and y_4 are known constants:

76. $-58y_{k+4} + 52y_{k+3} + 28y_{k+2} + -22y_{k+1} + 11y_k = -66$
77. $16y_{k+4} + -39y_{k+3} + -24y_{k+2} + 76y_{k+1} + -69y_k = 59$
78. $50y_{k+4} + 42y_{k+3} + 51y_{k+2} + 40y_{k+1} + 54y_k = -51$
79. $56y_{k+4} + -64y_{k+3} + 29y_{k+2} + 15y_{k+1} + 8y_k = 93$
80. $-73y_{k+4} + -14y_{k+3} + 21y_{k+2} + 43y_{k+1} + 3y_k = -4$
81. $-7y_{k+4} + 17y_{k+3} + -92y_{k+2} + 89y_{k+1} + -27y_k = 75$
82. $51y_{k+4} + 90y_{k+3} + -43y_{k+2} + 74y_{k+1} + -41y_k = 20$
83. $19y_{k+4} + 68y_{k+3} + -8y_{k+2} + 23y_{k+1} + 15y_k = 10$
84. $-11y_{k+4} + -86y_{k+3} + -24y_{k+2} + -82y_{k+1} + 97y_k = 52$
85. $-1y_{k+4} + -60y_{k+3} + -66y_{k+2} + -19y_{k+1} + -31y_k = 9$
86. $23y_{k+4} + -42y_{k+3} + -52y_{k+2} + -37y_{k+1} + 82y_k = 32$

87. $-22y_{k+4} + -94y_{k+3} + 51y_{k+2} + 43y_{k+1} + 90y_k = 22$
88. $-69y_{k+4} + 62y_{k+3} + 97y_{k+2} + -63y_{k+1} + 35y_k = 43$
89. $52y_{k+4} + -92y_{k+3} + 55y_{k+2} + 41y_{k+1} + -88y_k = -16$
90. $70y_{k+4} + -82y_{k+3} + -83y_{k+2} + -92y_{k+1} + -27y_k = -52$
91. $-19y_{k+4} + 78y_{k+3} + 9y_{k+2} + -12y_{k+1} + -31y_k = 4$
92. $-71y_{k+4} + 72y_{k+3} + -73y_{k+2} + 30y_{k+1} + -93y_k = 63$
93. $28y_{k+4} + 23y_{k+3} + -63y_{k+2} + -23y_{k+1} + 47y_k = -47$
94. $2y_{k+4} + -45y_{k+3} + -2y_{k+2} + 15y_{k+1} + 65y_k = 5$
95. $10y_{k+4} + 47y_{k+3} + 69y_{k+2} + 37y_{k+1} + 16y_k = -20$
96. $-63y_{k+4} + 67y_{k+3} + 43y_{k+2} + 77y_{k+1} + 5y_k = 80$
97. $39y_{k+4} + -6y_{k+3} + -85y_{k+2} + 57y_{k+1} + -46y_k = -30$
98. $18y_{k+4} + -87y_{k+3} + 16y_{k+2} + -10y_{k+1} + 71y_k = -46$
99. $-20y_{k+4} + 76y_{k+3} + -19y_{k+2} + -67y_{k+1} + 8y_k = -56$
100. $91y_{k+4} + -15y_{k+3} + 60y_{k+2} + 4y_{k+1} + -56y_k = 27$

101. Re-examine the solution to the clinic visit problem described in the last section of this chapter. Re-derive the solution after removing the constraint that s=1 is a root of the third-order, nonhomogeneous difference equation.
102. Prove that if

$$G(s) = \frac{1}{as^2 + bs + c} = \left[\frac{1}{As + B}\right]\left[\frac{1}{CS + D}\right] \tag{4.177}$$

then

$$G(s) \triangleright \left\{\frac{1}{BD}\left(\frac{-C}{D}\right)^k \sum_{j=0}^{k}\left(\frac{AD}{BC}\right)^j\right\} \tag{4.178}$$

5

Difference Equations: Variable Coefficients

5.1 Difficulty with the Generating Function Approach

As we have seen through the choice of the equation's order and homogeneity status, difference equations can be formulated to represent the dependence between the elements of a sequence. This flexibility in the collection of difference equation families can be further enriched by including equations whose coefficients are not constant, but are instead variable. Unfortunately, this generalization may lead to systems of equations for which there are no known general solutions. For example, the work of Chapter 4 reveals that the family of difference equations

$$y_{k+2} = 3y_{k+1} - 6y_k + k \tag{5.1}$$

is easily solved. However, the general family of difference equations

$$p_n(k)y_{k+n} + p_{n-1}(k)y_{k+n-1} + p_{n-2}(k)y_{k+n-2} + \cdots + p_0(k)y_k = R(k) \qquad (5.2)$$

in general cannot be solved when the functions $p_0(k)$, $p_1(k)$, $p_2(k)\ldots p_3(k)$ are functions of k. In the types of equations we have solved in Chapter 4, such as the family of third-order equations given by

$$270y_{k+4} - 129y_{k+3} - 95y_{k+2} - 129y_{k+1} + 3y_k = 11 \qquad (5.3)$$

the coefficients of sequence elements y_k to y_{k+4} are constants with respect to k. This allows a straightforward (although at times complicated) conversion of the infinite sequence of equations represented by equation (5.3) to one equation involving G(s) where $G(s) = \sum_{k=0}^{\infty} s^k y_k$. Having pointed this out, however, there are several interesting families of difference equations with variable coefficient that are of interest and can be explored with the mathematical tools developed so far in this text.

5.2 Bootstrapping the Generating Function

Having pointed out the difficulty with the generating function approach to the general solution of difference equations with nonconstant coefficients, there are some specific equations with nonconstant coefficients that have a solution. For example, consider, the family of difference equations described by

$$y_{k+1} = (-1)^k y_k \qquad (5.4)$$

for $k = 0,1,2,3,\ldots,\infty$; y_0 a known constant. Note the coefficient $(-1)^k$ is a function of k, and, as pointed out in the previous section, this will add complications to the general solution of the difference equation. Equation (5.4) is easy to solve iteratively as follows: $y_1 = y_0$. Then $y_2 = -y_0$, and $y_3 = -y_0$. The periodicity repeats as $y_4 = y_0$ and $y_5 = y_0$ and so on.. Note that the solution is a multiple of y_0 changing with every other value of k. We can write this solution succinctly through the use of indicator functions. A review of the sequence of the values of $y_1, y_2, y_3, \ldots$ reveals that the pattern is one of alteration of the sign preceding y_0 on every third entry.

The solution may be written succinctly using modulus, where the sign preceding y_0 is a function of k. Let $r\left(\frac{x}{y}\right)$ be the integer remainder or the quotient $\frac{x}{y}$. Then the value of y_k is a function of $r\left(\frac{k}{4}\right)$. If $r\left(\frac{k}{4}\right)$ is equal to 0 or 1, then $y_k = y_0$. If $r\left(\frac{k}{4}\right)$ is equal to 2 or 3, then $y_k = -y_0$. However, also note that k mod 4 = $r\left(\frac{k}{4}\right)$. Further define $I_{x\in A}$ to be the indicator function, equal to one when xεA, and equal to zero otherwise Then write the solution to equation (5.4) as

$$y_k = [I_{k=0 \bmod 4 \text{ or } k=1 \bmod 4} - I_{k=2 \bmod 4 \text{ or } k=3 \bmod 4}]y_0 \quad \text{for all } k = 0,1,2\ 3,\ldots,\infty$$

It would be interesting to attempt a solution to equation (5.4) using a generating function approach. Begin by defining $G(s) = \sum_{k=0}^{\infty} s^k y_k$

$$\begin{aligned} y_{k+1} &= (-1)^k y_k \\ s^k y_{k+1} &= s^k (-1)^k y_k \\ \sum_{k=0}^{\infty} s^k y_{k+1} &= \sum_{k=0}^{\infty} s^k (-1)^k y_k \end{aligned} \tag{5.5}$$

The next step that usually follows is to convert both sides of this equation to functions of G(s). The left side of course can be written as $s^{-1}[G(s) - y_0]$. However the right side poses a difficulty since it cannot be rewritten as a function of $G(s) = \sum_{k=0}^{\infty} s^k y_k$. In this example, the term $(-1)^k$ blocks our customary solution path. This therefore requires using a rather unusual approach. Begin with

$$\begin{aligned} &\sum_{k=0}^{\infty} s^k y_{k+1} = \sum_{k=0}^{\infty} s^k (-1)^k y_k \\ &\left[\sum_{k=0}^{\infty} s^k y_{k+1} - \sum_{k=0}^{\infty} s^k (-1)^k \right] y_k = 0 \end{aligned} \tag{5.6}$$

Proceeding

$$\begin{gathered}\sum_{k=0}^{\infty} s^k y_{k+1} - \sum_{k=0}^{\infty} s^k (-1)^k y_k = 0 \\ \sum_{k=0}^{\infty} \left[y_{k+1} - (-1)^k y_k \right] s^k = 0\end{gathered} \tag{5.7}$$

This may be written as

$$(y_1 - y_0) + (y_2 + y_1)s^1 + (y_3 - y_2)s^2 + (y_4 + y_3)s^3 + \ldots = 0 \tag{5.8}$$

For a fixed sequence of y_k's such that equation (5.8) is true for each and every value of $s \geq 0$, each coefficient of s^k must be individually equal to zero. Thus, start with $y_1 - y_0 = 0$ and find $y_1 = y_0$. Proceeding to the coefficient of s^2, find $y_2 + y_1 = 0$ revealing $y_2 = -y_1 = -y_0$. In this fashion, we can find the solution for any y_k in the sequence. It is debatable whether this is a generating function approach at all, since the solution has to be "built up" from the coefficients of the smaller to the coefficients of the larger powers of s.

5.3 Logarithmic Transformation

Another example of the application of generating functions in the solution of difference equations with nonconstant coefficients is the difference equation

$$y_{k+1} = a^k y_k \text{ for } k = 0 \text{ to } \infty \text{ for } a > 0 \tag{5.9}$$

of which a special case was evaluated in the previous section. Applying the generating function to this family of difference equations reveals

$$\begin{gathered}y_{k+1} = a^k y_k \\ s^k y_{k+1} = s^k a^k y_k \\ \sum_{k=0}^{\infty} s^k y_{k+1} = \sum_{k=0}^{\infty} s^k a^k y_k\end{gathered} \tag{5.10}$$

Now define $G(s) = \sum_{k=0}^{\infty} s^k y_k$, which allows the simplification of the left side of equation (5.10) to $s^{-1}[G(s) - y_0]$. However the right side of this equation does not allow us to express $\sum_{k=0}^{\infty} s^k a^k y_k$ as a function of G(s) – actually it is G(as). However, consider taking the logarithm of each side of equation (5.9).

$$\begin{aligned} \ln y_{k+1} &= \ln\left(a^k y_k\right) \\ \ln y_{k+1} &= \ln a^k + \ln y_k \\ \ln y_{k+1} &= k \ln a + \ln y_k \end{aligned} \tag{5.11}$$

Now, define $z_{k+1} = \ln y_{k+1}$, $z_k = \ln y_k$ and continue with the solution of equation (5.11).

$$z_{k+1} = k \ln a + z_k \tag{5.12}$$

This now looks like a family of difference equations that may be amenable to the generating function solution. Define

$$G(s) = \sum_{k=0}^{\infty} s^k z_k \tag{5.13}$$

Next apply the process of conversion, consolidation, and inversion to equation (5.12)

$$\begin{aligned} s^k z_{k+1} &= ks^k \ln a + s^k z_k \\ \sum_{k=0}^{\infty} s^k z_{k+1} &= \sum_{k=0}^{\infty} ks^k \ln a + \sum_{k=0}^{\infty} s^k z_k \\ s^{-1}\left[G(s) - z_0\right] &= \ln a \sum_{k=0}^{\infty} ks^k + G(s) \end{aligned} \tag{5.14}$$

From Chapter 2 and successive application of derivatives to the geometric series, write $\sum_{k=0}^{\infty} ks^k = \frac{s}{(1-s)^2}$. Proceed as follows:

$$s^{-1}\left[G(s)-z_0\right]=\ln a\frac{s}{(1-s)^2}+G(s)$$
$$G(s)-z_0=\ln a\frac{s^2}{(1-s)^2}+sG(s) \tag{5.15}$$

$$G(s)(1-s)=\ln a\frac{s^2}{(1-s)^2}+z_0$$
$$G(s)=\ln a\frac{s^2}{(1-s)^3}+\frac{z_0}{(1-s)} \tag{5.16}$$

Recall (from Chapter 2) that the successive applications of the derivatives and the sliding tool, reveals

$$\frac{s^2}{(1-s)^3} \triangleright \left\{\frac{k(k-1)}{2}\right\} \tag{5.17}$$

and we can conclude

$$z_k=\frac{k(k-1)}{2}\ln a+\ln y_0 \tag{5.18}$$

As a final step, convert z_k to y_k

$$y_k=y_0 a^{\frac{k(k-1)}{2}} \tag{5.19}$$

for the final solution.

As another example of this process, consider the equation:

$$y_{k+3}=a^k y_k \tag{5.20}$$

Equation (5.20) is a third-order family of difference equations. The logarithmic transformation may work here as well. Apply the same transformation $z_k = \ln y_k$

$$
\begin{aligned}
z_{k+3} &= k\ln a + z_k \\
s^k z_{k+3} &= (\ln a)ks^k + s^k z_k
\end{aligned}
\tag{5.21}
$$

Proceed

$$
\begin{aligned}
&\sum_{k=0}^{\infty} s^k z_{k+3} = (\ln a)\sum_{k=0}^{\infty} ks^k + \sum_{k=0}^{\infty} s^k z_k \\
&s^{-3}\left[G(s) - z_0 - sz_1 - s^2 z_2\right] = (\ln a)\frac{s}{(1-s)^2} + G(s) \\
&G(s) - z_0 - sz_1 - s^2 z_2 = (\ln a)\frac{s^4}{(1-s)^2} + s^3 G(s) \\
&G(s)\left(1-s^3\right) = z_0 + sz_1 + s^2 z_2 + (\ln a)\frac{s^4}{(1-s)^2}
\end{aligned}
\tag{5.22}
$$

Solve for G(s)

$$
G(s) = \frac{z_0}{1-s^3} + \frac{sz_1}{1-s^3} + \frac{s^2 z_2}{1-s^3} + (\ln a)\frac{s^4}{(1-s)^2\left(1-s^3\right)} \tag{5.23}
$$

The inversion uses the principles presented in Chapter 2. Begin by writing

$$
\begin{aligned}
\frac{1}{1-s^3} = \frac{1}{1-\left(s^2\right)s} &= 1 + s^2 s + \left(s^2\right)^2 s^2 + \left(s^2\right)^3 s^3 + \left(s^2\right)^4 s^4 \ldots \\
&= 1 + s^3 + s^6 + s^9 + s^{12} + \ldots \\
&\triangleright \left\{I_{K \bmod 3 = 0}\right\}
\end{aligned}
\tag{5.24}
$$

The inversion of the first three terms on the right side of equation (5.23) involves the straightforward application of the sliding principle (Chapter 2).

$$\frac{z_0}{1-s^3} \triangleright \{z_0 I_{K \bmod 3=0}\}$$
$$\frac{sz_1}{1-s^3} \triangleright \{z_1 I_{K \bmod 3=1}\} \quad (5.25)$$
$$\frac{s^2 z_2}{1-s^3} \triangleright \{z_2 I_{K \bmod 3=2}\}$$

We can now invert $\ln a \frac{s^4}{(1-s^3)(1-s)^2} = s^3 \ln a \left(\frac{1}{1-s^3}\right)\left(\frac{s}{(1-s)^2}\right)$. Since $\frac{1}{1-s^3} \triangleright \{I_{k \bmod 3=0}\}$, and (from Chapter 2) $\frac{s}{(1-s)^2} \triangleright \{k\}$, we can easily apply the convolution principle, revealing

$$\left(\frac{1}{1-s^3}\right)\left(\frac{s}{(1-s)^2}\right) \triangleright \left\{\sum_{j=0}^{k} jI_{(k-j) \bmod 3=0}\right\} \quad (5.26)$$

and by the sliding tool,

$$s^3\left(\frac{1}{1-s^3}\right)\left(\frac{s}{(1-s)^2}\right) \triangleright \left\{\sum_{j=0}^{k-3} jI_{(k-3-j) \bmod 3=0}\right\} \quad (5.27)$$

Thus,

$$G(s) = \frac{z_0}{1-s^3} + \frac{sz_1}{1-s^3} + \frac{s^2 z_2}{1-s^3} + (\ln a)\frac{s^4}{(1-s)^2(1-s^3)}$$
$$\triangleright \left\{z_0 I_{k \bmod 3=0} + z_1 I_{k \bmod 3=1} + z_2 I_{k \bmod 3=2} + \ln a \sum_{j=0}^{k-4} jI_{(k-3-j) \bmod 3=0}\right\} \quad (5.28)$$

Since $y_k = e^{z_k}$ we have

$$y_k = \left(y_0^{I_{k \bmod 3=0}}\right)\left(y_1^{I_{k \bmod 3=1}}\right)\left(y_2^{I_{k \bmod 3=2}}\right) a^{\sum_{j=0}^{k-4} j I_{(k-3-j) \bmod 3=0}} \tag{5.29}$$

5.4 Using Differential Equations to Solve Difference Equations

Consider the family of difference equations given by

$$y_{k+1} = \frac{y_k}{k+1} \tag{5.30}$$

for $k = 0$ to infinity and y_0 is a known constant. It is clear from a simple iterative approach what the solution to this equation is

$$\begin{aligned} y_1 &= y_0 \\ y_2 &= \frac{y_0}{2} \\ y_3 &= \frac{y_2}{3} = \frac{y_0}{3!} \\ y_4 &= \frac{y_3}{4} = \frac{y_0}{4!} \end{aligned} \tag{5.31}$$

and in general

$$y_k = \frac{y_0}{k!} \tag{5.32}$$

The family of difference equations in (5.30) that led to this solution is a family of difference equations with variable coefficients. However, they can be solved using the generating function procedure. Rewrite equation (5.30) as

$$(k+1)\, y_{k+1} = y_k \tag{5.33}$$

and define the generating function as

$$G(s) = \sum_{k=0}^{\infty} s^k y_k \tag{5.34}$$

and proceed by multiplying each side of equation (5.33) by s^k

$$\begin{aligned} (k+1)s^k y_{k+1} &= s^k y_k \\ \sum_{k=0}^{\infty}(k+1)s^k y_{k+1} &= \sum_{k=0}^{\infty} s^k y_k \end{aligned} \tag{5.35}$$

Note that $(k+1)s^k = \dfrac{d(s^{k+1})}{ds}$. Thus,

$$\sum_{k=0}^{\infty}(k+1)s^k y_{k+1} = \sum_{k=0}^{\infty}\frac{d(s^{k+1})}{ds} y_{k+1} = \frac{d\sum_{k=0}^{\infty} s^{k+1} y_{k+1}}{ds} = \frac{dG(s)}{ds} \tag{5.36}$$

leading to

$$\frac{dG(s)}{ds} = G(s) \tag{5.37}$$

Equation (5.37) is the one of the simplest differential equations to solve. Its solution is

$$G(s) = y_0 e^s \tag{5.38}$$

The inversion can continue by expanding e^s as

$$e^s = \sum_{k=0}^{\infty}\frac{s^k}{k!} \tag{5.39}$$

and the inversion is seen to be*

---

* This notion of converting a difference equation to a differential equation in s will be commonly employed in Chapters 11 and 12.

$$G(s) = y_0 e^s \triangleright \left\{ \frac{y_0}{k!} \right\} \tag{5.40}$$

5.5 Generalizing the Right Side

We can expand our ability to solve difference equations somewhat by including terms that are functions of k, not as coefficients of the terms from the sequence $\{y_k\}$ but as expressions on the right side of the equation. For example, consider the family of difference equations

$$y_{k+3} - y_k = 3^k - 2k + 1 \tag{5.41}$$

that holds for all value of $k = 0$ to infinity. The complication added in equation (5.41) are the terms 3^k and $-2k$ on the right side of the equation. Familiarity with generating function inversion will provide the guidance needed for the solution to this equation. Begin by defining

$$G(s) = \sum_{k=0}^{\infty} s^k y_k \tag{5.42}$$

and proceed with the conversion and consolidation of equation (5.41) as follows:

$$\begin{aligned} s^k y_{k+3} - s^k y_k &= s^k 3^k - 2s^k k + s^k \\ \sum_{k=0}^{\infty} s^k y_{k+3} - \sum_{k=0}^{\infty} s^k y_k &= \sum_{k=0}^{\infty} s^k 3^k - \sum_{k=0}^{\infty} 2s^k k + \sum_{k=0}^{\infty} s^k \end{aligned} \tag{5.43}$$

Equation (5.29) simplifies to

$$s^{-3}\left[G(s) - y_0 - sy_1 - s^2 y_2\right] - G(s) = \sum_{k=0}^{\infty} (3s)^k - 2\sum_{k=0}^{\infty} ks^k + \sum_{k=0}^{\infty} s^k \tag{5.44}$$

Proceed by noting that $\sum_{k=0}^{\infty} (3s)^k = \frac{1}{1-3s}$ and (from Chapter 2) $\sum_{k=0}^{\infty} ks^k = \frac{s}{(1-s)^2}$.

Thus, equation (5.44) becomes

$$
\begin{aligned}
&s^{-3}\left[G(s)-y_0-sy_1-s^2y_2\right]-G(s)=\frac{1}{1-3s}-\frac{2s}{(1-s)^2}+\frac{1}{1-s}\\
&G(s)-y_0-sy_1-s^2y_2-s^3G(s)=\frac{s^3}{1-3s}-\frac{2s^4}{(1-s)^2}+\frac{s^3}{1-s}\\
&G(s)\left(1-s^3\right)-y_0-sy_1-s^2y_2=\frac{s^3}{1-3s}-\frac{2s^4}{(1-s)^2}+\frac{s^3}{1-s}
\end{aligned}
\tag{5.45}
$$

Further simplification reveals

$$
\begin{aligned}
G(s)=&\frac{y_0}{1-s^3}+\frac{y_1s}{1-s^3}+\frac{y_2s^2}{1-s^3}\\
&+\frac{s^3}{(1-3s)\left(1-s^3\right)}-\frac{2s^4}{(1-s)^2\left(1-s^3\right)}+\frac{s^3}{(1-s)\left(1-s^3\right)}
\end{aligned}
\tag{5.46}
$$

The key to the inversion of G(s) involves inverting $\frac{1}{1-s^3}$. Write

$$
\begin{aligned}
\frac{1}{1-s^3}=\frac{1}{1-s^2s}&=1+s^2s+\left(s^2\right)^2s^2+\left(s^2\right)^3s^3+\left(s^2\right)^4s^4+\left(s^2\right)^5s^5\\
&=1+s^3+s^6+s^9+s^{12}+s^{15}\\
&\triangleright\left\{I_{k\bmod 3=0}\right\}
\end{aligned}
\tag{5.47}
$$

Proceeding

$$
\frac{y_0}{1-s^3}\triangleright\left\{y_0I_{k\bmod 3=0}\right\};\ \frac{y_1s}{1-s^3}\triangleright\left\{y_1I_{k\bmod 3=1}\right\};\ \frac{y_2s^2}{1-s^3}\triangleright\left\{y_2I_{k\bmod 3=2}\right\}
\tag{5.48}
$$

Similarly

$$\frac{2s^4}{(1-s)^2(1-s^3)} = s^3\left(\frac{2s}{(1-s)^2}\right)\left(\frac{1}{1-s^3}\right) \triangleright \left\{2\sum_{j=0}^{k-3} jI_{(k-3-j) \bmod 3=0}\right\} \tag{5.49}$$

$$\frac{s^3}{(1-3s)(1-s^3)} \triangleright \left\{\sum_{j=0}^{k-3} 3^j I_{(k-3-j) \bmod 3=0}\right\} \tag{5.50}$$

$$\frac{s^3}{(1-s)(1-s^3)} = s^3\left(\frac{1}{1-s}\right)\left(\frac{1}{1-s^3}\right) \triangleright \left\{\sum_{j=0}^{k-3} I_{j \bmod 3=0}\right\} \tag{5.51}$$

Continue with the inversion of G(s):

$$\begin{aligned} G(s) &= \frac{y_0}{1-s^3} + \frac{y_1 s}{1-s^3} + \frac{y_2 s^2}{1-s^3} \\ &+ \frac{s^3}{(1-3s)(1-s^3)} - \frac{2s^4}{(1-s)^2(1-s^3)} + \frac{s^3}{(1-s)(1-s^3)} \\ &\triangleright \left\{ \begin{aligned} & y_0 I_{k \bmod 3=0} + y_1 I_{k \bmod 3=1} + y_2 I_{k \bmod 3=2} \\ & + \sum_{j=0}^{k-3} 3^j I_{(k-3-j) \bmod 3=0} - 2\sum_{j=0}^{k-3} jI_{(k-3-j) \bmod 3=0} + \sum_{j=0}^{k-3} I_{j \bmod 3=0} \end{aligned} \right\} \end{aligned} \tag{5.52}$$

5.6 Powers of k on the Right Side

5.6.1 Example 1

A particularly illuminating example of the power of generating functions may be seen in the solution of

$$y_{k+2} = by_k + k^2 a^k \tag{5.53}$$

for k = 0 to infinity and y_0, y_1 known constants. For a negative value of the constant b, the equation can oscillate greatly. Define $G(s) = \sum_{k=0}^{\infty} s^k y_k$ and use the conversion and consolidation principles:

$$
\begin{aligned}
y_{k+2} &= by_k + k^2 a^k \\
s^k y_{k+2} &= bs^k y_k + k^2 a^k s^k \\
\sum_{k=0}^{\infty} s^k y_{k+2} &= \sum_{k=0}^{\infty} bs^k y_k + \sum_{k=0}^{\infty} k^2 a^k s^k \\
\sum_{k=0}^{\infty} s^k y_{k+2} &= b\sum_{k=0}^{\infty} s^k y_k + \sum_{k=0}^{\infty} k^2 (as)^k
\end{aligned}
\tag{5.54}
$$

$$
s^{-2}\left[G(s) - y_0 - y_1 s\right] = bG(s) + \sum_{k=0}^{\infty} k^2 (as)^k \tag{5.55}
$$

Recall from Chapter 2, equation (2.94) that the last term on the right side of equation (5.55) reveals $\frac{as + a^2s^2}{(1-as)^3} \triangleright \{k^2a^k\}$. Thus equation (5.55) may be rewritten as

$$
s^{-2}\left[G(s) - y_0 - y_1 s\right] = bG(s) + \frac{as + a^2s^2}{(1-as)^3} \tag{5.56}
$$

Solving for G(s)

$$
\begin{aligned}
s^{-2}\left[G(s) - y_0 - y_1 s\right] &= bG(s) + \frac{as + a^2s^2}{(1-as)^3} \\
G(s) &= \frac{y_0}{1-bs^2} + \frac{y_1 s}{1-bs^2} + \left(\frac{1}{1-bs^2}\right)\left(\frac{as^2 + a^2s^4}{(1-as)^3}\right)
\end{aligned}
\tag{5.57}
$$

Since we can write

$$
\frac{1}{1-bs^2} \triangleright \left\{ b^{\frac{k}{2}} I_{k \bmod 2 = 0} \right\} \tag{5.58}
$$

G(s) may be inverted directly to get

$$y_k = y_0 b^{\frac{k}{2}} I_{k \bmod 2=0} + y_1 b^{\frac{k-1}{2}} I_{k \bmod 2=1} + \sum_{j=0}^{k-2} j^2 a^j b^{\frac{k-2-j}{2}} I_{(k-2-j) \bmod 2=0} \tag{5.59}$$

5.6.2 Example 2

Consider the problem in the previous section, recall that the last term in equation (5.53) $k^2 a^k$ did not really require inversion in the process of solving for G(s). We may take advantage of this process by not transforming or inverting coefficients that do not multiply any member of $\{y_k\}$ sequence. Observe this process in operation by solving the family of difference equations

$$a y_{k+1} = y_k + (k+1)\log(k+1) + k^2 + c^k \tag{5.60}$$

We notice that the terms log(k+1), k, and c^k do not involve members of the sequence $\{y_k\}$ so we will not attempt to convert them to a function of G(s). Begin by multiplying each term in equation (5.60) by s^k and summing from k = zero to infinity.

$$a s^k y_{k+1} = s^k y_k + s^k (k+1)\log(k+1) + k^2 s^k + c^k s^k \tag{5.61}$$

$$\sum_{k=0}^{\infty} a s^k y_{k+1} = \sum_{k=0}^{\infty} s^k y_k + \sum_{k=0}^{\infty} s^k (k+1)\log(k+1) + \sum_{k=0}^{\infty} k^2 s^k + \sum_{k=0}^{\infty} c^k s^k \tag{5.62}$$

Defining $G(s) = \sum_{k=0}^{\infty} s^k y_k$ equation may be rewritten as

$$a s^{-1}\left[G(s) - y_0\right] = G(s) + \sum_{k=0}^{\infty} s^k (k+1)\log(k+1) + \sum_{k=0}^{\infty} k^2 s^k + \sum_{k=0}^{\infty} c^k s^k \tag{5.63}$$

and solving for G(s)

$$as^{-1}\left[G(s)-y_0\right]=G(s)+\sum_{k=0}^{\infty}s^k(k+1)\log(k+1)+\sum_{k=0}^{\infty}k^2s^k+\sum_{k=0}^{\infty}c^ks^k$$

$$G(s)=\frac{ay_0}{a-s}+\frac{\sum_{k=0}^{\infty}s^{k+1}(k+1)\log(k+1)}{a-s}+\frac{\sum_{k=0}^{\infty}k^2s^{k+1}}{a-s}+\frac{\sum_{k=0}^{\infty}c^ks^{k+1}}{a-s} \tag{5.64}$$

Each of these terms can be inverted using the tools of Chapter 2:

$$\frac{ay_0}{a-s}\triangleright\left\{\frac{y_0}{a^k}\right\} \tag{5.65}$$

$$\frac{\sum_{k=0}^{\infty}s^{k+1}(k+1)\log(k+1)}{a-s}\triangleright\left\{\sum_{j=1}^{k}j\log(j)\frac{1}{a^{k-j+1}}\right\} \tag{5.66}$$

$$\frac{\sum_{k=0}^{\infty}k^2s^{k+1}}{a-s}\triangleright\left\{\sum_{j=0}^{k}j^2\frac{1}{a^{k-j+1}}\right\} \tag{5.67}$$

$$\frac{\sum_{k=0}^{\infty}c^ks^{k+1}}{a-s}\triangleright\left\{\sum_{j=0}^{k-1}c^j\frac{1}{a^{k-j+1}}\right\} \tag{5.68}$$

and the solution is

$$y_k=\frac{y_0}{a^k}+\sum_{j=0}^{k}\frac{j\log(j)}{a^{k-j+1}}+\sum_{j=0}^{k}\frac{j^2}{a^{k-j+1}}+\sum_{j=0}^{k-1}\frac{c^j}{a^{k-j+1}} \tag{5.69}$$

The solution to equation (5.61) was straightforward because it was possible to carry the terms on the right side of that equation down to the final inversion of G(s).

5.7 Observations

As discussed in Chapter 1, the sequence $\{y_k\}$ that a difference equation represents is not a sequence whose members are all independent of one another; instead they are dependent in a very specific way. Difference equations represent a continued collection of intrasequence relationships, and the specifics of this dependency are provided in the difference equation. We have in fact taken advantage of the unvarying nature of this dependency by using the generating function, and have identified a generating function that is unique to the sequence in which the dependency is embedded. Although these interrelationships can be expanded by including families of difference equations with nonconstant coefficients, we have seen that in general there is no solution to these equations. However, in the previous sections of this chapter, we identified certain classes of difference equations that have variable coefficients but have general solutions through the generating function approach. The remainder of this chapter will focus on difference equations that are mixtures of other difference equations–an adaptation of the variable coefficient characterization.

5.8 Mixed Difference Equations

Consider the family of first-order homogeneous difference equations given by $y_{k+1} = ay_k$, $k = 0$ to ∞ y_0 a known constant. We know from Chapter 3 that the solution to this equation is $y_k=a^k y_0$. Rather than write the sequence $\{y_0, y_1, y_2, y_3, \ldots y_k, \ldots\}$, one might write the sequence as $\{y_0, ay_0, a^2y_0, a^3y_0, \ldots, a^ky_0, \ldots\}$. In this latter presentation, the relationship between the y_k's is clear; each succeeding y_k is obtained as a constant multiplier of the previous one, the constant being a. Now suppose we altered this sequence by changing the relationship between the y_k's beyond some fixed $k = k^*$. For example, consider the following alteration

$$y_0, ay_0, a^2y_0, a^3y_0, \ldots, a^{k^*}y_0, ba^{k^*}y_0, b^2a^{k^*}y_0, b^3a^{k^*}y_0, b^4a^{k^*}y_0, \ldots$$

The first portion of the sequence appears as it was first defined. However, later in the sequence, the relationship between the y_k's has changed. For the first part of the sequence the relationship was given by the difference equation $y_{k+1} = ay_k$. However, at some point in the sequence, this relationship was altered and a new relationship replaced it. We will denote such a sequence as a mix of sequences, and define the hinge point $k = k^*$, as the point in the sequence when the dependency change is specified, and we will assume that the location of the hinge point is known. These are difference equations with variable coefficients in the sense that the coefficients vary from one region of values of k to another.

Being able to manage this situation of sequence mixes will increase our ability to adapt circumstances in public health to the difference equation approach. We first saw an example of this sequence in Chapter 1 when we examined a possible relationship between the probability of ovulation and the day of the menstrual cycle. Relationships in sequence in real life are not static – but change over time. This changing relationship over time itself needs to be characterized. This chapter will adapt the generating function approach to these difference equations.

5.9 Applications of Generating Functions to Mixed Difference Equations

In order to develop this concept of difference equation mixtures, we will first need to represent the sequence mixture as a family of difference equations. The indicator function will be a useful tool in this effort. There are two circumstances we will consider. The first is demonstrated in the following difference equation formulation. Let $y_k = ay_{k-1}$ for $1 \leq k < 10$. Then for $k \geq 10$, the relationship changes to $y_k = by_{k-1}$. Using the indicator function write this relationship as

$$y_k = ay_{k-1}I_{k<10} + by_{k-1}I_{k\geq 10} \tag{5.70}$$

y_0 is a known constant. Note the role of the indicator function here. If $1 \leq k < 10$, the second summand on the right side of equation (5.70) is zero and the equation becomes $y_k = ay_{k-1}$. As k increases to ten or more, the equation switches to $y_k = by_{k-1}$. The indicator function shunts the equation to either ay_k or by_k depending on the value of k. This is "technically" a family of difference equations with variable coefficients. However, rather than the coefficients changing with each change in k, they change from one region of k (i.e., $1 \leq k < 10$) to another region ($k > 10$). Proceed as follows:

$$\begin{aligned} y_k &= ay_{k-1}I_{k<10} + by_{k-1}I_{k\geq 10} \\ s^k y_k &= as^k y_{k-1}I_{k<10} + bs^k y_{k-1}I_{k\geq 10} \\ \sum_{k=1}^{\infty} s^k y_k &= \sum_{k=1}^{\infty} as^k y_{k-1}I_{k<10} + \sum_{k=1}^{\infty} bs^k y_{k-1}I_{k\geq 10} \end{aligned} \tag{5.71}$$

leading to

$$\left[G(s) - y_0\right] = as\sum_{k=0}^{8} s^k y_k + bs\sum_{k=9}^{\infty} s^k y_k \tag{5.72}$$

The term $\sum_{k=1}^{\infty} s^k y_{k-1} I_{k<10}$ is just $s\sum_{k=0}^{8} s^k y_k$ because the indicator function assigns zero to all summands with $k \geq 9$. The last term in equation (5.72) is

$$b\sum_{k=9}^{\infty} s^k y_k = bs\left[G(s) - \sum_{k=0}^{8} s^k y_k\right] \tag{5.73}$$

Thus,

$$\begin{aligned} G(s) - y_0 &= as\sum_{k=0}^{8} s^k y_k + bs\left[G(s) - \sum_{k=0}^{8} s^k y_k\right] \\ G(s) - bsG(s) &= y_0 + (a-b)s\sum_{k=0}^{8} s^k y_k \end{aligned} \tag{5.74}$$

and

$$G(s) = \frac{y_0}{1-bs} + \frac{(a-b)s\sum_{k=0}^{8} s^k y_k}{1-bs} \tag{5.75}$$

The inversion of G(s) is straightforward. The second term on the right is a convolution that requires some additional attention. Using the convolution principle, write

$$\begin{aligned} &\frac{(a-b)s\sum_{k=0}^{8} s^k y_k}{1-bs} \\ &= (a-b)s\left(y_0 + y_1 s + y_2 s^2 + y_3 s^3 + \cdots + y_8 s^8\right)\left(1 + bs + b^2 s^2 + b^3 s^3 + \cdots\right) \\ &\triangleright \left\{(a-b)\sum_{j=0}^{\min(k-1,8)} y_j b^{k-j-1}\right\} \end{aligned} \tag{5.76}$$

So the complete inversion of G(s) is

$$G(s) \triangleright \left\{ y_0 b^k + (a-b) \sum_{j=0}^{\min(k-1,8)} y_j b^{k-j-1} \right\} \tag{5.77}$$

Obviously, this inversion requires knowledge of y_k for k = 0, 1, 2, ..., 8. The family of difference equations from which $y_1, y_2, y_3, \ldots, y_8$ are derived is a first-order homogeneous difference equation that is easily solvable. Therefore, in order to invert G(s) completely we must first solve the difference equation representing the first component of the mixed sequence.

We can proceed with the inversion of G(s) from equation (5.75) utilizing the knowledge gained from Chapter 3 for the solution of first-order homogeneous difference equations, that $y_k = a^k y_0$, for k = 1 to 8. This leads to the inversion

$$G(s) \triangleright \left\{ y_0 b^k + (a-b) y_0 \sum_{j=0}^{\min(k-1,8)} a^j b^{k-j-1} \right\} \tag{5.78}$$

This solution needs to be verified in each of the two regions, $k < 10$, and $k \geq 10$. For k=5,

$$\begin{aligned} y_5 &= y_0 b^5 + (a-b) y_0 \left[b^4 + ab^3 + a^2b^2 + a^3b + a^4 \right] \\ &= y_0 b^5 + y_0 ab^4 + y_0 a^2 b^3 + y_0 a^3 b^2 + y_0 a^4 b + y_0 a^5 \\ &- y_0 b^5 - y_0 ab^4 - y_0 a^2 b^3 - y_0 a^3 b^2 - y_0 a^4 b \\ &= y_0 a^5 \end{aligned} \tag{5.79}$$

A similar demonstration also shows that $y_{10} = a^9 b y_0$. G(s) is the solution to the mixed difference equation $y_k = a y_{k-1} I_{1 \leq k < 10} + b y_{k-1} I_{k \geq 10}$ for k =1 to ∞ but there is a complication to the generating function approach. We have to build up the solution by solving the first component of equation (5.70). This solution involves the identification of a related generating function and can be achieved with an additional step.

5.10 The Stepwise Approach

The solution for the elementary mixed difference equation revealed that the inversion of the generating function for the difference equation in the previous section requires the solution to the first difference equation $y_{k+1} = ay_k$, the first component equation in the mixture. This will be true in general for mixed difference equations. In order to get a general solution for the mixed family of difference equations that is composed of n distinct components of difference equations, we will need to solve the first n - 1 components separately, injecting the solution of each into the mixed generating function to complete the inversion. The stepwise approach demonstrated above is valid for only the cases where the region of k with the largest values is unbounded above. In this case, there is a separate equation in each of the regions that must be solved individually.

5.11 Solution to Mixtures of Three or More First-Order Difference Equations

5.11.1 Specific example

Apply the above idea next to understand the degree to which the generating functions that develop from the individual components of the difference equation mixture may be applied to find the solution to the specific mixed equation below:

$$y_k = ay_{k-1}I_{k<10} + by_{k-1}I_{10\le k<20} + cy_{k-1}I_{k\ge 20} \tag{5.80}$$

for k = 1 to ∞ and y_0 a known constant. Note this is a mixture involving three components and two hinge points (k = 10 and k = 20). Applying the procedure adopted in Chapter 4, note

$$\begin{aligned}
\left[G(s) - y_0\right] &= as\sum_{k=0}^{8} s^k y_k + bs\sum_{k=9}^{18} s^k y_k + cG(s) - c\sum_{k=0}^{18} s^k y_k \\
G(s)(1-cs) &= y_0 + (a-c)s\sum_{k=0}^{8} s^k y_k + (b-c)s\sum_{k=9}^{18} s^k y_k
\end{aligned} \tag{5.81}$$

As with the previous example, G(s) is easy to invert; however, the terms y_k for k = 1 to 19 are required. Given that $y_k = a^k y_0$ for $1 \le k < 9$ and $y_k = b^{k-9}a^9 y_0$ for k = 10 to 19, compute the solution to G(s) as

$$G(s) = \frac{y_0}{1-cs} + \frac{(a-c)s\sum_{k=0}^{8} s^k y_k}{1-cs} + \frac{(b-c)s\sum_{k=9}^{18} s^k y_k}{1-cs} \tag{5.82}$$

The inversion can be performed at once for

$$G(s) \triangleright \left\{ \begin{array}{l} y_0 c^k + (a-c) \sum_{j=0}^{\min(k-1,8)} y_j c^{k-j-1} \\ + (b-c) \sum_{j=9}^{\min(k-1,18)} y_j c^{k-j-1} \end{array} \right\} \tag{5.83}$$

Later in this chapter, we develop the general solution for a mixture of n first-order homogeneous difference equations. To provide some intuition for the overall solution, first solve a mixture of three homogeneous first-order difference equations. Once the pattern of the solution is observed for this set of equations, we will apply this pattern to a collection of n first-order homogeneous equations.

5.11.2 Mixtures of three first-order homogeneous equations

Consider the family of difference equations given by

$$y_k = a_1 y_{k-1} I_{k \le k_1} + a_2 y_{k-1} I_{k_1 < k < k_2} + a_3 y_{k-1} I_{k > k_2} \tag{5.84}$$

In this family of equations, there are three ranges of k and in each region a different first-order homogeneous difference equation holds. In the region $1 \le k \le k_1$, the difference equation $y_k = a_1 y_{k-1}$ holds. For the second region $k_1 < k \le k_2$, the equation $y_k = a_2 y_{k-1}$ is in force, and $y_k = a_3 y_{k-1}$ is true for the third region $k > k_2$. Proceeding with the solution by applying the customary generating function argument, begin with

$$G(s) = \sum_{k=0}^{\infty} s^k y_k \tag{5.85}$$

and multiply each side of equation (5.84) by s^k, and sum over each of the three distinct regions,

$$\begin{aligned} s^k y_k &= a_1 s^k y_{k-1} I_{k \le k_1} + a_2 s^k y_{k-1} I_{k_1 < k \le k_2} + a_3 s^k y_{k-1} I_{k > k_2} \\ \sum_{k=1}^{\infty} s^k y_k &= a_1 \sum_{k=1}^{k_1} s^k y_{k-1} + a_2 \sum_{k=k_1+1}^{k_2} s^k y_{k-1} + a_3 \sum_{k=k_2+1}^{\infty} s^k y_{k-1} \end{aligned} \tag{5.86}$$

to reveal

$$G(s)-y_0=a_1s\sum_{k=0}^{k_1-1}s^ky_k+a_2s\sum_{k=k_1}^{k_2-1}s^ky_k+a_3s\left[G(s)-\sum_{k=0}^{k_2-1}s^ky_k\right] \quad (5.87)$$

Some simplification is afforded by writing

$$G(s)-\sum_{k=0}^{k_2-1}s^ky_k=G(s)-\sum_{k=0}^{k_1-1}s^ky_k-\sum_{k=k_1}^{k_2-1}s^ky_k \quad (5.88)$$

And equation (5.88) becomes

$$G(s)-y_0=a_1s\sum_{k=0}^{k_1-1}s^ky_k+a_2s\sum_{k=k_1}^{k_2-1}s^ky_k+a_3s\left[G(s)-\sum_{k=0}^{k_1-1}s^ky_k-\sum_{k=k_1}^{k_2-1}s^ky_k\right] \quad (5.89)$$

Simplification reveals

$$G(s)-y_0=(a_1-a_3)s\sum_{k=0}^{k_1-1}s^ky_k+(a_2-a_3)s\sum_{k=k_1}^{k_2-1}s^ky_k+a_3sG(s) \quad (5.90)$$

This step is the key to the general solution. Regardless of how many equations are involved in the disjoint mixture, the summation over the last disjoint region can be obtained by writing it as G(s) minus a finite sum, and then distributing the finite sum over smaller finite sums that allow for its distribution across the other sums on the right side of equation (5.90). Continue with this development to find

$$G(s)-a_3sG(s)=y_0+(a_1-a_3)s\sum_{k=0}^{k_1-1}s^ky_k+(a_2-a_3)s\sum_{k=k_1}^{k_2-1}s^ky_k+$$

$$G(s)=\frac{y_0}{1-a_3s}+\frac{(a_1-a_3)\sum_{k=0}^{k_1-1}s^{k+1}y_k}{1-a_3s}+\frac{(a_2-a_3)\sum_{k=k_1}^{k_2-1}s^{k+1}y_k}{1-a_3s} \quad (5.91)$$

The inversion is a straightforward application of the sliding tool and convolution principles of Chapter 2.

$$G(s)=\frac{y_0}{1-a_3s}+\frac{(a_1-a_3)s\sum_{k=0}^{k_1-1}s^k y_k}{1-a_3s}+\frac{(a_2-a_3)s\sum_{k=k_1}^{k_2-1}s^k y_k}{1-a_3s} \tag{5.92}$$

$$\triangleright \left\{y_0a_3^k+(a_1-a_3)\sum_{j=0}^{\min(k-1,k_1-1)}y_ja_3^{k-j-1}+(a_2-a_3)\sum_{j=k_1}^{\min(k-1,k_2-1)}y_ja_3^{k-j-1}I_{k\geq k_1}\right\}$$

The demonstration that

$$\frac{(a_1-a_3)s\sum_{k=0}^{k_1-1}s^k y_k}{1-a_3s}\triangleright\left\{(a_1-a_3)\sum_{j=0}^{\min(k-1,k_1-1)}y_ja_3^{k-j-1}\right\} \tag{5.93}$$

and

$$\frac{(a_2-a_3)s\sum_{k=k_1}^{k_2-1}s^k y_k}{1-a_3s}\triangleright\left\{(a_2-a_3)\sum_{j=k_1}^{\min(k-1,k_2-1)}y_ja_3^{k-j-1}I_{k\geq k_1}\right\} \tag{5.94}$$

are posed as problems at the end of this chapter.

5.11.3 The general solution to mixtures of n first-order homogeneous difference equations

Consider the family of difference equations given by

$$y_k=a_1y_{k-1}I_{k\leq k_1}+\sum_{j=2}^{n-1}a_jy_{k-1}I_{k_{j-1}\leq k<k_j}+a_ny_{k-1}I_{k>k_n} \tag{5.95}$$

for $k \geq 1$. In this family of equations, there are now n contiguous, disjoint ranges of k, over each of which a different first-order difference equation determines the value of y_k. We proceed with this solution by applying our customary generating function argument. Begin with

$$G(s) = \sum_{k=0}^{\infty} s^k y_k \tag{5.96}$$

and multiply each side of equation (5.95) by s^k, and sum over each of the three distinct regions.

$$s^k y_k = a_1 s^k y_{k-1} I_{k \le k_1} + \sum_{j=2}^{n-1} a_j s^k y_{k-1} I_{k_{j-1} \le k < k_j} + a_n y_{k-1} s^k I_{k > k_{n-1}}$$

$$\sum_{k=1}^{\infty} s^k y_k = a_1 \sum_{k=1}^{k_1} s^k y_{k-1} + \sum_{j=2}^{n-1} \sum_{k=k_{j-1}+1}^{k_j} a_j s^k y_{k-1} + a_n s \left[G(s) - \sum_{k=0}^{k_{n-1}-1} s^k y_k \right] \tag{5.97}$$

Leading to

$$G(s) - y_0 = a_1 s \sum_{k=0}^{k_1-1} s^k y_k + \sum_{j=2}^{n-1} a_j s \sum_{k=k_{j-1}}^{k_j-1} s^k y_k + a_n s \left[G(s) - \sum_{k=0}^{k_{n-1}-1} s^k y_k \right] \tag{5.98}$$

simplify by writing

$$G(s) - \sum_{k=0}^{k_{n-1}-1} s^k y_k = G(s) - \sum_{k=0}^{k_1-1} s^k y_k - \sum_{j=2}^{n-1} \sum_{k=k_{j-1}}^{k_j-1} s^k y_k \tag{5.99}$$

Continue by substituting equation (5.99) into equation (5.97)

$$G(s) - y_0 = a_1 s \sum_{k=0}^{k_1-1} s^k y_k + \sum_{j=2}^{n-1} \sum_{k=k_{j-1}}^{k_j-1} a_j s^k y_k + a_n s \left[G(s) - \sum_{k=0}^{k_{n-1}-1} s^k y_k \right] \tag{5.100}$$

which, after some algebra becomes

$$G(s) - y_0 = (a_1 - a_n) s \sum_{k=0}^{k_1-1} s^k y_k + \sum_{j=2}^{n-1} (a_j - a_n) \sum_{k=k_{j-1}}^{k_j-1} s^k y_k + a_n s G(s) \tag{5.101}$$

Solving for G(s) reveals

$$G(s) = \frac{y_0}{1 - a_n s} + \frac{(a_1 - a_n) s \sum_{k=0}^{k_1-1} s^k y_k}{1 - a_n s} + \frac{\sum_{j=2}^{n-1} (a_j - a_n) s \sum_{k=k_{j-1}}^{k_j-1} s^k y_k}{1 - a_n s} \tag{5.102}$$

The inversion is straightforward

$$G(s) \triangleright \left\{ \begin{array}{l} y_0 a_n^k + (a_1 - a_n) \sum\limits_{i=0}^{\min(k-1,k_1-1)} y_i a_n^{k-i-1} \\ + \sum\limits_{j=2}^{n-1} (a_j - a_n) \sum\limits_{i=k_{j-1}}^{\min(k-1,k_j-1)} y_i a_n^{k-i-1} I_{k \geq k_{j-1}} \end{array} \right\} \tag{5.103}$$

5.12 Mixing Nonhomogeneous Difference Equations

As we pursue the issue of mixing difference equations, it is seen that a complicated system of equations can be constructed from mixtures of simpler systems. The strategy has consisted of first solving the smaller, simpler systems, then using the solutions to those equations to solve the more complex systems. We will demonstrate the utility of this approach by pursuing the general solution to a mixture of two first-order nonhomogeneous difference equations with constant coefficients. Write the system of equations as

$$y_k = [a y_{k-1} + b] I_{k<k^*} + [c y_{k-1} + d] I_{k \geq k^*} \tag{5.104}$$

for k = 1 to ∞: y_0 a known constant. Observing that this is a mixture of difference equations with one equation in operation for $k < k^*$, and a second equation in operation for $k \geq k^*$, we will need to solve the first component initially. Following the development from Chapter 2, recall that

if $y_k = a y_{k-1} + b$ then $y_k = y_0 a^k + b \sum_{j=0}^{k-1} a^j = y_0 a^k + b \frac{a^k - 1}{a - 1}$. Now proceed with the conversion:

$$s^k y_k = [ay_{k-1} + b]s^k I_{k<k*} + [cy_{k-1} + d]s^k I_{k \geq k*}$$

$$\sum_{k=1}^{\infty} s^k y_k = \sum_{k=1}^{\infty}[ay_{k-1} + b]s^k I_{k<k*} + \sum_{k=1}^{\infty}[cy_{k-1} + d]s^k I_{k \geq k*}$$

$$= as\sum_{k=0}^{k*-2} s^k y_k + cs\sum_{k=k*-1}^{\infty} s^k y_k + b\sum_{k=1}^{k*-1} s^k + d\sum_{k=k*}^{\infty} s^k$$

$$= as\sum_{k=0}^{k*-2} s^k y_k + cs\left(G(s) - \sum_{k=0}^{k*-2} s^k y_k\right) + b\sum_{k=1}^{k*-1} s^k \tag{5.105}$$

$$+ d\left(\frac{s}{1-s} - \sum_{k=1}^{k*-1} s^k y_k\right)$$

$$= (a-c)s\sum_{k=0}^{k*-2} s^k y_k + (b-d)\sum_{k=1}^{k*-1} s^k + csG(s) + d\frac{s}{1-s}$$

By simplifying, observe that

$$G(s) - csG(s) = y_0 + (a-c)s\sum_{k=0}^{k*-2} s^k y_k + (b-d)\sum_{k=1}^{k*-1} s^k + d\frac{s}{1-s}$$

$$G(s) = \frac{y_0}{1-cs} + \frac{(a-c)s\sum_{k=0}^{k*-2} s^k y_k}{1-cs} + \frac{(b-d)\sum_{k=1}^{k*-1} s^k}{1-cs} + \frac{ds}{(1-s)(1-cs)} \tag{5.106}$$

The inversion of $\dfrac{(a-c)s\sum_{k=0}^{k*-2} s^k y_k}{1-cs}$ has already been accomplished. This is already a partial generating function, requiring the individual values for y_k for k = 0 to k* - 2. However, we have only y_0 (from the boundary condition). From the solution of the first component of equation (5.104) we know that $y_k = y_0 a^k + b\dfrac{a^k - 1}{a-1}$ for k = 1 to k* - 1 from earlier in this section. We only need to apply this solution to the generating function to see that if

$$G(s) = \frac{y_0}{1-cs} + \frac{(a-c)s\sum_{i=0}^{k*-2} s^i y_i}{1-cs} + \frac{(b-d)\sum_{i=1}^{k*-1} s^i}{1-cs} + \frac{ds}{(1-s)(1-cs)} \tag{5.107}$$

then

$$G(s) \triangleright \begin{Bmatrix} y_0c^k + (a-c)\left(\sum_{i=0}^{\min(k-1,k^*-2)} y_i c^{k-i-1}\right) \\ +(b-d)\sum_{i=1}^{\min(k,k^*-1)} c^{k-i} + d\sum_{j=0}^{k-1} c^j \end{Bmatrix} \tag{5.108}$$

that simplifies to

$$G(s) \triangleright \begin{Bmatrix} y_0c^k + (a-c)\left(\sum_{i=0}^{\min(k-1,k^*-2)} y_i c^{k-i-1}\right) \\ +(b-d)\sum_{i=1}^{\min(k,k^*-1)} c^{k-i} + d\frac{c^k-1}{c-1} \end{Bmatrix} \tag{5.109}$$

The first and last terms of this solution are the solution to the first-order, nonhomogeneous equation $y_{k+1} = cy_k + d$. The middle terms reflect the influence of the mixture.

5.13 Overlapping Mixtures of First-Order Nonhomogeneous Difference Equations

So far the mixed difference equations considered had the feature of having one difference equation applicable in one region of integers, with another difference equation applicable over another realm. The two equations operated in disjoint regions. Another, more complex possibility is to allow both difference equations to operate in some overlapping region of k. Such mixtures will be called overlapping or nondisjoint. As an example consider the mixed difference equation

$$y_k = [9y_{k-1} + 7]I_{k\le 25} + [4y_{k-1} - 2]I_{k>20}: \tag{5.110}$$

where y_0 is a known constant. There are two ways to approach this. The first is to apply the generating function argument directly. Method 2 is to first see that this nondisjoint family of difference equations can be broken up into a mixed of three disjoint difference equations for $1 \leq k \leq 20; 21 \leq k \leq 25$ and $k > 25$.

5.13.1 Method 1

Here we will apply the difference equation from the top down, with no prior simplification of this nondisjoint equation. Thus

$$
\begin{aligned}
y_k &= [9y_{k-1} + 7]I_{k\leq 25} + [4y_{k-1} - 2]I_{k>20} \\
s^k y_k &= [9y_{k-1} + 7]s^k I_{k\leq 25} + [4y_{k-1} - 2]s^k I_{k>20} \\
\sum_{k=1}^{\infty} s^k y_k &= \sum_{k=1}^{\infty}[9y_{k-1} + 7]s^k I_{k\leq 25} + \sum_{k=1}^{\infty}[4y_{k-1} - 2]s^k I_{k>20}
\end{aligned}
\tag{5.111}
$$

and further simplification reveals

$$
G(s) - y_0 = 9s\sum_{k=0}^{24} s^k y_k + 4s\sum_{k=20}^{\infty} s^k y_k + 7\sum_{k=1}^{25} s^k - 2\sum_{k=21}^{\infty} s^k \tag{5.112}
$$

At this point we have to confront the overlap in the indices of the summations. Write

$$
\begin{aligned}
&9s\sum_{k=0}^{24} s^k y_k + 4s\sum_{k=20}^{\infty} s^k y_k = 9s\sum_{k=0}^{19} s^k y_k + 9s\sum_{k=20}^{24} s^k y_k + 4s\left(G(s) - \sum_{k=0}^{19} s^k y_k\right) \\
&= 4sG(s) + 5s\sum_{k=0}^{19} s^k y_k + 9s\sum_{k=20}^{24} s^k y_k
\end{aligned}
\tag{5.113}
$$

A similar evaluation shows

$$
\begin{aligned}
&7\sum_{k=1}^{25} s^k - 2\sum_{k=21}^{\infty} s^k = 7\sum_{k=1}^{20} s^k + 7\sum_{k=21}^{25} s^k - 2\left[\frac{s}{1-s} - \sum_{k=1}^{20} s^k\right] \\
&= 9\sum_{k=1}^{20} s^k + 7\sum_{k=21}^{25} s^k - \frac{2s}{1-s}
\end{aligned}
\tag{5.114}
$$

Returning to equation (5.112), we find

$$\left[G(s)-y_0\right] = 4sG(s)+5s\sum_{k=0}^{19} s^k y_k + 9s\sum_{k=20}^{24} s^k y_k + 9\sum_{k=1}^{20} s^k + 7\sum_{k=21}^{25} s^k - \frac{2s}{1-s} \tag{5.115}$$

Solving for G(s) reveals

$$G(s) = \frac{y_0}{1-4s} + \frac{5s\sum_{k=0}^{19} s^k y_k}{1-4s} + \frac{9s\sum_{k=20}^{24} s^k y_k}{1-4s} + \frac{5\sum_{k=21}^{25} s^k}{1-4s} - \frac{3\sum_{k=1}^{20} s^k}{1-4s} - \frac{2s}{(1-s)(1-4s)} \tag{5.116}$$

In order to proceed, we must have the value of the y_k's for k = 1 to 20 and for k = 20 to 24. For k = 1 to 20, we know that $y_k = 9y_{k-1} + 7$, for that the solution is

$$y_k = y_0 9^k - \frac{7}{8}\left(1-9^k\right) = \left[\frac{7+8y_0}{8}\right]9^k - \frac{7}{8} \tag{5.117}$$

For k = 20 to 24 we return to $y_k = \left[9y_{k-1}+7\right]I_{k\le 25} + \left[4y_{k-1}-2\right]I_{k>20} = 13y_{k-1}+5$ and in this range we know $y_k = 13^k\left[\frac{5+12y_{19}}{12}\right] - \frac{5}{12}$ where y_{19} comes from equation (4.6). An intermediate result that can be useful here and developed in the problems is

$$\text{if } G_1(s) = \sum_{j=k_1}^{k_2} a_j s^j \text{ and } G_2(s) = \sum_{j=0}^{\infty} b_j s^j$$
$$\text{then } G_1(s)G_2(s) \triangleright \left\{ \sum_{j=k_1}^{\min(k,k_2)} a_j b_{k-j} I_{k\ge k_1} \right\} \tag{5.118}$$

Proceed with the inversion:

$$G(s) = \frac{y_0}{1-4s} + \frac{5s\sum_{k=0}^{19} s^k y_k}{1-4s} + \frac{9s\sum_{k=20}^{24} s^k y_k}{1-4s} + \frac{7\sum_{k=21}^{25} s^k}{1-4s} + \frac{9\sum_{k=1}^{20} s^k}{1-4s} - \frac{2s}{(1-s)(1-4s)}$$

$$\triangleright \left\{ \begin{array}{l} y_0 4^k + 5 \sum_{i=0}^{\min(k-1,19)} \left(\left[\frac{7+8y_0}{8} \right] 9^i - \frac{7}{8} \right) 4^{k-i-1} \\ +9 \sum_{i=20}^{\min(k-1,24)} \left(13^i \left[\frac{5+12y_{20}}{12} \right] - \frac{5}{12} \right) 4^{k-i-1} I_{k \geq 21} \\ +7 \sum_{i=21}^{\min(k-1,25)} 4^{k-i-1} I_{k \geq 21} + 9 \sum_{i=1}^{\min(k-1,20)} 4^{k-i-1} \\ +\frac{2}{3}\left(4^k - 1\right) \end{array} \right\}$$

(5.119)

5.13.2 Method 2

A second method to solving mixtures of difference equations begins by breaking the index of the sequence $\{y_k\}$ into disjoint regions over which one equation applies. For $1 \leq k \leq 20$ $y_k = 9y_{k-1} + 7$ for y_0 known. For $20 < k \leq 25$, each equation is in force so $y_k = 9y_{k-1} + 7 + 4y_{k-1} - 2 = 13y_{k-1} + 5$ for y_{19} known. For $k > 25$, $y_k = 4y_{k-1} - 2$. Thus, the nondisjoint mixture of difference equations can be rewritten as a disjoint mixture of difference equations.

$$y_k = [9y_{k-1} + 7] I_{k \leq 20} + [13y_{k-1} + 5] I_{21 \leq k \leq 25} + [4y_{k-1} - 2] I_{k > 25} \qquad (5.120)$$

and we can apply generating function approach to this disjoint system.

$$y_k s^k = [9y_{k-1}+7]s^k I_{k\le 20} + [13y_{k-1}+5]s^k I_{21\le k\le 25} + [4y_{k-1}-2]s^k I_{k>25}$$

$$\sum_{k=1}^{\infty} y_k s^k = \sum_{k=1}^{\infty}[9y_{k-1}+7]s^k I_{k\le 20} + \sum_{k=1}^{\infty}[13y_{k-1}+5]s^k I_{21\le k\le 25} + \sum_{k=1}^{\infty}[4y_{k-1}-2]s^k I_{k>25}$$

$$G(s) - y_0 = 9s\sum_{k=0}^{19} y_k s^k + 7\sum_{k=1}^{20} s^k + 13s\sum_{k=20}^{24} y_k s^k + 5\sum_{k=21}^{25} s^k + 4\sum_{k=25}^{\infty} y_k s^k - 2\sum_{k=26}^{\infty} s^k$$

$$G(s) - y_0 = 9s\sum_{k=0}^{19} y_k s^k + 7\sum_{k=1}^{20} s^k + 13s\sum_{k=20}^{24} y_k s^k + 5\sum_{k=21}^{25} s^k$$
$$+ 4s\left[G(s) - \sum_{k=0}^{24} y_k s^k\right] - 2\left[\frac{s}{1-s} - \sum_{k=1}^{25} s^k\right] \tag{5.121}$$

Proceeding with the simplification reveals

$$G(s)(1-4s) = y_0 + 5s\sum_{k=0}^{19} y_k s^k + 9s\sum_{k=20}^{24} y_k s^k + 9\sum_{k=1}^{20} s^k + 7\sum_{k=21}^{25} s^k - \frac{2s}{1-s} \tag{5.122}$$

and inversion can take place:

$$G(s) = \frac{y_0}{1-4s} + \frac{5s\sum_{k=0}^{19} y_k s^k}{1-4s} + \frac{9s\sum_{k=20}^{24} y_k s^k}{1-4s} + \frac{9\sum_{k=1}^{20} s^k}{1-4s} + \frac{7\sum_{k=21}^{25} s^k}{1-4s} - \frac{2s}{(1-s)(1-4s)}$$

$$G(s) \quad \triangleright \quad \left\{ \begin{array}{l} y_0 4^k + 5\sum_{j=0}^{\min(k-1,19)} y_j 4^{k-j-1} + 9\sum_{j=20}^{\min(k-1,24)} y_j 4^{k-j-1} I_{k\ge 21} \\ + 9\sum_{j=1}^{\min(k-1,20)} 4^{k-j-1} + 7\sum_{j=21}^{\min(k-1,25)} 4^{k-j-1} I_{k\ge 21} - 2\sum_{j=0}^{k-1} 4^j \end{array} \right\} \tag{5.123}$$

where y_j for j=0 to 25 are identified from either the boundary condition or the first and second components of the disjoint system of equations, as in Method 1.

5.14 Mixtures of Two First-Order Nonhomogeneous Difference Equations

Families of second-order difference equations, whether homogeneous or nonhomogeneous, can be combined in mixes as well. We will consider disjoint mixes, since in the last section we demonstrated that nondisjoint mixtures of families of difference equations could be reconstructed into disjoint mixtures. The solution for the general case of disjoint mixtures of two families of second-order, nonhomogeneous difference equations will be identified. For $k \geq 0$, define

$$y_{k+2} = [ay_{k+1} + by_k + c]I_{k<k^*} + [dy_{k+1} + ey_k + f]I_{k \geq k^*} \tag{5.124}$$

as a mixture of two second-order nonhomogenous difference equations for k = zero to infinity. The hinge point at which the difference equations governing the values of the sequence $\{y_k\}$ is $k = k^*$. Proceeding as is the custom, with conversion and consolidation

$$\begin{aligned} y_{k+2} &= [ay_{k+1} + by_k + c]I_{k<k^*} + [dy_{k+1} + ey_k + f]I_{k \geq k^*} \\ s^k y_{k+2} &= [ay_{k+1} + by_k + c]s^k I_{k<k^*} + [dy_{k+1} + ey_k + f]s^k I_{k \geq k^*} \\ \sum_{k=0}^{\infty} s^k y_{k+2} &= \sum_{k=0}^{\infty} [ay_{k+1} + by_k + c]s^k I_{k<k^*} + \sum_{k=0}^{\infty} [dy_{k+1} + ey_k + f]s^k I_{k \geq k^*} \end{aligned} \tag{5.125}$$

leading to

$$\begin{aligned} \sum_{k=0}^{\infty} s^k y_{k+2} = {} & a\sum_{k=0}^{\infty} s^k y_{k+1} I_{k<k^*} + b\sum_{k=0}^{\infty} s^k y_k I_{k<k^*} + c\sum_{k=0}^{\infty} s^k I_{k<k^*} \\ & + d\sum_{k=0}^{\infty} s^k y_{k+1} I_{k \geq k^*} + e\sum_{k=0}^{\infty} s^k y_k I_{k \geq k^*} + f\sum_{k=0}^{\infty} s^k I_{k \geq k^*} \end{aligned} \tag{5.126}$$

The left side of the equation (5.126) is $s^{-2}(G(s) - y_0 - y_1 s)$. The remaining six sums are evaluated independently.

$$a\sum_{k=0}^{\infty} s^k y_{k+1} I_{k<k^*} = a\sum_{k=0}^{k^*-1} s^k y_{k+1} = as^{-1}\sum_{k=0}^{k^*-1} s^{k+1} y_{k+1} = as^{-1}\sum_{k=1}^{k^*} s^k y_k \tag{5.127}$$

$$b\sum_{k=0}^{\infty} s^k y_k I_{k<k^*} = b\sum_{k=0}^{k^*-1} s^k y_k \tag{5.128}$$

$$c\sum_{k=0}^{\infty} s^k I_{k<k^*} = c\sum_{k=0}^{k^*-1} s^k \tag{5.129}$$

$$\begin{aligned} d\sum_{k=0}^{\infty} s^k y_{k+1} I_{k\geq k^*} &= ds^{-1}\sum_{k=k^*}^{\infty} s^{k+1} y_{k+1} \\ &= ds^{-1}\sum_{k=k^*+1}^{\infty} s^k y_k = ds^{-1}\left[G(s) - \sum_{k=0}^{k^*} s^k y_k\right] \end{aligned} \tag{5.130}$$

$$e\sum_{k=0}^{\infty} s^k y_k I_{k\geq k^*} = e\sum_{k=k^*}^{\infty} s^k y_k = e\left[G(s) - \sum_{k=0}^{k^*-1} s^k y_k\right] \tag{5.131}$$

$$f\sum_{k=0}^{\infty} s^k I_{k\geq k^*} = f\sum_{k=k^*}^{\infty} s^k = f\left[\frac{1}{1-s} - \frac{1-s^{k^*}}{1-s}\right] = f\frac{s^{k^*}}{1-s} \tag{5.132}$$

We are now ready to proceed with further simplification.

$$\begin{aligned} s^{-2}\left[G(s) - y_0 - y_1 s\right] &= as^{-1}\sum_{k=1}^{k^*} s^k y_k + b\sum_{k=0}^{k^*-1} s^k y_k + c\sum_{k=0}^{k^*-1} s^k \\ &+ ds^{-1}\left[G(s) - \sum_{k=0}^{k^*} s^k y_k\right] + e\left[G(s) - \sum_{k=0}^{k^*-1} s^k y_k\right] + f\frac{s^{k^*}}{1-s} \end{aligned} \tag{5.133}$$

Multiplying each side of equation (5.133) by s^{-2} reveals

$$\begin{aligned} G(s) - y_0 - y_1 s &= as\sum_{k=1}^{k^*} s^k y_k + bs^2\sum_{k=0}^{k^*-1} s^k y_k + cs^2\sum_{k=0}^{k^*-1} s^k \\ &+ ds\left[G(s) - \sum_{k=0}^{k^*} s^k y_k\right] + es^2\left[G(s) - \sum_{k=0}^{k^*-1} s^k y_k\right] + f\frac{s^{k^*+2}}{1-s} \end{aligned} \tag{5.134}$$

Adjusting the summands for the factors of s leads to

Gathering like terms and isolating G(s) results in

$$G(s)\left[1-ds-es^{2}\right]=y_{0}+y_{1}s+as\sum_{k=1}^{k^{*}}s^{k}y_{k}+bs^{2}\sum_{k=0}^{k^{*}-1}s^{k}y_{k}+cs^{2}\sum_{k=0}^{k^{*}-1}s^{k}$$
$$-ds\sum_{k=0}^{k^{*}}s^{k}y_{k}-es^{2}\sum_{k=0}^{k^{*}-1}s^{k}y_{k}+f\frac{s^{k^{*}+2}}{1-s} \tag{5.135}$$

and the expression for G(s) is

$$G(s)=\frac{y_{0}}{1-ds-es^{2}}+\frac{y_{1}s}{1-ds-es^{2}}+\frac{as\sum_{k=1}^{k^{*}}s^{k}y_{k}}{1-ds-es^{2}}+\frac{bs^{2}\sum_{k=0}^{k^{*}-1}s^{k}y_{k}}{1-ds-es^{2}}$$
$$+\frac{cs^{2}\sum_{k=0}^{k^{*}-1}s^{k}}{1-ds-es^{2}}-\frac{ds\sum_{k=0}^{k^{*}}s^{k}y_{k}}{1-ds-es^{2}}-\frac{es^{2}\sum_{k=0}^{k^{*}-1}s^{k}y_{k}}{1-ds-es^{2}}+\frac{fs^{k^{*}+2}}{(1-s)\left(1-ds-es^{2}\right)} \tag{5.136}$$

Proceed with the inversion by recognizing (from Chapter 2) that

$$\frac{1}{1-ds-es^{2}}\triangleright\left\{\sum_{m=0}^{k}\sum_{j=0}^{m}\binom{m}{j}e^{j}d^{m-j}I_{m+j=k}\right\}=\{A(k)\} \tag{5.137}$$

and see

$$G(s)\triangleright\left\{\begin{array}{l}y_{0}A(k)+\ y_{1}A(k-1)+a\sum_{i=1}^{\min(k-1,k^{*})}y_{i}A(k-i-1)\\+b\sum_{i=0}^{\min(k-2,k^{*}-1)}y_{i}A(k-i-2)\ +c\sum_{i=0}^{\min(k-2,k^{*}-1)}A(k-i-2)\\-d\sum_{i=0}^{\min(k-1,k^{*})}A(k-i-1)-e\sum_{i=0}^{\min(k-2,k^{*}-1)}A(k-i-2)+fI_{k\geq k^{*}+2}\sum_{i=0}^{k-k^{*}-2}A(i)\end{array}\right\} \tag{5.138}$$

where y_k for $0 < k < k^*$ is the solution to the equation $y_{k+2}=ay_{k+1}+by_{k}+c$.

5.15 Example: Tracking Drug Use in a Changing Environment

5.15.1 Federal Food and Drug Administration review process

One of the goals of pharmaceutical companies, in concert with the federal Food and Drug Administration (FDA), is to ensure that the compounds that are produced and marketed are both safe and effective for the condition for which they are indicated. In order to have some measure of predictability and order in the drug approval process, it is important for both sponsors of new drugs, biologics (naturally occurring substances produced by manufacturers to make up deficiencies or enhance body function) or devices and the FDA, to have a mutual understanding of what is expected for the FDA to give its approval.

From the perspective of the drug manufacturer, this mutual understanding allows the planning of drug development and decision making with a reasonable prospect of success if all works out as hoped. From the perspective of the agency (the FDA), this joint involvement allows them to approve a drug submitted based upon standards understood by both sides. This introduces a fairness into the process.

5.15.2 The issue

Consider a compound produced by a sponsor (i.e., pharmaceutical company) that has been approved by the FDA for use by patients. The sponsor is interested in tracking sales (number of units) of this drug. Let y_k be the number of units of a particular drug in the kth month. Assuming that the sales of this drug in any month is a moving average of the sales of this drug for the previous three months, y_k is governed by the difference equation

$$y_k = \frac{y_{k-1} + y_{k-2} + y_k}{3} \tag{5.139}$$

where y_0, y_1, and y_2 are known and $k \geq 3$.

As currently approved, the drug requires a prescription, so patients must see a physician in order to receive the drug. The sponsoring drug company anticipates that they will be able to convince the regulatory agencies to remove the requirement for a prescription. This means that patients will be able to buy the drug "OTC" or over the counter, without first seeing a physician. Removing the requirement of a physician is expected to change sales of the drug since neither the expense nor the inconvenience of the physician visit will be borne by the patient. The sponsor expects approval of the OTC status after the drug has been available for 36 months. Once OTC approval is obtained, they expect drug sales to be governed by the difference equation

$$y_k = \frac{y_{k-1} + y_{k-2}}{2} \tag{5.140}$$

The sponsor is interested in projecting drug sales over five years. These five-year projections span a period of time in equation (5.139) and equation (5.140) are in force. Write this as a disjoint mix of two families of difference equations:

$$y_k = \frac{y_{k-1} + y_{k-2} + y_{k-3}}{3} I_{k<36} + \frac{y_{k-1} + y_{k-2}}{2} I_{k\geq 36} \tag{5.141}$$

for $k = 0$ to ∞, and y_0, y_1, and y_2 are known. This is a mixture of a third-order family and a second-order family of homogeneous difference equations. Apply the generating function transformation, beginning with conversion and consolidation

$$s^k y_k = s^k \frac{y_{k-1} + y_{k-2} + y_{k-3}}{3} I_{k<36} + s^k \frac{y_{k-1} + y_{k-2}}{2} I_{k\geq 36} \tag{5.142}$$

Simplification reveals

$$s^k y_k = s^k \frac{y_{k-1}}{3} I_{k<36} + s^k \frac{y_{k-2}}{3} I_{k<36} + s^k \frac{y_{k-3}}{3} I_{k<36}$$
$$+ s^k \frac{y_{k-1}}{2} I_{k\geq 36} + s^k \frac{y_{k-2}}{2} I_{k\geq 36} \tag{5.143}$$

and summing each expression in equation (5.143) from $k = 3$ to infinity to find

$$\sum_{k=3}^{\infty} s^k y_k = \sum_{k=3}^{\infty} s^k \frac{y_{k-1}}{3} I_{k<36} + \sum_{k=3}^{\infty} s^k \frac{y_{k-2}}{3} I_{k<36} + \sum_{k=3}^{\infty} s^k \frac{y_{k-3}}{3} I_{k<36}$$
$$+ \sum_{k=3}^{\infty} s^k \frac{y_{k-1}}{2} I_{k\geq 36} + \sum_{k=3}^{\infty} s^k \frac{y_{k-2}}{2} I_{k\geq 36}$$

(5.144)

The first level of simplification adjusts the bounds on the summation signs

$$\sum_{k=3}^{\infty} s^k y_k = \sum_{k=3}^{35} s^k \frac{y_{k-1}}{3} + \sum_{k=3}^{35} s^k \frac{y_{k-2}}{3} + \sum_{k=3}^{35} s^k \frac{y_{k-3}}{3}$$
$$+\sum_{k=36}^{\infty} s^k \frac{y_{k-1}}{2} + \sum_{k=36}^{\infty} s^k \frac{y_{k-2}}{2} \tag{5.145}$$

The next adjustment synchrnizes the power of s with the index of y in each of the summands

$$\sum_{k=3}^{\infty} s^k y_k = s\sum_{k=3}^{35} s^{k-1} \frac{y_{k-1}}{3} + s^2\sum_{k=3}^{35} s^{k-2} \frac{y_{k-2}}{3} + s^3\sum_{k=3}^{35} s^{k-3} \frac{y_{k-3}}{3}$$
$$+s\sum_{k=36}^{\infty} s^{k-1} \frac{y_{k-1}}{2} + s^2\sum_{k=36}^{\infty} s^{k-2} \frac{y_{k-2}}{2} \tag{5.146}$$

Simplification can proceed

$$\sum_{k=3}^{\infty} s^k y_k = \frac{s}{3}\sum_{k=2}^{34} s^k y_k + \frac{s^2}{3}\sum_{k=1}^{33} s^k y_k + \frac{s^3}{3}\sum_{k=0}^{32} s^k y_k$$
$$+\frac{s}{2}\sum_{k=35}^{\infty} s^k y_k + \frac{s^2}{2}\sum_{k=34}^{\infty} s^k y_k \tag{5.147}$$

Now observe

$$\sum_{k=3}^{\infty} s^k y_k = G(s) - y_0 - y_1 s - y_2 s^2 \tag{5.148}$$

$$\frac{s}{3}\sum_{k=2}^{34} s^k y_k + \frac{s^2}{3}\sum_{k=1}^{33} s^k y_k + \frac{s^3}{3}\sum_{k=0}^{32} s^k y_k$$
$$= \frac{1}{3}\left[\begin{array}{l} s\sum_{k=2}^{34} s^k y_k + s^2\left(\sum_{k=2}^{34} s^k y_k + y_1 s - y_{34}s^{34}\right) \\ +s^3\left(\sum_{k=2}^{34} s^k y_k + y_0 + y_1 s - y_{33}s^{33} - y_{34}s^{34}\right) \end{array}\right] \tag{5.149}$$

And observe that

$$\frac{s}{2}\sum_{k=35}^{\infty}s^k y_k+\frac{s^2}{2}\sum_{k=34}^{\infty}s^k y_k$$
$$=\frac{1}{2}G(s)\left(s+s^2\right)-\frac{s}{2}\left(\sum_{k=2}^{34}s^k y_k+y_0+y_1 s\right)-\frac{s^2}{2}\left(\sum_{k=2}^{34}s^k y_k+y_0+y_1 s-y_{34}s^{34}\right) \tag{5.150}$$

Which leads to the reexpression of equation (5.150) as

$$G(s)\left[1-\frac{s+s^2}{2}\right]-y_0-y_1 s-y_2 s^2=\frac{1}{6}\sum_{k=2}^{34}s^k y_k\left[2s+2s^2+2s^3-3s-3s^2\right]$$
$$+y_0\left[\frac{s^3}{3}-\frac{s}{2}-\frac{s^2}{2}\right]+y_1\left[\frac{s^3}{3}+\frac{s^4}{3}-\frac{s^2}{2}-\frac{s^3}{2}\right] \tag{5.151}$$
$$-\frac{y_{34}}{3}s^{36}-\frac{y_{33}}{3}s^{36}+\frac{y_{34}}{2}s^{36}-\frac{y_{34}}{3}s^{37}$$

Further simplication reveals

$$G(s)(1-s)(2+s)=2\left[y_0+y_1 s+y_2 s^2\right]+\frac{1}{3}s\left[2s^2-s-1\right]\sum_{k=2}^{34}s^k y_k$$
$$+\frac{y_0}{3}\left[2s^3-3s^2-3s\right]+\frac{1}{3}y_1\left[2s^4-s^3-3s^2\right]+\frac{1}{3}\left[y_{34}s^{36}-2y_{33}s^{36}-2y_{34}s^{37}\right] \tag{5.152}$$

It now remains to gather like powers of s

$$G(s)=\frac{2y_0}{(1-s)(2+s)}+\frac{(2y_1-y_0)s}{(1-s)(2+s)}-\frac{(y_0+y_1-2y_2)s^2}{(1-s)(2+s)}$$
$$+\frac{(2y_0-y_1)s^3}{3(1-s)(2+s)}+\frac{2}{3}\frac{y_1 s^4}{(1-s)(2+s)}+\frac{1}{3}\frac{(y_{34}-2y_{33})s^{36}}{(1-s)(2+s)}$$
$$-\frac{2}{3}\frac{y_{34}s^{37}}{(1-s)(2+s)}-\frac{s}{3}\frac{(2s+1)}{2+s}\sum_{k=2}^{34}y_k s^k \tag{5.153}$$

Recall that $\frac{1}{2+s} \triangleright \left\{(-1)^k \left[\frac{1}{2}\right]^{k+1}\right\}$, and a simple convolution argument demonstrating

$$\frac{1}{(1-s)(2+s)} \triangleright \left\{\sum_{j=0}^{k} \frac{(-1)^j}{2^{j+1}}\right\} = \{A(k)\} \tag{5.154}$$

to allow the inversion of equation (5.153) to proceed, revealing

$$\begin{aligned}
y_k &= 2y_0A(k) + (2y_1 - y_0)A(k-1) - (y_0 + y_1 - 2y_2)A(k-2) \\
&+ \frac{2y_0 - y_1}{3}A(k-3) + \frac{2}{3}y_1A(k-4) + \frac{1}{3}(y_{34} - 2y_{33})A(k-36) \\
&- \frac{2}{3}y_{34}A(k-37) - \frac{2}{3}\sum_{j=2}^{\min(k-2,34)} y_j(-1)^{k-2-j}\left(\frac{1}{2}\right)^{k-1-j} \\
&- \frac{1}{3}\sum_{j=2}^{\min(k-1,34)} y_j(-1)^{k-1-j}\left(\frac{1}{2}\right)^{k-j}
\end{aligned} \tag{5.155}$$

Problems

1. Prove the following:

If $G_1(s) = \sum_{j=k_1}^{k_2} a_j s^j$ and $G_2(s) = \sum_{j=0}^{\infty} b_j s^j$

then $G_1(s)G_2(s) \triangleright \left\{\sum_{j=\min(k,k_1)}^{\min(k,k_2)} a_j b_{k-j} I_{k \geq k_1}\right\}$.

2. Consider a mixture of three difference equations in three disjoint but contiguous regions.

Show $\dfrac{(a_1 - a_3)s\sum_{k=0}^{k_1-1} s^k y_k}{1 - a_3 s} \triangleright \left\{(a_1 - a_3)\sum_{j=0}^{\min(k-1,k_1-1)} y_j a_3^{k-j-1}\right\}$

3. Show $\dfrac{(a_2 - a_3)s\sum_{k=k_1}^{k_2-1} s^k y_k}{1 - a_3 s} \triangleright (a_2 - a_3)\sum_{j=\min(k-1,k_1)}^{\min(k-1,k_2-1)} y_j a_3^{k-j-1} I_{k \ge k_1}$

4. Show $\dfrac{1}{d - s - s^2} \triangleright \left\{\sum_{m=0}^{k} \frac{1}{d^{m+1}} \sum_{i=0}^{m} \binom{m}{j} I_{m+j=k}\right\}$ where d is a known constant.

Solve the following families of difference equations with variable coefficients using a generating function approach:

5. $3y_{k+1} = by_k + k^2 a^k$
6. $6y_{k+3} = by_k + a\log(k-1)$
7. $-y_{k+4} = by_{k+3} + ak - bk^2$
8. $7y_{k+2} = by_{k+1} + a^k(k+1)$
9. $-4y_{k+1} = by_k + \cos(k)a^k c^{-k-1}$ where k is in radians

Solve the following difference equation mixtures using a generating function argument:

10. $y_{k+1} = y_k I_{k<10} - y_k I_{k\ge 10}$
11. $3y_{k+2} = 6y_{k+1} I_{k<15} - 4y_k I_{k\ge 15}$
12. $4y_{k+2} = (y_{k+1} - 3y_k) I_{k<22} - (y_k + 1) I_{k\ge 22}$
13. $-y_{k+2} = (4y_{k+1} + 5y_k) I_{k<13} - (4y_{k+1} + 5y_k) I_{k\ge 13}$
14. $6y_{k+2} = (4y_{k+1} + 5y_k) I_{k<23} - y_k I_{23\le k\le 30} - y_k (4y_{k+1} + 5y_k) I_{k>30}$
15. $2y_{k+2} = (y_{k+1} - 11y_k) I_{k<10} - (4y_k + 1) I_{5\le k\le 40} - (9y_{k+1} - 7y_k) I_{k>30}$

6

Difference Equations: Run Theory

6.1 Introduction

Probability models have proven to be very useful in the quantitative study of health care phenomena but their successful implementation requires the user to understand the underlying event whose probability is to be obtained. The researcher must often examine the event from a different and new perspective, while simultaneously grasping the nature of the available probability models. Only in this way can the worker adapt the underlying event and mold the probability model to generate relevant probabilities for events of interest. It is this joint process that makes the application of probability theory somewhat of an art. A useful tool for modeling public health problems of interest is the use of the theory of runs. After a brief review of run theory, this chapter will build a bridge to run theory with difference equations. This work will span two chapters, and introduce two models, the $\mathbf{R}_{i,k}(n)$ model and the $\mathbf{T}_{[K,L]}(n)$ model. Each of these models will be valuable in using difference equations to generate probability distributions of interest to public health issues that can be considered as a "run" of specific events.

The $\mathbf{R}_{i,k}(n)$ model, which is the simpler of the two models to be introduced, will be developed and solved first. During this solution, we will identify the telescoping difference equation, a potential obstacle to the generating function approach to its solution, and demonstrate how to solve this difficult family of equations. The more complicated $\mathbf{T}_{[K,L]}(n)$ will be developed in full in Chapter 7 when the use of difference equations to provide order statistics is discussed.

6.2 Review of Run Theory

6.2.1 Probability of consecutive Bernoulli trials

Classic run theory begins with an understanding of Bernoulli trials. A Bernoulli trial (introduced in Chapter 2) is an experiment that results in one of two outcomes: success or failure. If x denotes the result of the experiment, then we can say that if the experiment results in a success, then $x = 1$. Analogously, if a failure occurs, then $x = 0$. The probability of a success is denoted by $P[x = 1] = p$, and the $P[x = 0] = q = 1 - p$. A sequence of independent, identically distributed Bernoulli trials is a sequence of such experiments where 1) the probability of success remains the same from experiment to experiment and 2) the sequence of experiments are independent, i.e., the occurrence of a success on one experiment does not influence the outcome of any future experiment.

A simple example of such a sequence is consecutive flips of a coin for which the probability of a head is p. Each flip (experiment) results in either a success (heads) or failure (tails). The probability of success is the same from experiment to experiment. Also, the independence assumption is applicable in this sequence of experiments. For example, knowledge that the third flip of the coin resulted in a head does not affect the probability that subsequent flips will be heads. Thus, this operation of consecutive flips of a fair coin represents a sequence of Bernoulli trials.

One can easily apply the binomial probability model introduced in Chapter 2 to compute useful probabilities e.g., the probability that there are exactly three tails in four flips of the coin. One intuitive way to solve this problem is to count the number of ways three tails can occur in four flips. They are

HTTT, THTT, TTHT, TTTH

The probability for each of these sequences is obtained by multiplying the probability associated with each occurrence H or T. For the sequence **HTTT** the probability is $p \cdot q \cdot q \cdot q = pq^3$. The key to the use of multiplication is that a head *and* a tail *and* a tail *and* a tail must occur. The occurrence of these simultaneous events requires that the probabilities be multiplied. Similarly, the probability of **THTT** is qpq^2, the probability of **TTHT** is q^2pq, and the probability of **TTTH** is q^3p. Each of these probabilities is equal to pq^3.

To compute the probability of three tails in four flips of the coin, the probability of the occurrence of **HTTT,** *or* **THTT,** *or* **TTHT,** *or* **TTTH** is needed. The occurrence of either of these circumstances satisfies the probability of interest (three tails in four flips of a fair coin). In this case we add the probabilities. Thus P[three tails in four flips of a fair coin] is P[**HTTT**] + P[**THTT**] + P[**TTHT**] +P[**TTTH**] = $4pq^3$.

However, this sequence of Bernoulli trials can be the basis of more complicated events of interest. For example, if the coin is flipped four times, how likely is it that there will only be three heads, and that these heads will occur consecutively (i.e., three heads in a row)? How likely is it that heads and tails will alternate throughout the sequence? Although the underlying experiments represent Bernoulli trials, these last two events have introduced a new level of complexity.

6.2.2 Definition of a run

In his classic textbook on probability, Feller [1] provides a succinct definition of a run of Bernoulli trials. Consider a sequence of Bernoulli trials of known length. Within this sequence, Feller [1] states that we should count the length of each subsequence of the same element. The occurrence of these lengths is the distribution of run lengths in the original sequence. For example, consider the following sequence resulting from six consecutive flips of the same coin.

HHTTTH

Runs are always separated by meetings of unlike neighbors. This sequence of six consecutive Bernoulli trials resulted in one run of length two, one run of length three, and one run of length one. If we focused on the occurrence of failure runs, we would say that this sequence contains one run failure run of length three. If instead the focus was on the occurrence of runs of success, we would characterize this sequence as having a success run of length two and an additional success run of length one.

There should be some other way to consider the distribution of run lengths than by counting them, and several esteemed probabilists have devoted attention to this problem. In Chapter XIII, Feller [1] provides the classic generating function approach to finding the probability that a run of length r occurs at the nth trial and the probability runs of length r occurs for the first time on the nth trial. A calculation of derivatives involving generating functions reveals the mean and variance of the recurrence times of runs of length r. Feller [1] also gives some weak convergence results on the distribution of the number of runs using the central limit theorem.

6.2.3 Retrospective run theory

A major contribution to run theory is provided by A. M. Mood [2], who after a review of the literature on the subject, provides an in-depth examination of the theory of the distribution of runs in a sequence of Bernoulli trials. The first sections of his paper derive the probability distribution obtained from random arrangements of a fixed number of each kind of element (success or failure) in a sequence of Bernoulli trials. The most common use of this one result of Mood's has been to compute the distribution of the number of runs in a sequence of n Bernoulli trials if it is known that there have been n_1 successes and n_2 failures. In this formulation, we assume that we have already observed the results of a sequence of Bernoulli trials, and use the results (in terms of n_1 and n_2) to compute the probability of the occurrence of runs of different lengths. This problem is also discussed in Brownlee [3], Stevens [4], and Wald and Wolfowitz [5]. The cumulative distribution of the number of runs is tabulated in Swed and Eisenhart [6]. As these computations are historically cumbersome, and the tabulations of Swed and Eisenhart limited, much work has been undertaken to compute an approximation for the probability distribution of the number of runs. These approximations, given in Stevens [4], Wald and Wolfowitz [5], and Wallis [7] have been useful to quality control engineers.

6.2.4 Prospective run theory

A difficulty with the retrospective approach is that it presupposes knowledge of n_1 and n_2, the total number of successes and failures in the sequence of Bernoulli trials. Thus, one must wait until the sequence has occurred and knowledge of n_1 and n_2 is available before the probability distribution of the total number of runs in the sequence is computed. This methodology is only of limited use in predicting the future behavior of Bernoulli sequences, since one can only guess n_1 and n_2 for the sequence of Bernoulli sequences yet to be observed.

The second portion of Mood's work [2] speaks to this issue, for it is here that he examines the distribution of runs of elements from a binomial or multinomial population without knowing specifically the values of n_1 and n_2. That is, instead of having the number of elements of each kind fixed, he now supposes that they are randomly drawn from a binomial or multinomial population. We term this investigation the prospective approach to predicting the future occurrence of runs, since predictions are made without knowing the exact makeup of the sequence.

In Mood's approach to this prospective delineation, the number of successes and failures, n_1 and n_2 now become random variables subject only to the restriction that their sum equals the sample size n. From this Mood has identified the probability that there are exactly i runs of length k. Unfortunately, this perspective on the prospective approach is impeded by a final expression, that is provided in terms of n_1 and n_2. Mood acknowledges this "dependence among the arguments" and indicates that it may be removed by summing over n_1 and n_2 to obtain the distributions for the r's alone. However, he states that the results of these summations are quite cumbersome, and in some cases cannot be explicitly identified. Thus, although Mood solves the prospective run probability distribution, the solution is left in the framework of the retrospective approach, i.e., summations or expectations of n_1 and n_2. He does, however, go on to identify the moments of the run length probability distributions.

6.2.5 Cumulative sum approaches

Another approach to run theory is the examination of cumulative sums (CUSUMS). Returning to the use of x in describing the results of a sequence of Bernoulli trials ($x_i = 1$ if the trial results in a success, and $x_i = 0$ if the trial results in a failure), one can consider accumulating the sums of these results (a binomial random variable) and focus attention on the manner in which this sum may provide information about the occurrence of runs. Page [8] has produced a pivotal paper in discussing cumulative sum analysis. An important example of this methodology's utility is in industry, where this approach is used to identify when the quality of a massed produced product begins to deviate sharply from expected. The relationship between cumulative sum analysis and run theory lies in the fact that a change in the cumulative sum is directly related to a change in the distribution of the average run length. The sampling schemes and hypothesis tests that were produced from this approach were seen to be based on the average run length of the sequence of interest.

This paper prompted much theoretical discussion of the average run length of successes in a sequence of Bernoulli trials, crucial to an understanding of the behavior of the cumulative sum. Woodall [9] carried Page's results one step further by constructing a Markov chain representation of the two-sided CUSUM procedure based on integer-valued cumulative sums. The general problem is that one observes successive independent random variables x_1, x_2, x_3,... generated by a sequence of probability mass functions $f(x_i;0)$, $i = 1, 2, 3, \ldots$. The CUSUM procedure leads to a formal statistical hypothesis test to detect a shift in either direction of the sequence of observed occurrences. The run length of the procedure is the length of time over which the shift is detected. However, in order to determine the run length, the author was forced to identify the state space, which consists of all possible combinations of values for the two cumulative sums. This requirement led to the development of an approximation of the run length distribution using the theory of Markov chains.

More recently O'Brien [10] proposed a procedure for testing the hypothesis that Bernoulli trials are independent with common probability of success. This is a test of the hypothesis that a fixed number of successes and failures fall randomly in the sequence, and is based on the generation of weighted linear combinations of the variances of run lengths' successes and failures. The null hypothesis is rejected if the total number of runs is too small (a clustered arrangement) or too large (a systematic arrangement). However, this is a retrospective approach, requiring knowledge of the total number of successes and failures *a priori*.

Janson [11] proves convergence series for cumulative sums of Bernoulli sequences and applies them to run theory. Also, Pittel [12] has discussed convergence results of the distribution of the number of runs in selected circumstances. Shaunesy [13] presents computational procedures that can be used to calculate critical values for statistical tests based on the total number of runs.

Possible applications to this general run theory are numerous. Schwager [14] notes applications in DNA sequencing, psychology (successful people experience runs or strings of successes), sociology, and radar astronomy.

6.3 Motivation for the $R_{iK}(n)$ Model

6.3.1 The definition of $R_{ik}(n)$

We begin by defining $\mathbf{R}_{ik}(n)$ as the probability that there are exactly i failure runs of length k in n Bernoulli trials. $\mathbf{R}_{ik}(n)$ may be regarded as the probability distribution of the number of failure runs of a given length in n trials. Using this definition once n is known, it is evident that the maximum number of failure runs that may occur is dependent on both n and k. Reflecting on the possible failure run lengths, we see that the shortest failure run length is zero. Let I be the largest number of runs possible of length k in n trials. Then

$$I = \left[\frac{n+1}{k+1} \right] \tag{6.1}$$

where the bracket indicates the greatest integer contained in the fraction. For example, the largest number of failure runs of length 2 in 6 trials is [7/3] = [2.6667] = 2. As another example, there can be [15/3] = 5 failure runs of length 2 in 14 trials.

The plan is–once q, k, and n are provided–to compute a difference equation for $\mathbf{R}_{ik}(n)$. Once we identify the probability distribution of the number of failure runs of length k in n trials, for fixed k and n, we will be able to compute other quantities of interest. For example, defining the expected number of failure runs of a given length k as $E_k(n)$ then if follows that

$$E_k(n) = \sum_{i=1}^{I} iR_{ik}(n) \tag{6.2}$$

6.4 Development of the $R_{ik}(n)$ Model

6.4.1 The $R_{0,k}(n)$ model

The $\mathbf{R}_{ik}(n)$ model is a family of difference equations in two variables. For fixed i, the model represents a difference equation in n. However we will see that the model can also be approached as a difference equation in i. We will proceed with identifying the solution for $\mathbf{R}_{ik}(n)$ by first identifying $\mathbf{R}_{0,k}(n)$. From this solution, we will proceed to the solution for $\mathbf{R}_{1k}(n)$, ending with $\mathbf{R}_{ik}(n)$ for $i > 1$. $\mathbf{R}_{0,k}(n)$ is the probability that there are exactly 0 runs of length k. This is the probability of, in a sequence of n Bernoulli trials, runs of length 1, 2, 3, 4, ..., k - 1, k + 1, k +2, k + 3,..., n. The several boundary conditions are

$$\begin{aligned} R_{0,k}(n) &= 1 \qquad \text{for } n < k \\ R_{0,k}(n) &= 0 \qquad \text{for } n < 0 \\ R_{0,k}(0) &\equiv 1 \end{aligned} \tag{6.3}$$

Some examination of these boundary conditions is in order. There can of course be no runs of length k for $n < k$, so $\mathbf{R}_{0,k}(n) = 0$ in this circumstance. It is useful to assert that the probability of no runs of length k is zero when n is negative. Also define $\mathbf{R}_{0,k}(0) = 1$. Using these boundary conditions provided in the relationships laid out in (6.3), the general difference equation for the probability of exactly no runs of length k in n trials when $n \geq k$ may be written as

$$\begin{aligned} R_{0,k}(n) = {} & pR_{0,k}(n-1) + qpR_{0,k}(n-2) + q^2pR_{0,k}(n-3) + q^3pR_{0,k}(n-4) \\ & + \ldots + q^{k-1}pR_{0,k}(n-k) + q^{k+1}pR_{0,k}(n-k-2) \\ & + \ldots + q^{n-1}pR_{0,k}(0) + q^nI_{n \neq k} \end{aligned} \tag{6.4}$$

Each term on the right side of equation (6.4) permits a failure run length of a specified length. The first term includes a failure run length of zero (i.e., p =

q^0p). The next term permits a failure run of length one, the following term permits a run length of two, etc. Every possible failure run length is permitted except a run of length k, a requirement of $\mathbf{R}_{0,k}(n)$. Thus, $\mathbf{R}_{0,k}(n)$ is exactly the probability of permitting all possible failure run lengths except a failure run of length k. Insight into the equation (6.4) is provided in section 6.5. However, we first complete the specification of $\mathbf{R}_{i,k}(n)$ for all nonnegative integer values of i for completeness.

6.4.2 The $R_{1,k}(n)$ and $R_{i,k}(n)$, for i > 1

A similar set of equations may be defined to compute the probability of exactly one failure run of length k in n trials, $\mathbf{R}_{1k}(n)$, as follows

$$
\begin{aligned}
&R_{1,k}(n) = 0 \text{ for } n < k \\
&R_{1,k}(n) = pR_{1,k}(n-1) + qpR_{1,k}(n-2) + q^2pR_{1,k}(n-3) + q^3pR_{1,k}(n-3) + \cdots \\
&\qquad + q^{k-1}pR_{1,k}(n-k) + q^{k+1}pR_{1,k}(n-k-2) + \cdots \\
&\qquad + q^{n-1}pR_{1,k}(0) + q^kpR_{0,k}(n-k-1)
\end{aligned}
\tag{6.5}
$$

There is one additional observation we must make at this point. For $i > 1$, there is a minimum number of observations n_i, such that for $n < n_i$ $\mathbf{R}_{ik}(n) = 0$. It can be easily shown that $n_i = i(k + 1) - 1$. Also,

$$R_{i,k}(n_i) = q^k \left(pq^k\right)^{i-1} \tag{6.6}$$

for $i > 1$. For $i > 1$ find

$$
\begin{aligned}
&R_{i,k}(n) = 0 \text{ for } n \le n_i \\
&R_{i,k}(n) = pR_{i,k}(n-1) + qpR_{i,k}(n-2) + q^2pR_{i,k}(n-3) + q^3pR_{i,k}(n-3) + \cdots \\
&\qquad + q^{k-1}pR_{i,k}(n-k) + q^{k+1}pR_{i,k}(n-k-2) + \cdots \\
&\qquad + q^{n-1}pR_{i,k}(0) + q^kpR_{i-1,k}(n-k-1)
\end{aligned}
\tag{6.7}
$$

Note the last term in equation (6.5). It permits a failure run of length k if i the remaining n – k - 1 trials, there are no runs of length k. In order to compute $\mathbf{R}_{ik}(n)$ for a fixed run length k for all possible values of i and n, we will need to solve each of equations (6.4), (6.5), and (6.7). The initial strategy will be to combine both the generating function and the iterative approaches. The generating function approach will provide the solution for all values of $n \geq k$. The iterative approach will be used to solve the difference equation for i.

6.5 $R_{0,k}(n)$ Examples and Solution

6.5.1 Intuition and background for $R_{0,k}(n)$

Before initiating the generating function approach for the solution of $R_{i,k}(n)$, we will provide some simple examples of its operation. The formulation for $\mathbf{R}_{0,k}(n)$ is

$$\begin{aligned}R_{0,k}(n) &= pR_{0,k}(n-1)+qpR_{0,k}(n-2)+q^2pR_{0,k}(n-3)+q^3pR_{0,k}(n-4)+\cdots\\&+q^{k-1}pR_{0,k}(n-k)+q^{k+1}pR_{0,k}(n-k-2)+\ldots+q^{n-1}pR_{0,k}(0)+q^nI_{n\neq k}\end{aligned} \tag{6.8}$$

While equation (6.8) appears complicated, it is actually made up of several repeating, intuitive components. The relationship is recursive in n, and assumes that both k and q (of course $p = 1 - q$) are fixed. Equation (6.8) permits failure run lengths of 0, 1, 2, 3, ..., $k - 1$, $k + 1$, $k + 2$, $k + 3$,...,n. Thus, every failure run length except for length k is permitted. As noted before, $\mathbf{R}_{0,k}(n)$ is 1 for $n < k$, since it is impossible to have a failure run length of k in less than k trials. $\mathbf{R}_{0,k}(n) = 0$ for negative values of n. However, define $\mathbf{R}_{0,k}(0) = 1$ without any loss of generality.

As an example of the use of equation (6.8), consider computing the probability that there will be no failure runs of length k in k trials. For this simple example, we need not turn to a difference equation for the solution. A moment's reflection reveals that there is only one way to get a failure run length of k in k trials i.e., each of the k trials results in a failure. The probability of this event is q^k. Thus, the probability of no runs of length k in k trials is $1 - q^k$. To verify the accuracy of equation (6.8) compute

$$\begin{aligned}R_{0,k}(k) &= pR_{0,k}(k-1)+qpR_{0,k}(k-2)+q^2pR_{0,k}(k-3)\\&\quad+q^3pR_{0,k}(k-4)+\ldots+q^{k-1}pR_{0,k}(0)\\&= p+qp+q^2p+q^3p+q^4p+q^5p+\ldots+q^{k-1}p\\&= p\left(1+q+q^3+q^4+q^5+q^6+\ldots+q^{k-1}\right)\end{aligned} \tag{6.9}$$

and simplify to see that

$$R_{0,k}(k) = p\left[\frac{1-q^k}{1-q}\right] = 1-q^k \tag{6.10}$$

confirming our intuitive solution. As another example, consider the solution for $\mathbf{R}_{0,k}(k+1)$. Again, a moment's reflection reveals that there are only two ways to obtain 1 or more runs of length k . The first event is that the first trial is a success, and the second event is that only the last of the n + 1 trials is a success. Each of these two events occurs with probability pq^k . Since the desired solution is the absence of the occurences of these events, compute $\mathbf{R}_{0,k}(k+1) = 1 - 2pq^k$. Can we find the same result through the application of equation (6.8)?

$$\begin{aligned} R_{0,k}(k+1) &= pR_{0,k}(k) + qpR_{0,k}(k-1) + q^2pR_{0,k}(k-2) \\ &\quad + q^3pR_{0,k}(k-3) + \ldots + q^{k-1}pR_{0,k}(1) + q^{k+1} \\ &= p(1-q^k) + qp + q^2p + q^3p + q^4p + q^5p + \ldots + q^{k-1}p + q^{k+1} \\ &= p(1-q^k) + qp\left(1 + q + q^3 + q^4 + q^5 + q^6 + \ldots + q^{k-2}\right) + q^{k+1} \end{aligned} \tag{6.11}$$

Additional simplification reveals

$$\begin{aligned} R_{0,k}(k+1) &= p(1-q^k) + qp\left[\frac{1-q^{k-1}}{1-q}\right] + q^{k+1} = p - pq^k + q - q^k + q^{k+1} \\ &= 1 - pq^k - q^k(1-q) = 1 - 2pq^k \end{aligned} \tag{6.12}$$

For these simple examples, it is of course more efficient to find the solution by implementing some elementary probability considerations than to use a difference equation. However, these excursions do provide some experience with the reasoning behind $\mathbf{R}_{0,k}(n)$ as well as providing some experience in the computations involved.

6.5.2 Solution of $R_{0,k}(n)$: The complications of conversion

An initial attempt to solve $\mathbf{R}_{0,k}(n)$ will be to apply the generating function approach. Begin with rewriting equation (6.8)

$$
\begin{aligned}
R_{0,k}(n) = {} & pR_{0,k}(n-1) + qpR_{0,k}(n-2) + q^2pR_{0K}(n-3) \\
& + q^3pR_{0,k}(n-4) + \ldots + q^{k-1}pR_{0K}(n-k) \\
& + q^{k+1}pR_{0,k}(n-k-2) + \ldots + q^{n-1}pR_{0,k}(0) + q^nI_{n\neq k}
\end{aligned}
\tag{6.13}
$$

and see that this can be summarized as

$$
R_{0,k}(n) = \sum_{j=0}^{k-1} pq^jR_{0,k}(n-j-1) + \sum_{j=k+1}^{n-1} pq^jR_{0,k}(n-j-1) + q^nI_{n\neq k} \tag{6.14}
$$

Note that the second summation has $n-1$ as an upper bound. This term points to an unusual feature in a difference equation; the number of terms in the equation is not fixed but is instead a function of n. For example

$$
\begin{aligned}
R_{0,k}(1) &= pR_{0,k}(0) + q \\
R_{0,k}(2) &= pR_{0,k}(1) + pqR_{0,k}(0) + q^2 \\
R_{0,k}(3) &= pR_{0,k}(2) + pqR_{0,k}(1) + pq^2R_{0,k}(0) + q^3 \\
R_{0,k}(4) &= pR_{0,k}(3) + pqR_{0,k}(2) + pq^2R_{0,k}(1) + pq^3R_{0,k}(0) + q^4
\end{aligned}
\tag{6.15}
$$

This is a telescoping difference equation. As n increases, the number of terms in the difference equation increases as well. This is somewhat unusual for us and it will have important implications as we move through the conversion process. Proceeding in the customary fashion, define G(s) as

$$
G(s) = \sum_{n=0}^{\infty} s^nR_{0,k}(n) \tag{6.16}
$$

and proceed by multiplying each side of equation (6.13) by s^n to find

$$
\begin{aligned}
s^nR_{0,k}(n) = {} & s^n\sum_{j=0}^{k-1} pq^jR_{0,k}(n-j-1) + s^n\sum_{j=k+1}^{n-1} pq^jR_{0,k}(n-j-1) \\
& + s^nq^nI_{n\neq k}
\end{aligned}
\tag{6.17}
$$

Summing over the range of n for $k > 0$

$$\sum_{n=0}^{\infty} s^n R_{0,k}(n) = \sum_{n=0}^{\infty} s^n \sum_{j=0}^{k-1} pq^j R_{0,k}(n-j-1) + \sum_{n=0}^{\infty} s^n \sum_{j=k+1}^{n-1} pq^j R_{0,k}(n-j-1) \tag{6.18}$$
$$+ \sum_{n=0}^{\infty} s^n q^n I_{n \neq k}$$

The left side of equation (6.18) is G(s). Evaluate each of the terms on the right side of equation (6.18) term by term, beginning with $\sum_{n=0}^{\infty} s^n \sum_{j=0}^{k-1} pq^j R_{0,k}(n-j-1)$ as follows:

$$\sum_{n=0}^{\infty} s^n \sum_{j=0}^{k-1} pq^j R_{0,k}(n-j-1) = \sum_{j=0}^{k-1} \sum_{n=0}^{\infty} s^n pq^j R_{0,k}(n-j-1)$$
$$= \sum_{j=0}^{k-1} pq^j \sum_{n=0}^{\infty} s^n R_{0,k}(n-j-1) = \sum_{j=0}^{k-1} pq^j s^{j+1} \sum_{n=0}^{\infty} s^{n-j-1} R_{0,k}(n-j-1) \tag{6.19}$$
$$= \sum_{j=0}^{k-1} pq^j s^{j+1} G(s)$$

6.6 Continued Consolidation of $R_{0,k}(n)$

The next term in the expression for G(s) from the second term of the right side of equation (6.18) requires careful attention. This expression

$$\sum_{n=0}^{\infty} s^n \sum_{j=k+1}^{n-1} pq^j R_{0,k}(n-j-1) \tag{6.20}$$

involves n in each of the two summations, a complication requiring careful attention. The difficulty is that as n increases, the number of terms in the difference equation increases as well. We will be able to isolate G(s), but it will take some additional work.

Begin the process by observing that for $n < k + 2$, $pq^j R_{0,k}(n-j-1)$ will be zero, since $\mathbf{R}_{0,k}(n) = 0$ for $n < 0$, $k > 0$. Thus, rewrite equation (6.20) as

$\sum_{n=k+2}^{\infty}\sum_{j=k+1}^{n-1} pq^j s^n R_{0,k}(n-j-1)$. We will gather the terms of each of these summands. Beginning with the outer summand and letting n = k + 2, we observe that the inner summation consists of only one term, $pq^{k+1}s^{k+2}R_{0,k}(0)$. For n = k + 3, the inner summation consists of the sum of two expressions, $pq^{k+1}s^{k+3}R_{0,k}(1)$ + $pq^{k+2}s^{k+3}R_{0,k}(0)$. We can continue to build terms this way visualizing the following:

$$
\begin{aligned}
&\sum_{n=k+2}^{\infty} s^n \sum_{j=k+1}^{n-1} pq^j R_{0,k}(n-j-1) = \\
&pq^{k+1}s^{k+2}R_{0,k}(0) + \\
&pq^{k+1}s^{k+3}R_{0,k}(1) + pq^{k+2}s^{k+3}R_{0,k}(0) + \\
&pq^{k+1}s^{k+4}R_{0,k}(2) + pq^{k+2}s^{k+4}R_{0,k}(1) + pq^{k+3}s^{k+4}R_{0,k}(0) + \\
&pq^{k+1}s^{k+5}R_{0,k}(3) + pq^{k+2}s^{k+5}R_{0,k}(3) + pq^{k+3}s^{k+5}R_{0,k}(1) + pq^{k+4}s^{k+5}R_{0,k}(0) + \\
&\vdots
\end{aligned}
\tag{6.21}
$$

The double summation is carried out, not row by row, but column by column. The first column of equation (6.21) is

$$\sum_{m=0}^{\infty} pq^{k+1}s^{k+2+m}R_{0,k}(m) \tag{6.22}$$

that may be simplified to

$$\sum_{m=0}^{\infty} pq^{k+1}s^{k+2+m}R_{0,k}(m) = pq^{k+1}s^{k+2}\sum_{m=0}^{\infty} s^m R_{0,k}(m) = pq^{k+1}s^{k+2}G(s) \tag{6.23}$$

Similarly for the second column,

$$\sum_{m=0}^{\infty} pq^{k+2}s^{k+3+m}R_{0,k}(m) = pq^{k+2}s^{k+3}\sum_{m=0}^{\infty} s^m R_{0,k}(m) = pq^{k+2}s^{k+3}G(s) \tag{6.24}$$

Proceeding across the infinite number of columns in expression (6.21) we may write

$$\sum_{n=0}^{\infty} s^n \sum_{j=k+1}^{n-1} pq^j R_{0,k}(n-j-1) = \sum_{j=0}^{\infty} pq^{k+j+1} s^{k+j+2} G(s)$$
$$= pq^{k+1} s^{k+2} G(s) \sum_{j=0}^{\infty} q^j s^j = \frac{pq^{k+1} s^{k+2} G(s)}{1-qs} \tag{6.25}$$

revealing

$$\sum_{n=0}^{\infty} s^n \sum_{j=k+1}^{n-1} pq^j R_{0,k}(n-j-1) = \frac{pq^{k+1} s^{k+2} G(s)}{1-qs} \tag{6.26}$$

We can now proceed with the consolidation of G(s). From the previous section

$$\sum_{n=0}^{\infty} s^n R_{0,k}(n) = \sum_{n=0}^{\infty} s^n \sum_{j=0}^{k-1} pq^j R_{0,k}(n-j-1)$$
$$+ \sum_{n=0}^{\infty} s^n \sum_{j=k+1}^{n-1} pq^j R_{0,k}(n-j-1) + \sum_{n=0}^{\infty} s^n q^n I_{n \neq k} \tag{6.27}$$

can now be written as

$$G(s) = \sum_{j=0}^{k-1} pq^j s^{j+1} G(s) + \frac{pq^{k+1} s^{k+2} G(s)}{1-qs} + \frac{1}{1-qs} - (qs)^k \tag{6.28}$$

The task is to now solve for G(s)

$$G(s)\left[1 - \sum_{j=0}^{k-1} pq^j s^{j+1} - \frac{pq^{k+1} s^{k+2}}{1-qs}\right] = \frac{1}{1-qs} - (qs)^k$$
$$G(s) = \frac{\dfrac{1}{1-qs} - (qs)^k}{1 - \sum_{j=0}^{k-1} pq^j s^{j+1} - \dfrac{pq^{k+1} s^{k+2}}{1-qs}} \tag{6.29}$$

Equation (6.29) may be reformulated as

$$
\begin{aligned}
G(s) &= \frac{1}{1-qs-(1-qs)\sum_{j=0}^{k-1} pq^j s^{j+1} - pq^{k+1}s^{k+2}} \\
&- \frac{(1-qs)(qs)^k}{1-qs-(1-qs)\sum_{j=0}^{k-1} pq^j s^{j+1} - pq^{k+1}s^{k+2}}
\end{aligned}
\tag{6.30}
$$

Further expansion of the first term on the right hand side of equation (6.30) reveals

$$
\begin{aligned}
\frac{1}{d} &= \frac{1}{1-qs-(1-qs)\sum_{j=0}^{k-1} pq^j s^{j+1} - pq^{k+1}s^{k+2}} \\
&= \frac{1}{1-qs-\sum_{j=0}^{k-1} pq^j s^{j+1} + \sum_{j=0}^{k-1} pq^{j+1} s^{j+2} - pq^{k+1}s^{k+2}}
\end{aligned}
\tag{6.31}
$$

canceling terms in the denominator leads to

$$
\begin{aligned}
\frac{1}{d} = \frac{1}{1-qs+pq^k s^{k+1} - ps - pq^{k+1}s^{k+2}} &= \frac{1}{1-s+pq^k s^{k+1} - pq^{k+1}s^{k+2}} \\
= \frac{1}{1-\left[1-pq^k s^k(1-qs)\right]s} &\triangleright_s \left\{\left[1-pq^k s^k(1-qs)\right]^n\right\}
\end{aligned}
\tag{6.32}
$$

6.7 Inverting G(s) for $R_{0,k}(n)$

Attempt an inversion

$$\frac{1}{d}=\frac{1}{1-\left[1-pq^ks^k(1-qs)\right]s} >_s \left\{\left[1-pq^ks^k+pq^{k+1}s^{k+1}\right]^n\right\} \tag{6.33}$$

To continue, apply the multinomial to $\left[1-pq^ks^k+pq^{k+1}s^{k+1}\right]^n$ to find

$$\begin{aligned}\left[1-pq^ks^k+pq^{k+1}s^{k+1}\right]^n &= \sum_{j=0}^{n}\sum_{h=0}^{n-j}\binom{n}{j\ \ h}(-1)^j\left(pq^ks^k\right)^j\left(pq^{k+1}s^{k+1}\right)^h \\ &= \sum_{j=0}^{n}\sum_{h=0}^{n-j}\binom{n}{j\ \ h}(-1)^j p^{j+h}q^{k(j+h)+h}s^{k(j+h)+h}\end{aligned} \tag{6.34}$$

Thus the inversion of 1/d requires the identification of the cofficient of s^n from the infinite series whose nth term is $\sum_{j=0}^{n}\sum_{h=0}^{n-j}\binom{n}{j\ \ h}(-1)^j p^{j+h}q^{k(j+h)+h}s^{n+k(j+h)+h}$.
Using the method of coefficient collection from Chapter 3, complete the inversion

$$\begin{aligned}\frac{1}{d} &= \frac{1}{1-\left[1-pq^ks^k(1-qs)\right]s} \\ &> \left\{\sum_{m=0}^{n}\sum_{j=0}^{m}\sum_{h=0}^{m-j}\binom{m}{j\ \ h}(-1)^j p^{j+h}q^{k(j+h)+h}I_{m+k(j+h)+h=n}\right\}\end{aligned} \tag{6.35}$$

Now that the inversion of d is complete, recall that

$$G(s)=\frac{1}{d}-\frac{(1-qs)(qs)^k}{d}=\frac{1}{d}-\frac{q^ks^k}{d}+\frac{q^{k+1}s^{k+1}}{d} \tag{6.36}$$

and apply the addition principle, the scaling tool and the sliding tool to find

$$G(s) \triangleright \left\{ \begin{array}{l} \sum_{m=0}^{n}\sum_{j=0}^{m}\sum_{h=0}^{m-j}\binom{m}{j\;\;h}(-1)^{j}p^{j+h}q^{k(j+h)+h}I_{m+k(j+h)+h=n} \\ -q^{k}\sum_{m=0}^{n-k}\sum_{j=0}^{m}\sum_{h=0}^{m-j}\binom{m}{j\;\;h}(-1)^{j}p^{j+h}q^{k(j+h)+h}I_{m+k(j+h)+h=n-k} \\ +q^{k+1}\sum_{m=0}^{n-k-1}\sum_{j=0}^{m}\sum_{h=0}^{m-j}\binom{m}{j\;\;h}(-1)^{j}p^{j+h}q^{k(j+h)+h}I_{m+k(j+h)+h=n-k-1} \end{array} \right\} \tag{6.37}$$

Solutions for $R_{1,k}(n)$ and $R_{i,k}(n)$ will be explored in the problems.

6.8 General Background and History of Drought Prediction

6.8.1 Background

The term "drought" is of course very general, and has different interpretations in various parts of the world. In Bali, a period of six days without rain is considered by its denizens as a drought. In parts of Libya, droughts are acknowledged only after a period of two years without rain. In Egypt, any year the Nile River does not flood its banks is a drought year, regardless of rainfall.

In the 18th and 19th centuries, two distinct approaches were used in the analysis of hydrologic and water resources. Each was motivated by the belief that the availability of water resources was cyclical. The first was a direct examination of random fluctuations in annual values of precipitation and runoff by statistical methods. This would hopefully lead to the discovery of periodicity in precipitation across time. The second method involved an analysis of the major periodic variables (e.g, the seasons) in determining and contributing to precipitation and runoff. Once these variables were identified, if was hoped, a parent, cyclical behavior would be revealed.

The development of these two determined perspectives for drought prediction paralleled the growth in population. Since modern societies have grown in size and interdependency, these societies are less able to accept the conventional, sometime catastrophic risks of drought. Hydrologists are therefore encouraged to provide the most accurate and far ranging estimates of drought occurrences. The discussion in this chapter examines the role of difference equations in the examination of the public health problem of drought prediction.

Hydrologists have been able to formulate the problem of drought likelihood in fairly complicated manners, (e.g., Cramer [15]), but not only are the computations complicated, there was sustained difficulty in translating the abstract terms in which these solutions are formulated to a specific hydrologic context. It should therefore come as no surprise that a review of the drought literature reveals that the application of run theory to drought prediction is restricted.

Run theory has been identified as useful to hydrologists in the assessment of drought likelihood [16]. However, results of importance to hydrologists that have been based on the established theory of runs, rests on either simplistic probability models or require familiarity with asymptotic behavior of run length. Beginning with the definition of the occurrence of adequate rainfall in a sequence of years as a sequence of Bernoulli trials, then typically, the hydrologist has access to q, the probability of failure and also k, the defined drought length. This information in the Bernoulli format is sufficient to obtain estimates of drought likelihood. The approach offered here permits this and is therefore of interest from a number of perspectives. First, it offers an exact solution for the prediction of the future occurrence of runs of an arbitrary length in a sequence of Bernoulli trials with no approximation required. Secondly, as will be seen in models to be developed in the next chapter, the solution is in terms of parameters with which the hydrologist is familiar and has direct access. Thus starting from estimates of familiar parameters and using the presented model, the practicing hydrologist will gain a pertinent, accurate assessment of the likelihood of droughts in the region of interest.

6.8.2 Early run theory and drought prediction

The first suggestion that the theory of runs might be applied to predictions concerning droughts was made by Cochran [17]. However, it was Yevjevich [16, 18] who attempted a specific formulation of a model for drought recurrence time based on modern run theory. He believed that it should be possible to determine the probability of occurrence of a run of given length based on monthly measurements of precipitation. If one plotted the available water over a time period, and chose a point y_0 on the ordinate as reflecting the minimum amount of water required before a water shortage was declared, then the sequence of runs of points below y_0 reflect the state of water insufficiency. This examination presumes the value of y_0 as a known constant.

Computational simplicity is a major attraction of the Bernoulli model. Let the Bernoulli trial be the occurrence of a time period (for example one year) in which there are inadequate water resources, and let the occurrence of this event be independent from one year to the next*. To continue the occurrence of failure runs of length n, it can be easily shown that for a discrete series, we need only obtain the probability that available water resources are greater than y_0, denoted

* We will discuss the tenability of this assumption later in this chapter.

as p, and q = 1 - p. Yevjevich defined P(n) = P[a new drought begins next year and lasts for n-1 years] and demonstrated using the geometric distribution†

$$P(n) = pq^{n-1} \tag{6.38}$$

that has the expected drought length as $\frac{1}{q}$ and variance $\frac{p}{q^2}$. Yevjevich then proceeded to use the method of moments to estimate p. No cohesive application of run theory to drought precipitation has been attempted since this study.

6.9 Drought Prediction

6.9.1 Introduction

The $\mathbf{R}_{i,k}(n)$ model provides the necessary foundation for the computation of the distribution of number of failure runs of a specified length, and, by extension, for the number of droughts of a given length. The form of its output (expected number of droughts of a given length) allows the comparison of observed findings with expected findings, and provides at least an initial impression of the ability of a complicated, albeit Bernoulli trial based model to compute the expected number of droughts of a specified length in a given time interval.

6.9.2 Plan to assess observed versus predicted drought lengths

The Bernoulli model first investigated by Yevjevich is a useful tool in the investigation of drought occurrence. However, the underlying assumption of the Bernoulli model must be examined. As was discussed in the previous section, the Bernoulli model assumes independence of rainfall over consecutive years. This assumption may (Friedman [19]) or may not (Tannehill [20]) be valid. In order to justify the independence assumption underlying Bernoulli model, it would be useful to compare the expected occurrence of drought lengths derived from the run theory model with the run lengths actually observed.

6.9.3 Texas data

In order to critique the performance of $\mathbf{R}_{ik}(n)$ in drought prediction, rainfall data was obtained from the clinatological records for Texas [21]. The methodology for this comparison procedure has been described by Moyé and Kapadia [22].

---

† The geometric distribution was developed in the last section of Chapter 2.

Data were available for the 93 years (from 1892 to 1984) for the entire state as well as for several clinatological regions of the state. The Texas Almanac defined a drought as a period when annual precipitation was less than 78% of the thirty-year norm (1931-1960 period).

In order to calculate the expected drought frequencies using the $\mathbf{R}_{i,k}(n)$ model, we must 1) estimate q, the probability of inadequate water resources 2) determine n, the number of years for which the prediction is to be made, and also decide on the drought length of interest, k, 3) compute the run length probabilities based on q, n, and k and 4) compute the expected number of runs for each drought length of interest from equation. This process was followed for each of the computations.

Texas is divided into ten climactic regions. An assessment of the $\mathbf{R}_{ik}(n)$ model's predictive ability is provided for two of them. In each of these two regions, a value for q, the probability of inadequate rainfall in a given year is required. Following the guidance of Yevjevich [16], this value is computed using the maximum likelihood approach, that, simply put, is the proportion of total years observed for which there is inadequate rainfall. This criteria for these two regions were met when precipitation was less than 75% of the normal by the duration of the period of record. For the Upper Coast and Southern Climatic regions of Texas, these estimates of q were respectively, 0.13 and 0.14.

6.9.4 Regional results

For drought lengths between one and four years (k=1 to 4), $\mathbf{R}_{i,k}(n)$ was computed. From these probabilities, the expected number of failure runs of length k were computed using equation (6.2), and the following tables report the expected and observed frequencies of droughts of various lengths, from one to five years, during the period from 1892 to 1984. Table 6.1 provides the estimates from the model for the Upper Coast Climactic Division in Texas, and the findings for the Southern Climactic Region are provided in Table 2. Tables 6.1 and 6.2 demonstrate a remarkable correspondence between the observed drought length distribution and the number of droughts predicted by the model. In each of these regions, the expected number of long droughts was low, accurately predicting the absence of these droughts. In each of the Upper Coast

Table 6.1 Comparison of Observed vs. Predicted Drought Distribution: Upper Coast Climatic Division, Texas

Drought Length	Observed Droughts	Predicted Droughts
1	8	9.13
2	2	1.17
3	0	0.15
4	0	0.02
5	0	0

Table 6.2 Comparison of Observed vs. Predicted Drought Distribution: Southern Climatic Division, Texas

Drought Length	Observed Droughts	Predicted Droughts
1	8	9.65
2	2	1.34
3	0	0.18
4	0	0.03
5	0	0.06

and Southern Climactic Regions, the model had a slight inclination to overestimate the occurrence of droughts of one year, with a consequent small propensity to underestimate droughts of two years. However, in each of these two regions the predictions from the $\mathbf{R}_{i,k}(n)$ model were reasonably good.

Table 6.3 Texas Average Annual Rainfall 1892-1984

Year	cm	Year	cm	Year	cm
1892	26.32	1923	37.24	1954	19.30
1893	18.50	1924	22.32	1955	23.59
1894	25.61	1925	25.37	1956	16.17
1895	29.83	1926	32.97	1957	36.93
1896	25.15	1927	24.32	1958	32.71
1897	24.21	1928	27.56	1959	31.29
1898	24.56	1929	29.47	1960	33.78
1899	27.57	1930	28.44	1961	30.20
1900	36.87	1931	28.37	1962	24.50
1901	20.13	1932	32.76	1963	20.95
1902	28.28	1933	26.15	1964	24.11
1903	29.64	1934	25.59	1965	28.68
1904	26.78	1935	35.80	1966	28.68
1905	35.98	1936	30.32	1967	28.44
1906	29.19	1937	25.89	1968	34.54
1907	28.51	1938	25.25	1969	29.85
1908	29.06	1939	23.52	1970	26.36
1909	21.58	1940	32.70	1971	29.58
1910	19.52	1941	42.62	1972	27.73
1911	26.83	1942	30.68	1973	38.37
1912	24.92	1943	24.28	1974	32.78
1913	35.25	1944	34.80	1975	28.70
1914	35.19	1945	30.60	1976	33.37
1915	28.79	1946	35.16	1977	24.40
1916	23.05	1947	24.75	1978	27.00
1917	14.30	1948	21.79	1979	31.43
1918	26.02	1949	35.80	1980	24.49
1919	42.15	1950	24.48	1981	32.65
1920	29.90	1951	21.99	1982	26.97
1921	25.18	1952	23.27	1983	25.75
1922	29.83	1953	24.76	1984	26.80

6.9.5 *Statewide results*

An important issue, however, is based on the observation that each of these computations were based on one definition of a drought (a period when annual precipitation was less than 75% of the 1931-1960 time period); it would be useful to consider alternative drought definitions to investigate the robustness of the $\mathbf{R}_{i,k}(n)$ model predictions. A drought could, alternatively, be defined as having a minimum length of two years. This application is illustrated using the entire state of Texas annual average precipitation for 1892-1984 (Table 6.3). Each entry in Table 6.3 represents the average rainfall (cm) in Texas for the specified year. From these data, the model's ability to predict the frequency of droughts of various lengths may be examined (Table 6.4) since drought definition could be made arbitrarily.

In order to carry out this computation, the period for which q is predicted requires some consideration. In order to further examine the robustness of the model, q was computed from a period of time different then that for the predictions. The mean precipitation rate and q were computed for the first thirty years of data. After q was computed, the $\mathbf{R}_{ik}(n)$ model was used to make drought predictions for the next 69 years. Thus, the model was placed in a position of making prospective predictions of the occurrences of droughts. Since the precipitation record for these 69 years was known, it was possible to make a direct comparison of the predictions based on the $\mathbf{R}_{ik}(n)$ model vs. what was actually observed for this time period.

Table 6.4 uses a drought definition of any year with less than 90% of the 1931-1960 statewide average. When this definition is used, the value of q is computed to be 0.34. We see the correspondence between the observed and predicted drought lengths is a good one. There was one drought observe to be six years in length. Although the model predicted only 0.05 droughts of this length, the probability of a drought at least as long as four years from the $\mathbf{R}_{ik}(n)$ is 0.50 + 0.016 + 0.05 + ... = 0.74.

If the definition of a drought is expanded from one year to two consecutive years with less than normal (based on the 1892-1922). A comparison of observed and expected droughts using this definition is provided in Table 6.5. For this drought definition, q=0.525. The correspondence between the observed and predicted drought lengths is remarkable for drought lengths up to four years in duration. The expected number of droughts greater than four years in duration is 0.61 + 0.32 + ... = 1.27, and exactly one drought was observed.

Table 6.4 Comparison of Observed vs. Predicted Drought Distribution: Statewide Data (drought = 1 years of inadequate water)

Drought Length	Observed Droughts	Predicted Droughts
2	5	4.35
3	2	1.51
4	0	0.50
5	0	0.16
6	1	0.05
7	0	0.02
8	0	0.01
9	0	0.00
10	0	0.00

6.9.6 Conclusions for $R_{i,k}(n)$ model

Although it has been recognized that run theory would be useful in drought prediction, it has been difficult to provide useful assessment thus far. Mood and Schwager have developed expressions that theoretically predict future run behavior. However, these results are not left in the form most suitable to hydrologists, i.e., the offered solutions are functions of parameters to which hydrologists have no access. The hydrologist has access to q, the probability of inadequate annual rainfall in the past, n, the number of years for which the predictions are to be made, and k, the defined drought length period. Using the $\mathbf{R}_{i,k}(n)$, this is all that is required. The difference equation perspective on run theory and its application to drought prediction is of interest from several perspectives. First, it offers an exact solution for the prediction of the future occurrences of runs of an arbitrary length in a sequence of Bernoulli trials, with no required approximations. Second, the resulting model provides a solution that is in terms of parameters that the hydrologist understands. Third, the outcome of the model (expected number of droughts of an arbitrary length) are measures of hydrologic importance.

Table 6.5 Comparison of Observed vs. Predicted Drought Distribution: Statewide Data (drought = 2 years of inadequate water)

Drought Length	Observed Droughts	Predicted Droughts
2	4	4.03
3	2	2.14
4	1	1.14
5	0	0.61
6	0	0.32
7	1	0.17
8	0	0.09
9	0	0.05
10	0	0.03

However useful as this appears to be, the $R_{ik}(n)$ does not provide information about the expected maximum drought length, that researchers (Sen) have suggested are of greater importance. Having established that the Bernoulli model performs reasonably well in predicting droughts in various regions of Texas, we can proceed to an evaluation of order statistics.

Problems

1. Show that the largest number of runs of length k in n trials is $i = \left[\frac{n+1}{k+1}\right]$ where the bracket indicates greatest integer function.

2. Show that

$$R_{i,k}(n_i) = q^k \left(pq^k\right)^{i-1} \qquad \text{for } n_i = i(k+1) - 1$$

3. Formulate the generating function argument to solve for $R_{1,k}(n)$ in terms of $R_{0,K}(n)$.

4. Formulate the generating function argument to solve for $\mathbf{R}_{i,k}(n)$ in terms of $\mathbf{R}_{i-1,k}(n)$.

5. Write the family of difference equations for the probability distribution of the number of success runs of length k.

6. Find $\mathbf{R}_{0,4,}(7)$ using the difference equation approach, and check this solution through enumeration.

7. Compute $\mathbf{R}_{1,4}(7)$ in terms of $\mathbf{R}_{0,4,}(7)$, and check this solution through enumeration.

8. Based on the development of $R_{i,k}(n)$, find a difference equation for the probability of the minimum failure run length in a sequence of Bernoulli trials.

9. Prove or Disprove: $R_{i,k}(n) \geq R_{i-1,k}(n)$ for $i > 0$, $k > 0$ and the conditions of problem 1.

10. Prove or Disprove: $R_{i,k+1}(n) \geq R_{i,k}(n)$ for $i > 0$, $k > 0$ and the conditions of problem 1.

References

1. Feller W. An Introduction to Probability Theory and Its Applications. New York. John Wiley and Sons. 1968.
2. Mood A.M. The distribution theory of runs. *Annals of Mathematical Statistics* 11:367-392. 1940.
3. Brownlee K.A. Statistical Theory and Methodology in Science and Engineering. Second Edition. New York. John Wiley and Sons. 1965.
4. Stevens W.L. Distribution of groups in a sequence of alternatives. *Annals of Eugenics* 9:10-17. 1939.
5. Wald A., Wolfowitz J. On a test whether two samples are from the same population. *Annals of Mathematical Statistics* 11:147-62. 1940.
6. Swed R.S., Eisenhart C. Tables for testing randomness of groupings in a sequence of alternatives. *Annals of Mathematical Statistics.* 14:66-87. 1941.
7. Wallis W., and Allen M. Rough and ready statistical tests. *Industrial Quality Control* 8: 35-40. 1952.
8. Page E.S. Continuous inspection schemes. *Biometrika* 41:111-115. 1954.
9. Woodall W.H. On the Markov Chain Approach to the two sampled CUSUM procedure. *Technometrics* 26:41-46. 1984.
10. O'Brien P.C., Dyck P.J. A runs test based on run lengths. *Biometrics* 41:237-44. 1985.
11. Janson S. Runs in m-dependent sequences. *The Annals of Probability* 12:237-244. 1984.
12. Pittel B.G. A process of runs and its convergence to Brownian motion. *Stochastic Processes Application* 10:33-48. 1980.

13. Shaunesy P.W. Multiple runs distributions; recurrence and critical values *Journal of the American Statistical Association* 76: 732-736. 1981.
14. Schwager S.J. Run probabilities in a sequence on markov-dependent trials. *Journal of the American Statistical Association* 78: 168-175. 1983.
15. Cramer H.L., Leadbetter M.R. Stationary and related stochastic processes. John Wiley and Sons. New York. 1967.
16. Yevjevich V. An objective approach to definitions and investigations or continental hydrologic droughts. Hydrology Papers Colorado State University. 1967.
17. Cochran W.G. An extension of Gold's method of examining the apparent persistence of one type of weather. *Quarterly Journal of the Royal Meterological Society*. 64:631-634. 1938.
18. Yevjevich V. Stochastic Processes in Hydrology. Water Resource Publication. Colorado State University. Fort Collins, Colorado. 1972.
19. Friedman D.C. The prediction of long-continuing drought in South and Southwest Texas. Occasional Research Paper in Meteorology No 1. The Weather Research Center. Travelers Insurance Company. Hartford Connecticut. 1957.
20. Tannehil I.R. Drought, Its Causes and Effects. Princeton, New Jersey. Princeton University Press, 1947.
21. Kingston M. ed. Texas Almanac and State Industrial Guide. AH Belo Corp. Communications Center, Dallas, Texas. 1986-1987.
22. Moyé L.A., Kapadia A.S., Cech I.M., Hardy R.J. The theory of runs with application to drought prediction. *Journal of Hydrology* 103:127-137. 1988.

7

Difference Equations: Drought Prediction

7.1 Introduction

Chapter 6 served as a brief review of the theory of runs, or the consecutive occurrence of events in a sequence of Bernoulli trials. During this process we developed a family of difference equations denoted by $\mathbf{R}_{i,k}(n)$. From this family of equations, we were able to compute the probability distribution of the number of failure runs of a given length. With access to hydrology data from the state of Texas, we were able to verify that the distribution of drought lengths observed in Texas in the early to middle 20th century was what we might predict from the $\mathbf{R}_{i,k}(n)$ model. This was a gratifying finding since we were particularly concerned about the applicability of the assumption of applying independence across Bernoulli trials to the model.

The purpose of this chapter is to further elaborate the difference equation approach to run theory by providing a specific application to the prediction of droughts. The $T_{[K,L]}(n)$ model mentioned in the previous chapter will be the focus of the mathematical developments of this chapter. In order to confirm the ability of a difference equation model to predict droughts, this system of difference equations must be motivated and solved.

7.2 Introduction to the $T_{[K,L]}(n)$ Model

One might wonder why, after the derivation of the complicated $R_{0,k}(n)$ model in the previous chapter, an additional difference family should be considered. The reason is that although the distribution of the number of runs of specified lengths is useful, there are other, more useful quantities specified by the order statistics (for example the minimum and maximum run lengths). Knowledge of the likelihood of occurrence of the worst run length and the best run length help workers to plan for the worst and best contingencies. However, the $R_{i,k}(n)$ model does not directly reveal the probability distribution of these order. In this section we will derive a model based on the use of difference equations that will predict the probability distribution of shortest and longest run lengths using the prospective perspective. That is, we will identify the probability distribution of the minimum and maximum run lengths before the sequence has occurred, and therefore make these predictions without knowledge of the number of successes and failures.

7.2.1 Definition and assumptions of the $T_{[K,L]}(n)$ model

We will assume that we will observe a sequence of independent Bernoulli trials with fixed probability of success p. However we are not interested in the total number of runs that occurs in this sequence, but instead concentrate on the failure runs. Thus, we are interested in addressing the following question. What is the probability that in a future occurrence of n Bernoulli trials, each of the minimum failure run length and the maximum failure run length occur are between length K and length L when $0 \leq K \leq L \leq n$?

Before we proceed with this, however, some statements on order statistics are required. Order statistics are measures of the relative ranks of a collection of observations. For example, consider a collection of random variables, e.g., the annual rainfall in a region. Let the annual rainfall in the ith year be x_i, for $i = 1$ to n. With only this knowledge, it is possible to index the rainfall as $x_1, x_2, x_3, \ldots, x_n$. However, it is also possible to order them using a different framework. One alternative is to order these observations from the smallest to the largest, creating the new sequence $x_{[1]}, x_{[2]}, x_{[3]}, \ldots, x_{[n]}$, where $x_{[1]} \leq x_{[2]} \leq x_{[3]} \leq \ldots \leq x_{[n]}$. These ranks are called the order statistics of the sample. Identify $x_{[1]}$ as the minimum or smallest observation in the sequence, and $x_{[n]}$ is the maximum or largest observation.

Events involving these order statistics require careful thought and attention. For example, if the minimum observation of a sequence of random variables is greater than a constant c then all of the observations are greater than c. Since all the observations are greater than c and the minimum is the smallest of these observations, then the minimum must be greater than c. Setting a lower bound for the minimum sets a lower bound for the entire set of observations.

A similar line of reasoning applies to the maximum of the observations. If the maximum of an observation is less than a constant c then all of the observations must be less than c. Thus if we know that for a collection of observations the minimum is greater than or equal to K and the maximum is less than or equal to L, then all of the observations are "trapped" between K and L. Define $\mathbf{T}_{[K,L]}(n)$ as

> $\mathbf{T}_{[K,L]}(\mathbf{n})$ = P[the minimum and maximum failure run lengths are between length K and length L respectively in n trials]

$\mathbf{T}_{[K,L]}(\mathbf{n})$ is the probability that all failure run lengths are trapped between length K and length L in a sequence of n Bernoulli trials.* For a simple example, $\mathbf{T}_{[5,7]}(9)$ is the probability that in n Bernoulli trials all failure runs have lengths of either 5, 6, or 7. If consists of the following enumerated events

FFFFFHHHH	FFFFFFHHH	FFFFFFFHH
HFFFFFHHH	HFFFFFFHH	HFFFFFFFH
HHFFFFFHH	HHFFFFFFH	HHFFFFFFF
HHHFFFFFH	HHHFFFFFF	
HHHH**FFFFF**		

Note that all sequences with failure run lengths of one, two, three, four, eight, and nine have been excluded. The probability of the sequences depicted above is $\mathbf{T}_{[5,7]}(9)$. Notice that in the above sequences, the sequence of all heads (no failures) is not considered.

Although each of the $\mathbf{R}_{ik}(n)$ and $\mathbf{T}_{[K,L]}(n)$ models have as their common foundation, the Bernoulli model, they compute two different probabilities. The $\mathbf{R}_{ik}(n)$ model focuses on a failure run of a specific length, and computes the probability of the number of failure runs of that length. The $\mathbf{T}_{[K,L]}(n)$ focuses on the probability not of a single failure run length, but whether all failure runs are in a certain run length interval. It is useful to think of $\mathbf{T}_{[K,L]}(n)$ as trapping failure run lengths within intervals.

---

* A Bernoulli trial was defined in Chapter 2.

7.2.2 Motivation for the $T_{[K,L]}(n)$ model

The goal of this chapter is to formulate $\mathbf{T}_{[K,L]}(n)$ in terms of K, L, q, and n. By completing this computation, the user has access to the following probabilities for n consecutive Bernoulli trials

$\mathbf{T}_{[K,K]}(n)$ = P[all failure runs have the same run length (= K)]
$\mathbf{T}_{[0,L]}(n)$ = P[all failure run lengths are $\leq$ L in length]
$1 - \mathbf{T}_{[0,L]}(n)$ = P[at least one failure run length is greater than length L]

Further investigation reveals that this model provides important information involving order statistics. Observe that $\mathbf{T}_{[K,n]}(n)$ is the probability that the minimum failure run length is greater than or equal to K.

Similarly, $\mathbf{T}_{[0,L]}(n)$, (that is the probability that all failure run lengths are less than or equal to length L) is the probability that the maximum failure run length is $\leq$ L. $\mathbf{T}_{[0,L-1]}(n)$ is the probability that the maximum failure run length is less than or equal to L - 1 in length. Thus $\mathbf{T}_{[0,L]}(n) - \mathbf{T}_{[0,L-1]}(n)$ is the probability that the maximum failure run length is exactly of length L. With this formulation, the expected maximum failure run length $E[M_F(n)]$ and its variance $V(\mathbf{M}_F(n))$ can be computed as

$$E\left[M_F(n)\right]=\sum_{L=0}^{n} L\left[T_{[0,L]}(n)-T_{[0,L-1]}(n)\right] \tag{7.1}$$

$$\mathrm{Var}\left[M_F(n)\right]=\sum_{L=0}^{n} L^2\left[T_{[0,L]}(n)-T_{[0,L-1]}(n)\right] \;-\; E^2\left[M_F(n)\right]$$

An analogous computation for the minimum failure run length is available. Observe that $\mathbf{T}_{[K,n]}(n)$ is the probability that the minimum run length is $\geq$ K. Then $\mathbf{T}_{[K+1,n]}(n)$ is the probability that the minimum failure run length is $\geq$ K+1, and $\mathbf{T}_{[K,n]}(n) - \mathbf{T}_{[K+1,n]}(n)$ is the probability that the minimum run length is exactly K. We can therefore compute the expected value of the minimum failure run length $m_F(n)$, $E[m_F n)]$, and its variance $Var[m_F(n)]$

$$E\left[m_F(n)\right]=\sum_{K=0}^{n} K\left[T_{[K,n]}(n)-T_{[K+1,n]}(n)\right] \tag{7.2}$$

$$\mathrm{Var}\left[m_F(n)\right]=\sum_{K=0}^{n} K^2\left[T_{[K,n]}(n)-T_{[K+1,n]}(n)\right] \;-\; E^2\left[m_F(n)\right]$$

7.2.3 An application in hydrology

Consider the following formulation of a problem in water management. Workers in hydrology have long suspected that run theory with a basis in Bernoulli trials may be an appropriate foundation from which to predict sequences of hydrologic events. The hydrologist Yevjevich [1] was among the first at attempting a prediction of properties of droughts using the geometric probability distribution, defining a drought of k years as k consecutive years when there are not adequate water resources. In this development, the hallmark of events of hydrologic significance is consecutive occurrences of the event. We define a drought (negative run) as a sequence of consecutive years of inadequate water resources. We will also assume that the year to year availability of adequate water resources can be approximated by a sequence of Bernoulli trials in which only one of two possible outcomes (success with probability p or failure with probability q) may occur. These probabilities remain the same from year to year, and knowledge of the result of a previous year provides no information for the water resource availability of any following year. This model has been employed in modeling hydrologic phenomena [2].

If q is the probability of inadequate water resources, then $\mathbf{T}_{[K,L]}(n)$ is the probability that all droughts that occur in n trials are between K and L years in length. If $\mathbf{T}_{[K,L]}(n)$ can be formulated in terms of K, L, q, and n, quantities to which the hydrologist has access, the following probabilities for n consecutive years may be identified:

$\mathbf{T}_{[K,L]}(n)$ = P[all drought lengths are between K and L years long, inclusive]
$1 - \mathbf{T}_{[0,L]}(n)$ = P[at least one drought > L years in length]

Probabilities for the order statistics involving the maximum and minimum drought length are particularly noteworthy.

P[minimum drought length $\geq$ K years] = $\mathbf{T}_{[K,n]}(n)$
P[maximum drought length is $\leq$ L years] = $\mathbf{T}_{[0,L]}(n)$

Access to these probabilities provides useful information for predicting extremes in drought lengths. The notion of order statistics can be further developed by observing that $\mathbf{T}_{[0,L]}(n)$ is the probability that all droughts in n years are $\leq$ L years long, i.e., the maximum drought length is $\leq$ L years in length. $\mathbf{T}_{[0,L-1]}(n)$ is the probability that the maximum drought length is $\leq$ L-1 years in length. Thus $\mathbf{T}_{[0,L]}(n) - \mathbf{T}_{[0,L-1]}(n)$ is the probability that the maximum drought length is exactly L years in length. The expected maximum drought length $E[\mathbf{M}_D(n)]$ and its variance $V(\mathbf{M}_D(n))$ can be computed from equation (7.1)

7.3 Formulation of the $T_{[K,L]}(n)$ and $T_{0,[K,L]}(n)$ Models

7.3.1 $T_{[K,L]}(n)$ versus $T_{0,[K,L]}(n)$

Recall that the modeling goal here is to formulate $\mathbf{T}_{[K,L]}(n)$ as a family of difference equations and then to apply generating function arguments to solve them. In this process we will consecutively work through the procedures of conversion, consolidation, and inversion as introduced in Chapter 2 and applied in Chapters 4 and 5.

We begin the general derivation of the $\mathbf{T}_{[K,L]}(n)$ model by first defining a necessary offshoot of this model, $\mathbf{T}_{0,[K,L]}(n)$. Let K and L be integers such that $0 < K < L$. Let us start with the assertion that if both the minimum failure run length is in [K, L] and the maximum failure run length is in [K, L], then all failure run lengths are in [K, L]. Then $\mathbf{T}_{0,[K,L]}(n)$ is the probability that either the maximum failure run length is zero, or the minimum failure run length and the maximum failure run lengths are each in [K, L]. $\mathbf{T}_{0,[K,L]}(n)$, in addition to the events included in the calculation of $T_{[K,L]}(n)$, also allows for the possibility of a sequence of n successes.

As with $\mathbf{T}_{[K,L]}(n)$, we define as the probability that both the minimun and maximum run lengths are in [K, L]. In addition the $\mathbf{T}_{0,[K,L]}(n)$ also computes the probability that the sequence of n Bernoulli trials has absolutely no failures at all. The desired quantity $\mathbf{T}_{[K,L]}(n)$ is

$$T_{[K,L]} = T_{0,[K,L]}(n) - p^n \tag{7.3}$$

The family of difference equations will be derived for $\mathbf{T}_{0,[K,L]}(n)$. $\mathbf{T}_{[K,L]}(n)$ will then be computed from equation (7.3).

7.3.2 Examples and verifications of the $T_{0,[K,L]}(n)$

As an introductory example, compute $T_{0,[1,2]}(n)$ and $T_{[1,2]}(n)$ for increasing values of n. We can begin by just defining $T_{0,[1,2]}(0) = 1$. Enumeration reveals $T_{0,[1,2]}(1) = q + p$, and $T_{[1,2]}(1) = q$. Analogously

$$T_{0,[1,2]}(2) = p^2 + pq + qp + q^2 = (p+q)^2 = 1 \tag{7.4}$$

and $\mathbf{T}_{[1,2]}(2) = 1 - p^2$. For $T_{0,[1,2]}(3)$, note that the probability of either no failures in three trials, or, when failures occur, they occur as length 1 or length 2 is

$$
\begin{aligned}
T_{0,[1,2]}(3) &= p^3 + qpp + pqp + ppq + qpq + qqp + pqq \\
&= p^3 + 3p^2q + 3pq^2 = p^3 + 3pq(p+q) \\
&= p^3 + 3pq
\end{aligned}
\tag{7.5}
$$

Note also

$$
\begin{aligned}
T_{[1,2]}(3) &= qpp + pqp + ppq + qpq + qqp + pqq \\
&= 3p^2q + 3pq^2 = 3pq(p+q) \\
&= 3pq
\end{aligned}
\tag{7.6}
$$

And note that

$$
T_{[1,2]}(n) = T_{0,[1,2]}(n) - p^3 \tag{7.7}
$$

7.3.3 The $T_{0,[K,L]}$ family of difference equations

How can we model this process? Define $T_{0,[K,L]}(0) \equiv 1$, and $T_{0,[K,L]}(n) = 0$ for negative values of n. Then we may write the difference equation for $T_{0,[K,L]}(n)$ as

$$
\begin{aligned}
T_{0,[K,L]}(n) &= pT_{0,[K,L]}(n-1) + q^K pT_{0,[K,L]}(n-K-1) + q^{K+1}pT_{0,[K,L]}(n-K-2) \\
&\quad + q^{K+2}pT_{0,[K,L]}(n-K-3) + K + q^L pT_{0,[K,L]}(n-L-1) + q^n I_{K \le n \le L}
\end{aligned}
\tag{7.8}
$$

When this equation is considered for the special case of K = 1 and L = 2, equation (7.8) simplifies to

$$
T_{0,[1,2]}(n) = pT_{0,[1,2]}(n-1) + qpT_{0,[1,2]}(n-2) + q^2pT_{0,[1,2]}(n-3) + q^n I_{n=1 \text{ or } n=2} \tag{7.9}
$$

We can now use the results from equation (7.9) to check the enumerated results for $T_{[1,2]}(n)$, for n = 1, 2, and 3. We compute from equation (7.9)

$$
T_{0,[1,2]}(1) = pT_{0,[1,2]}(0) + qpT_{0,[1,2]}(-1) + q^2pT_{0,[1,2]}(-2) + q = p + q \tag{7.10}
$$

that matches our solution through enumeration. We can proceed with $T_{0,[1,2]}(2)$ as

$$\begin{aligned} T_{0,[1,2]}(2) &= pT_{0,[1,2]}(1) + qpT_{0,[1,2]}(0) + q^2pT_{0,[1,2]}(-1) + q^2 \\ &= p(p+q) + qp + q^2 = p^2 + 2pq + q^2 \end{aligned} \tag{7.11}$$

that again matches our enumerated result. Finally, for $T_{0,[1,2]}(3)$ we have

$$\begin{aligned} T_{0,[1,2]}(3) &= pT_{0,[1,2]}(2) + qpT_{0,[1,2]}(1) + q^2pT_{0,[1,2]}(0) \\ &= p(p^2 + 2pq + q^2) + qp(p+q) + q^2p(1) \\ &= p^3 + 2p^2q + pq^2 + p^2q + pq^2 + pq^2 \\ &= p^3 + 3p^2q + 3pq^2 \end{aligned} \tag{7.12}$$

Notice that the term p^3, necessary for the solution of $\mathbf{T}_{0,[1,2]}(3)$ is propagated from the term $p\mathbf{T}_{0,[1,2]}(2)$. It is this term that distinguishes $\mathbf{T}_{0,[1,2]}(3)$ from $\mathbf{T}_{[1,2]}(3)$.

As a final example of increased complexity, we will verify the solution for $T_{0,[1,2]}(4)$. To compute this probability through enumeration, we count the sequences of Bernoulli trials that have the probability that the maximum failure run length is 0 or both the minimum and maximum failure run lengths are in [1,2] as

pppp	qpqp	qpqq
qppp	pqpq	qqpq
pqpp	pqqp	
ppqp	qqpp	
pppq	ppqq	

These probabilities sum to $p^4 + 2pq^3 + 5p^2q^2 + 4p^3q$. Using equation (7.9) we write:

$$\begin{aligned} T_{0,[1,2]}(4) &= pT_{0,[1,2]}(3) + qpT_{0,[1,2]}(2) + q^2pT_{0,[1,2]}(1) \\ &= p(p^3 + 3p^2q + 3q^2p) + qp(p^2 + 2pq + q^2) + q^2p(p+q) \\ &= p^4 + 3p^3q + 3p^2q^2 + p^3q + 2p^2q^2 + pq^3 + p^2q^2 + pq^3 \\ &= p^4 + 2pq^3 + 5p^2q^2 + 4p^3q \end{aligned} \tag{7.13}$$

In general, the boundary conditions for $\mathbf{T}_{0,[K,L]}(n)$ are

$\mathbf{T}_{0,[K,L]}(n) = 0$ for all $n < 0$: $\mathbf{T}_{0,[K,L]}(n) = p^n$ for $0 < n < K$
$\mathbf{T}_{0,[K,L]}(n) = 1$ for $n = 0$:

Using the indicator function, we may write the recursive relationship for $\mathbf{T}_{0,[K,L]}(n)$ for $0 < K \leq \min(L, n)$ as

$$\begin{aligned} T_{0,[K,L]}(n) &= pT_{0,[K,L]}(n-1) + q^K pT_{0,[K,L]}(n-K-1) \\ &+ q^{K+1} pT_{0,[K,L]}(n-K-2) + q^{K+2} pT_{0,[K,L]}(n-K-3) \\ &+ \ldots + q^L pT_{0,[K,L]}(n-L-1) + q^n I_{K \leq n \leq L} \end{aligned} \tag{7.14}$$

7.3.4 Examination of the $T_{0,[K,L]}(n)$ model's structure

The $\mathbf{T}_{0,[K,L]}(n)$ model may be written as

$$T_{0,[K,L]}(n) = pT_{0,[K,L]}(n-1) + \sum_{j=K}^{L} q^j pT_{0,[K,L]}(n-j-1) + q^n I_{K \leq n \leq L} \tag{7.15}$$

This is a L + 1th order difference equation. We assume that the run length interval bounds K and L are known. We will also assume that the probability of failure, q is known, and that p = 1 - q. Equation (7.15) is a new family of difference equations and requires some development and motivation before we proceed with its solution. It is easiest if we examine this relationship through a term by term examination of the component pieces of this equation. Each component in the equation describes a way by which a failure run might occur whose length is between K and L, or a special condition when $0 < n \leq L$.

The term $\sum_{j=K}^{L} q^j pT_{0,[K,L]}(n-j-1)$ is the heart of the recursive relationship. This finite sum spans the failure run lengths for the designated run interval from K to L. Each failure run length (q^j) is terminated by a success ($q^j p$). After any of these sequences of length j+1 has occurred, there are only n – j - 1 trials left, in which only failure run lengths of between K and L are permitted. The last term in equation (7.15), $q^n I_{K \leq n \leq L}$ is necessary since when the length of the sequence n is in the permitted run length (i.e., $K \leq n \leq L$) a failure run of length n is consistent with the definition of $\mathbf{T}_{0,[K,L]}(n)$.

The right side of equation (7.15) completely captures all possibilities for a sequence that is permitted to have failure run lengths of length zero or between lengths K and L.

7.3.5 Further practice with the $T_{0,[K,L]}(n)$

To provide some intuition for the underlying reasoning behind this difference equation as elaborated in equation (7.15), we examine its performance in some specific instances. Assume there is interest in computing $T_{0,[3,6]}(n)$, for n = 1 to 10, i.e., use equation (7.15) and its boundary conditions to compute $T_{0,[3,6]}(1)$, $T_{0,[3,6]}(2)$, $T_{0,[3,6]}(3)$,..., $T_{0,[3,6]}(10)$. Observe at once that $T_{0,[3,6]}(1) = p$ and $T_{0,[3,6]}(2) = p^2$. Using $T_{[3,6]}(n) = T_{0,[3,6]}(n) - p^n$, see that $T_{[3,6]}(1) = T_{[3,6]}(2) = 0$. This satisfies our intuition, which is that these probabilities must be zero since it is impossible to have failure run lengths of 3 or more with less than three trials. Returning to equation (7.15) to compute $T_{0,[3,6]}(3)$ as

$$\begin{aligned} T_{0,[3,6]}(3) &= pT_{0,[3,6]}(2) + q^3pT_{0,[3,6]}(-1) + q^4pT_{0,[3,6]}(-2) \\ &\quad + q^5pT_{0,[3,6]}(-3) + q^6pT_{0,[3,6]}(-4) + q^3 \\ &= p^3 + q^3 \end{aligned} \tag{7.16}$$

The middle terms involving failure runs of lengths three, four, five, and six are each zero because $T_{0,[K,L]}(n) = 0$ for $n < 0$. Thus, we find $T_{[3,6]}(3) = T_{0,[3,6]}(3) - p^3 = q^3$.

Move on to $T_{0,[3,6]}(4)$, computing

$$\begin{aligned} T_{0,[3,6]}(4) &= pT_{0,[3,6]}(3) + q^3pT_{0,[3,6]}(0) + q^4pT_{0,[3,6]}(-1) + q^5pT_{0,[3,6]}(-2) \\ &\quad + q^6pT_{0,[3,6]}(-3) + q^4 \\ &= p\left(p^3 + q^3\right) + q^3p + q^4 \\ &= p^4 + pq^3 + pq^3 + q^4 \\ &= p^4 + 2pq^3 + q^4 \end{aligned} \tag{7.17}$$

and $T_{[3,6]}(4) = T_{0,[3,6]}(4) - p^4 = 2pq^3 + q^4$. Consider how in a sequence of four Bernoulli trials there are only failure runs of length three through six and observe that the evaluation must include only failure runs of lengths three or four. There are two ways to obtain a failure run of length three, each occurring with probability pq^3, and of course, the probability of a failure run length of four is q^4. Thus equation (7.15) returns the correct solution. It is useful to examine the mechanics of the performance of equation equation (7.15) in this example. Since $T_{0,[3,6]}(0) = 1$ by definition, the term $q^3pT_{0,[3,6]}(0)$ makes a contribution to this computation. Also, the necessity of the last term in equation equation (7.15) is clear now since a failure run length equal to the duration of the sequence is necessary when the sequence length is between K and L.

Proceeding to $T_{0,[3,6]}(5)$, we find

$$\begin{aligned}
T_{0,[3,6]}(5) &= pT_{0,[3,6]}(4) + q^3pT_{0,[3,6]}(1) + q^4pT_{0,[3,6]}(0) \\
&\quad + q^5pT_{0,[3,6]}(-1) + q^3p(p) + q^6pT_{0,[3,6]}(-2) + q^5 \\
&= p\left(p^4 + 2pq^3 + q^4\right) + p^2q^3 + q^4p + q^5 \\
&= p^5 + 2p^2q^3 + pq^4 + p^2q^3 + q^4p + q^5 \\
&= p^5 + 3p^2q^3 + 2pq^4 + q^5
\end{aligned} \tag{7.18}$$

and

$$T_{[3,6]}(5) = 3p^2q^3 + 2pq^4 + q^5 \tag{7.19}$$

We now compute $T_{0,[3,6]}(6)$

$$\begin{aligned}
T_{0,[3,6]}(6) &= pT_{0,[3,6]}(5) + q^3pT_{0,[3,6]}(2) + q^4pT_{0,[3,6]}(1) \\
&\quad + q^5pT_{0,[3,6]}(0) + q^6pT_{0,[3,6]}(-1) + q^6 \\
&= p\left(p^5 + 3p^2q^3 + 2pq^4 + q^5\right) + q^3p(p^2) + q^4p(p) + q^5p + q^6 \\
&= p^6 + 3p^3q^3 + 2p^2q^4 + pq^5 + p^3q^3 + p^2q^4 + pq^5 + q^6 \\
&= p^6 + 4p^3q^3 + 3p^2q^4 + 2pq^5 + q^6
\end{aligned} \tag{7.20}$$

and

$$T_{[3,6]}(6) = 4p^3q^3 + 3p^2q^4 + 2pq^5 + q^6 \tag{7.21}$$

For the computation of $T_{0,[3,6]}(7)$ we will lose the last term since $L < 7$ to find

$$
\begin{aligned}
T_{0,[3,6]}(7) &= pT_{0,[3,6]}(6)+q^3pT_{0,[3,6]}(3)+q^4pT_{0,[3,6]}(2)+q^5pT_{0,[3,6]}(1) \\
&\quad +q^6pT_{0,[3,6]}(0) \\
&= p\left(p^6+4p^3q^3+3p^2q^4+2pq^5+q^6\right)+q^3p\left(p^3+q^3\right)+q^4p(p^2) \\
&\quad +q^5p(p)+q^6p
\end{aligned} \tag{7.22}
$$

Continuing, find

$$
\begin{aligned}
T_{0,[3,6]}(7) &= p^7+4p^4q^3+3p^3q^4+2p^2q^5+pq^6+p^4q^3+pq^6+p^3q^4 \\
&\quad +p^2q^5+pq^6 \\
&= p^7+5p^4q^3+4p^3q^4+3p^2q^5+2pq^6
\end{aligned} \tag{7.23}
$$

and

$$
T_{[3,6]}(7) = 5p^4q^3+4p^3q^4+3p^2q^5+2pq^6 \tag{7.24}
$$

The computations for $T_{0,[3,6]}(8)$, and $T_{0,[3,6]}(9)$, proceed analogously.

$$
\begin{aligned}
T_{0,[3,6]}(8) &= pT_{0,[3,6]}(7)+q^3pT_{0,[3,6]}(4)+q^4pT_{0,[3,6]}(3)+q^5pT_{0,[3,6]}(2) \\
&\quad +q^6pT_{0,[3,6]}(1) \\
&= p\left(p^7+5p^4q^3+4p^3q^4+3p^2q^5+3pq^6\right)+q^3p\left(p^4+2pq^3+q^4\right) \\
&\quad +q^4p\left(p^3+q^3\right)+q^5p(p^2)+q^6p(p) \\
&= p^8+5p^5q^3+4p^4q^4+3p^3q^5+3p^2q^6+p^5q^3+2p^2q^6+pq^7+p^4q^4 \\
&\quad +pq^7+p^3q^8+p^2q^6 \\
&= p^8+6p^5q^3+5p^4q^4+4p^3q^5+6p^2q^6+2pq^7
\end{aligned} \tag{7.25}
$$

and

$$
T_{[3,6]}(8) = 6p^5q^3+5p^4q^4+4p^3q^5+6p^2q^6+2pq^7 \tag{7.26}
$$

$$
\begin{aligned}
T_{0,[3,6]}(9) &= pT_{0,[3,6]}(8)+q^3pT_{0,[3,6]}(5)+q^4pT_{0,[3,6]}(4)+q^5pT_{0,[3,6]}(3)+q^6pT_{0,[3,6]}(2) \\
&= p\left(p^8+6p^5q^3+5p^4q^4+4p^3q^5+6p^2q^6+2pq^7\right)+q^3p\left(p^5+3p^2q^3+2pq^4+q^5\right) \\
&\quad +q^4p\left(p^4+2pq^3+q^4\right)+q^5p\left(p^3+q^3\right)+q^6p(p^2) \\
&= p^9+6p^6q^3+5p^5q^4+4p^4q^5+6p^3q^6+2p^2q^7+p^6q^3+3p^3q^6+2p^2q^7+pq^8 \\
&\quad +p^5q^4+2p^2q^7+pq^8+p^4q^5+pq^8+p^3q^6 \\
&= p^9+7p^6q^3+6p^5q^4+5p^4q^5+10p^3q^6+6p^2q^7+3pq^8
\end{aligned}
\tag{7.27}
$$

and

$$
T_{[3,6]}(9) = 7p^6q^3+6p^5q^4+5p^4q^5+10p^3q^6+6p^2q^7+3pq^8 \tag{7.28}
$$

Note that the terms in the final solution that contain powers of q greater than 6 do not denote failure runs lengths greater than six, but multiple failure runs each of which has a length between 3 and 6.

7.4 Solution to $T_{0,[K,L]}(n)$ I : Conversion and Consolidation

7.4.1 First steps toward the solution

Having motivated and provided intuition for $T_{0,[K,L]}(n)$ family of difference equations, we will now undertake the solution to the general family of difference equations. That family can be represented as

$$
T_{0,[K,L]}(n) = pT_{0,[K,L]}(n-1)+\sum_{j=K}^{L} q^j pT_{0,[K,L]}(n-j-1)+q^n I_{K\le n\le L} \tag{7.29}
$$

Subject to the boundary conditions

$T_{[0,[K,L]}(n) = 0$ for all $n < 0$:
$T_{0,[K,L]}(n) = 1$ for $n = 0$:
$T_{0,[K,L]}(n) = 0$ for $K > L$:
$T_{0,[K,L]}(n) = p^n$ for $0 < n < K$

As noted in the previous section, our final goal is to compute $T_{[K,L]}(n)$, the probability that all failure run lengths are between K and L. If we instead solve for $T_{0,[K,L]}(n)$, we can compute $T_{0,[K,L]}(n)$ by $T_{[K,L]}(n) = T_{0,[K,L]}(n) - p^n$. Thus, our intermediate goal will be to compute $T_{0,[K,L]}(n)$. The procedures that we will follow here will be those we have developed in Chapters 2 through 4. Note that this is a difference equation indexed not by k, that has been our custom so far, but by n. In this family of difference equations, we assume that p, q, K, and L are known constants.

7.4.2 Conversion of $T_{0,[K,L]}(n)$

Before we begin with the conversion, we define G(s) as

$$G(s) = \sum_{n=0}^{\infty} s^n T_{0,[K,L]}(n) \tag{7.30}$$

and thereby declare that the probabilities we seek will be coefficients of powers of s. The process will start as presented in Chapter 3, multiplying each term in equation (7.29) by s^n and then take the sums over the range from n = 0 to infinity, revealing

$$s^n T_{0,[K,L]}(n) = ps^n T_{0,[K,L]}(n-1) + \sum_{j=K}^{L} s^n q^j p T_{0,[K,L]}(n-j-1) + s^n q^n I_{K \le n \le L} \tag{7.31}$$

and taking the sum of each term in equation (7.31) as n goes from zero to infinity reveals

$$\begin{aligned}\sum_{n=0}^{\infty} s^n T_{0,[K,L]}(n) &= \sum_{n=0}^{\infty} ps^n T_{0,[K,L]}(n-1) + \sum_{n=0}^{\infty}\sum_{j=K}^{L} s^n q^j p T_{0,[K,L]}(n-j-1) \\ &\quad + \sum_{n=0}^{\infty} s^n q^n I_{K \le n \le L}\end{aligned} \tag{7.32}$$

Taking these terms individually, we see that the expression on the left of the equation sign in equation (7.32) is G(s). We take the terms on the right side of the equation term by term. The first expression is evaluated as

$$\sum_{n=0}^{\infty} ps^n T_{0,[K,L]}(n-1) = \sum_{n=0}^{K-1} ps^n T_{0,[K,L]}(n-1) + \sum_{n=K}^{\infty} ps^n T_{0,[K,L]}(n-1) \tag{7.33}$$

The first term on the right side of equation (7.33) is

$$\sum_{n=0}^{K-1} ps^n T_{0,[K,L]}(n-1) = \sum_{n=0}^{K-1} ps^n p^{n-1} = \sum_{n=0}^{K-1} s^n p^n \tag{7.34}$$

since for $0 \leq n \leq K - 1$, $T_{0,[K,L]}(n-1) = p^{n-1}$. The second term on the right side of equation (7.33) may be simplified as follows

$$\begin{aligned} p\sum_{n=K}^{\infty} s^n T_{0,[K,L]}(n-1) &= ps\sum_{n=K}^{\infty} s^{n-1} T_{0,[K,L]}(n-1) = ps \sum_{n=K-1}^{\infty} s^n T_{0,[K,L]}(n) \\ &= ps\left(G(s) - \sum_{n=0}^{K-2} s^n T_{0,[K,L]}(n) \right) \end{aligned} \tag{7.35}$$

The middle term in equation (7.32) will require the most work, but is easy to simplify since the inner summation is a finite sum over the permissible failure run lengths.

$$\begin{aligned} &\sum_{n=0}^{\infty}\sum_{j=K}^{L} s^n q^j p T_{0,[K,L]}(n-j-1) = \sum_{n=j+1}^{\infty}\sum_{j=K}^{L} s^n q^j p T_{0,[K,L]}(n-j-1) \\ &= \sum_{j=K}^{L}\sum_{n=j+1}^{\infty} s^n q^j p T_{0,[K,L]}(n-j-1) \end{aligned} \tag{7.36}$$

In equation (7.36), $T_{0,[K,L]}(n) = 0$ for $n < 0$. To proceed fix j for some value between K and L, then evaluate the inner sum as

$$\begin{aligned} &\sum_{n=j+1}^{\infty} s^n q^j p T_{0,[K,L]}(n-j-1) \\ &= q^j p \sum_{n=j+1}^{\infty} s^n T_{0,[K,L]}(n-j-1) = q^j ps^{j+1} \sum_{n=j+1}^{\infty} s^{n-j-1} T_{0,[K,L]}(n-j-1) \\ &= q^j ps^{j+1} \sum_{n=0}^{\infty} s^n T_{0,[K,L]}(n) = q^j ps^{j+1} G(s) \end{aligned} \tag{7.37}$$

and substituting the result of equation (7.37) into equation (7.36) we find that

$$\sum_{n=0}^{\infty}\sum_{j=K}^{L} s^n q^j p T_{0,[K,L]}(n-j-1) = G(s)\sum_{j=K}^{L} q^j p s^{j+1} \tag{7.38}$$

Finally, we quickly simplify the last term on the right side of equation (7.32):

$$\sum_{n=0}^{\infty} s^n q^n I_{K \le n \le L} = \sum_{n=K}^{L} s^n q^n \tag{7.39}$$

We can now simplify equation (7.32) by writing

$$\begin{aligned} G(s) &= \sum_{n=0}^{K-1} s^n p^n + ps\left(G(s) - \sum_{n=0}^{K-2} s^n T_{0,[K,L]}(n) \right) + G(s)\sum_{j=K}^{L} q^j p s^{j+1} + \sum_{n=K}^{L} s^n q^n \\ &= ps\left(G(s) - \sum_{n=0}^{K-2} s^n T_{0,[K,L]}(n) \right) + G(s)\sum_{j=K}^{L} q^j p s^{j+1} + \sum_{n=0}^{K-1} s^n p^n + \sum_{n=K}^{L} s^n q^n \end{aligned} \tag{7.40}$$

We are now in a position to consolidate the G(s) terms. However, we can see that, in the final expression for the solution to G(s) prior to inversion, the last two summation in equation (7.40) will most likely reside in the numerator of G(s), and will involve only the use of the sliding tool for inversion. Our major effort in the solution of equation (7.40) will be in the first two terms of the right side of the equality.

7.4.3 Consolidation of $T_{0,[K,L]}(n)$

The consolidation of equation (7.40) is actually straightforward. Combining all terms involving G(s) to the left side of the equation reveals

$$\begin{aligned} G(s) = {} & ps\left(G(s) - \sum_{n=0}^{K-2} s^n T_{0,[K,L]}(n) \right) + G(s)\sum_{j=K}^{L} q^j p s^{j+1} \\ & + \sum_{n=0}^{K-1} s^n p^n + \sum_{n=K}^{L} s^n q^n \end{aligned} \tag{7.41}$$

Distributing the multiplication and removing the parenthesis leads to

$$G(s)-psG(s)-\sum_{j=K}^{L}q^{j}ps^{j+1}G(s)=\sum_{n=0}^{K-1}s^{n}p^{n}+\sum_{n=K}^{L}s^{n}q^{n}-ps\sum_{n=0}^{K-2}s^{n}T_{0,[K,L]}(n) \quad (7.42)$$

Solving for G(s) reveals

$$G(s)\left[1-ps-\sum_{j=K}^{L}q^{j}ps^{j+1}\right]=\sum_{n=0}^{K-1}s^{n}p^{n}+\sum_{n=K}^{L}s^{n}q^{n}-ps\sum_{n=0}^{K-2}s^{n}T_{0,[K,L]}(n) \quad (7.43)$$

Fortunately, there is substantial simplification that can take place on the right side of the equation in the sums involving $T_{0,[K,L]}(n)$.

$$\begin{aligned} ps\sum_{n=0}^{K-2}s^{n}T_{0,[K,L]}(n) &= ps\left[1+ps+p^{2}s^{2}+\ldots+p^{K-2}s^{K-2}\right] \\ &= ps\left[\frac{1-p^{K-1}s^{K-1}}{1-ps}\right] \end{aligned} \quad (7.44)$$

since $T_{0,[K,L]}(0) = 1$ and $T_{0,[K,L]}(n) = p^{n}$ for $1 \le n \le K$.

Using these simplifications and solving for G(s) from equation (7.43) we find

$$G(s)=\frac{\sum_{n=0}^{K-1}s^{n}p^{n}}{1-ps-\sum_{j=K}^{L}q^{j}ps^{j+1}}+\frac{\sum_{n=K}^{L}s^{n}q^{n}}{1-ps-\sum_{j=K}^{L}q^{j}ps^{j+1}}-\frac{ps\left[\frac{1-p^{K-1}s^{K-1}}{1-ps}\right]}{1-ps-\sum_{j=K}^{L}q^{j}ps^{j+1}} \quad (7.45)$$

Conversion and consolidation of the difference equation family is now complete, and we can proceed with the inversion.

7.5 The Inversion and Solution of the $T_{0,[K,L]}(n)$ Model

An examination of the denominator of G(s) where

$$G(s)=\frac{\sum_{j=0}^{K-1}s^jp^j}{1-ps-\sum_{j=K}^{L}q^jps^{j+1}}+\frac{\sum_{j=K}^{L}s^jq^j}{1-ps-\sum_{j=K}^{L}q^jps^{j+1}}-\frac{ps\left[\frac{1-p^{k-1}s^{k-1}}{1-ps}\right]}{1-ps-\sum_{j=K}^{L}q^jps^{j+1}} \tag{7.46}$$

reveals the major work in the inversion is concentrated in the denominator. We will therefore focus first on the inversion of

$$H(s)=\frac{1}{1-ps-\sum_{j=K}^{L}q^jps^{j+1}} \tag{7.47}$$

H(s) is a generating function with polynomial (in s) in its denominator. In order to get a sense of how to proceed, we might begin by using specific examples of K and L to provide some guidance. If we were to choose, for example, K = 5 and L = 7, then H(s) becomes

$$H(s)=\frac{1}{1-ps-\sum_{j=5}^{7}q^jps^{j+1}}=\frac{1}{1-ps-pq^5s^6-pq^6s^7-pq^7s^8} \tag{7.48}$$

In general, the roots of the polynomial in s that reside in the denominator of H(s) in equation (7.48) are neither simple nor obvious. We cannot therefore appeal directly to either the method of partial fractions or the convolution principle of Chapter 2. We therefore focus our efforts instead on the method of coefficient collection, planning to invoke the multinomial theorem. We begin with the inversion of H(s) by writing

$$\frac{1}{1-ps-pq^5s^6-pq^6s^7-pq^7s^8}=\frac{1}{1-p\left(1+q^5s^5+q^6s^6+q^7s^7\right)s} \tag{7.49}$$

Thus $H(s) \triangleright_s \left\{p^k\left(1+q^5s^5+q^6s^6+q^7s^7\right)^k\right\}$ and we utilize the multinomial theorem to assess the middle term.

$$\left(1+q^5s^5+q^6s^6+q^7s^7\right)^k = \sum_{i=0}^{k}\sum_{j=0}^{k-i}\sum_{h=0}^{k-i-j}\binom{k}{i\ \ j\ \ h}\left(q^7s^7\right)^i\left(q^6s^6\right)^j\left(q^5s^5\right)^h$$
$$= \sum_{i=0}^{k}\sum_{j=0}^{k-i}\sum_{h=0}^{k-i-j}\binom{k}{i\ \ j\ \ h}q^{7i+6j+5h}s^{7i+6j+5h} \tag{7.50}$$

Thus

$$p^k\left(1+q^5s^5+q^6s^6+q^7s^7\right)^k s^k = p^k\sum_{i=0}^{k}\sum_{j=0}^{k-i}\sum_{h=0}^{k-i-j}\binom{k}{i\ \ j\ \ h}q^{7i+6j+5h}s^{k+7i+6j+5h} \tag{7.51}$$

The process of coefficient collection can now proceed as developed in Chapter 3 to reveal

$$H(s) = \frac{1}{1-ps-pq^5s^6-pq^6s^7-pq^7s^8}$$
$$\triangleright \left\{\sum_{m=0}^{k}p^m\sum_{i=0}^{m}\sum_{j=0}^{m-i}\sum_{h=0}^{m-i-j}\binom{m}{i\ \ j\ \ h}q^{7i+6j+5h}I_{m+7i+6j+5h=k}\right\} \tag{7.52}$$

The evaluation of this simple example for K = 5 and L =7 of H(s) has provided guidance on how to proceed with the inversion of H(s) in equation (7.47) for any K, L such that $0 \le K \le L \le n$. The only complication is the range of the permissible failure run lengths. The greater this range, the greater the number of summands in the final inversion. We proceed as follows

$$H(s) = \frac{1}{1-ps-\sum_{j=K}^{L}q^jps^{j+1}} = \frac{1}{1-\left[p\sum_{j=K}^{L}q^js^j\right]s} \tag{7.53}$$

Recalling that we can now write

$$H(s) \triangleright_s \left\{p^n\left(\sum_{j=K}^{L}q^js^j\right)^n s^n\right\} \tag{7.54}$$

we require the nth term of the series $p^n\left(\sum_{j=K}^{L} q^j s^j\right)^n s^n$ written so we may see the contributions of each power of s. We now apply the multinomial theorem to the expression $\left(\sum_{j=K}^{L} q^j s^j\right)^n$

$$\left(\sum_{j=K}^{L} q^j s^j\right)^n = \sum_{n_K=0}^{n} \sum_{n_{K+1}=0}^{n-n_k} \sum_{n_{K+2}=0}^{n-n_K-n_{K+1}} \cdots \sum_{n_L}^{n-n_K-n_{K+1}-n_{K+2}-\ldots n_{L-1}} \binom{n}{n_K\ n_{K+1}\ n_{K+2} \cdots n_L} q^{\sum_{h=K}^{L} hn_h} s^{\sum_{h=K}^{L} hn_h} \tag{7.55}$$

where $\sum_{j=K}^{L} n_j = n$. This leads to

$$p^n\left(\sum_{j=K}^{L} q^j s^j\right)^n s^n$$
$$= p^n \sum_{n_K=0}^{n} \sum_{n_{K+1}=0}^{n-n_k} \sum_{n_{K+2}=0}^{n-n_K-n_{K+1}} \cdots \sum_{n_L}^{n-n_K-n_{K+1}-n_{K+2}-\ldots n_{L-1}} \binom{n}{n_K\ n_{K+1}\ n_{K+2} \cdots n_L} q^{\sum_{h=K}^{L} hn_h} s^{n+\sum_{h=K}^{L} hn_h} \tag{7.56}$$

It only remains to collect coefficients of powers of s to invert H

$$H(s) = \frac{1}{1-ps-\sum_{j=K}^{L} q^j p s^{j+1}}$$
$$\triangleright \left\{ \sum_{m=0}^{n} p^m \sum_{n_K=0}^{m} \sum_{n_{K+1}=0}^{m-n_K} \cdots \sum_{n_L}^{m-n_K-n_{K+1}-n_{K+2}-\ldots n_{L-1}} \binom{m}{n_K\ n_{K+1}\ n_{K+2} \cdots n_L} q^{\sum_{h=K}^{L} hn_h} I_{m+\sum_{h=K}^{L} hn_h = n} \right\} \tag{7.57}$$

Although this looks daunting, it is at its root merely sums of powers of q.

Having completed the inversion of H(s) we can return to the original generating function G(s) to see that, if

$$G(s) = \frac{\sum_{j=0}^{K-1} s^j p^j}{1 - ps - \sum_{j=K}^{L} q^j ps^{j+1}} + \frac{\sum_{j=K}^{L} s^j q^j}{1 - ps - \sum_{j=K}^{L} q^j ps^{j+1}} - \frac{ps\left[\frac{1 - p^{K-1}s^{K-1}}{1 - ps}\right]}{1 - ps - \sum_{j=K}^{L} q^j ps^{j+1}} \tag{7.58}$$

then

$$G(s) \triangleright \left\{ \sum_{j=0}^{\min(K-1,n)} p^j A(n-j) + \sum_{j=K}^{\min(L,n)} q^j A(n-j) I_{n \geq K} - p \sum_{j=0}^{n-1} p^j A(n-1-j) + p^K \sum_{j=0}^{n-K} p^j A(n-K-j) \right\} \tag{7.59}$$

where

$$A(n) = \sum_{m=0}^{n} p^m \sum_{n_K=0}^{m} \sum_{n_{K+1}=0}^{m-n_K} \cdots \sum_{n_L}^{m-n_K-n_{K+1}-n_{K+2}-\ldots n_{L-1}} \binom{m}{n_K \; n_{K+1} \; n_{K+2} \cdots n_L} q^{\sum_{h=K}^{L} h n_h} I_{m + \sum_{h=K}^{L} h n_h = n} \tag{7.60}$$

An examination of the behavior of this solution as applied to problems in public health will follow in the next sections.

7.6 Order Statistics and Drought Prediction

7.6.1 Motivation

Application of the $\mathbf{R}_{ik}(n)$ to the prediction of average drought lengths carried out in Chapter 6, revealed that the results of this model correspond closely to the findings of recorded precipitation history. However, some have been critical of this approach to drought prediction. In ensuing correspondence, Şen [3] was critical of these predictions, particularly with regard to the use of averages (e.g., average drought length), a natural offshoot of the $\mathbf{R}_{ik}(n)$ model. This criticism was anchored in the view that hydrologists should try to find solutions for critical values rather than averages of lengths of hydrologic phenomena. An interesting enumerative approach for maximum drought lengths was identified by Şen [4] using a Markovian model.

The difference equation work to follow continues the advances of formal stochastic modeling based using difference equations for the examination of hydrologic phenomena. By integrating the notion of sequential occurrences of events with extreme order statistics (minimums and maximum event lengths), the model extends the application of formal probability modeling to hydrologic events of interest. As with the work of the previous section, the solutions are expressed in parameters that are readily available to workers in the field.

Having documented the ability of a Bernoulli trial based model to adequately predict drought lengths in one region of the United States, it would be interesting to examine the implications of this model. For this, we return to the $\mathbf{T}_{0,[K,L]}(n)$ model developed in here, demonstrating that $\mathbf{T}_{0,[K,L]}(n)$ can be formulated in terms of K, L, q, and n. Also, by solving this system, the user has access to the order statistics and can examine the impact of varying these parameters on the examined maximum drought length is easily computed by the authors [5].

7.6.2 Prediction of Drought Length Intervals

We can begin this process by making a prediction for the future drought length in the next 20 years as a function of the probability of inadequate water resources, and provide some interesting and direct assessments of the relationship between the yearly probability of inadequate water resources and the probability of drought (Table 7.1).

In this table, the probability of a drought is provided as a function of the probability of inadequate water resources (q). Three different drought lengths are considered; short drought lengths (0-5 years), medium drought lengths (6-10 years), and long drought lengths (16-20 years). For example, from Table 7.1, the probability that in the next twenty years, all drought lengths will be less than or equal to five years in duration is 0.99 if the probability of inadequate water resources is 0.10.

There is an interesting phenomena predicted by the model that may not be so self evident. The probability of a short drought actually decreases as the probability of inadequate water resources increases. The probability of shorter drought lengths is very high for smaller probabilities of inadequate rainfall, but gradually, the likelihood of only short drought lengths declines as

Table 7.1 Predictions of Drought Length in 20 Years as a Function of Yearly Probability of Inadequate Water Resources (q)

		Yearly Probability of Inadequate Water Resources				
		0.30	0.50	0.70	0.90	0.99
Drought Length (years)	0 to 5	0.99	0.88	0.46	0.03	0.00
	6 to 15	0.00	0.00	0.01	0.15	0.07
	15 to 20	0.00	0.00	0.00	0.15	0.84

the probability of inadequate water increases. The probability that drought lengths will be restricted to between 6 and 15 years decreases as the probability of inadequate water resources increases, rising to a maximum of 0.15 for q = 0.90. It is notable that this probability then decreases as q increases further, falling to 0.07 for the maximum value of q examined (0.99). This somewhat paradoxical behavior is explained by the fact that 15 years is not the maximum possible drought length in 20 years (Figure 7.1). As q increases from 0.10 to 0.99, the probability that the only droughts that occur are long droughts (16-20 years in length) increases slowly at first for small values of q, then increasingly dramatically as q increases to 0.99. Thus, the prevalence of droughts changes as a function of the probability of inadequate water resources, as revealed in Figure 7.1.

7.6.3 Maximum Drought Length Predictions

Our discussion thus far has focused on the occurrence of droughts of lengths chosen arbitrarily. We now turn our attention to the occurrence of maximum and minimum drought lengths. Equation (7.2) provides for the computation of expected maximum drought lengths based on the $T_{0,[K,L]}(n)$ model. Table 7.2 examines the expected lengths of maximum drought lengths as a function of q. Table 7.2 reveals that the probabilities for longer maximum drought lengths increase as q increases. For example, in a 20 year period, the probability the maximum drought length is greater than 4 years increases from 0.004 for q = 0.20 to 0.731 for q = 0.70. However, even for large values of q, the probability

Table 7.2 Predictions of Maximum Drought Length Probabilities in 20 Years as a Function of Yearly Probability of Inadequate Water Resources (q)

		Yearly Probability of Inadequate Water Resources				
		0.20	0.40	0.50	0.60	0.70
Maximum	> 2 yrs	0.11	0.56	0.79	0.93	0.99
Drought	> 4 yrs	0.00	0.10	0.25	0.48	0.73
Length	> 6 yrs	0.00	0.01	0.06	0.17	0.38
	> 8 yrs	0.00	0.00	0.01	0.05	0.17
	> 10 yrs	0.00	0.00	0.00	0.02	0.07

Table 7.3 Predictions of Expected Maximum Drought Length with Standard Deviation for 20 Years as a Function of the Probability of Inadequate Water Resources (q)

	Yearly Probability of Inadequate Water Resources				
	0.20	0.40	0.50	0.60	0.70
Expected Length	1.63	2.90	3.73	4.80	6.28
Standard Deviation	0.76	1.26	1.59	2.03	2.65
Lower Bound	0.14	0.43	0.61	0.82	1.09
Upper Bound	3.12	5.37	6.85	8.78	11.47

An evaluation of the behavior of the expected maximum drought length as a function of q provides important new information for droughts over the next twenty years. For each yearly probability of inadequate water resources, Table 7.8 provides the expected duration of the worst drought (maximum drought length), and its standard deviation. In addition, the lower and upper bounds of a 95% confidence interval are provided based on the expected length and its standard deviation. Thus, if the probability of a inadequate water resources is 0.40, the length of the worst drought is most likely to be 2.9 years. However, the worst drought could be as short as 0.43 years and as long as 5.3 years. This table demonstrates that the expected maximum drought length increases as the yearly probability of inadequate water resources increases. In addition, the standard deviation increases, making prediction of the length of the worst drought increasingly hazardous. Thus, for a high probability of inadequate water resources of 0.70, the expected maximum drought length is 6.28, but the standard deviation of 2.65 makes the upper bound on the maximum drought length of 11.47 years.

Despite the advances in stochastic hydrology, an important hurdle for its application remained the incomplete examination of "critical drought length," or extreme order statistics, as mentioned by Şen [3]. The emphasis on maximum values would provide a firm basis for decisions concerning the performance of engineering structures at their most extreme risk. A prospectively based run theory model permits this examination of critical drought length through the use of order statistics. The model allows prediction of droughts of arbitrary length, and is easily converted to allow the computation of the probabilities of different bounty lengths.

A criticism of this model is the use of Bernoulli trials. However, a prospective run theory model based on independent Bernoulli trials has been demonstrated to work sufficiently well in predicting rainfall in Texas. Thus, although there is correlation from year to year in rainfall amounts, the

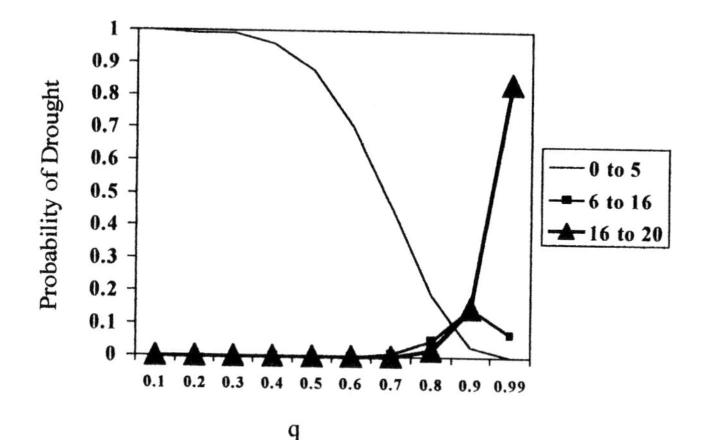

Figure 7.1 Probability of droughts of three different lengths as a function of q.

correlation is of such a nature as to allow the independence assumption as a reasonable approximation.

The model allows prediction of droughts of arbitrary length, and is easily converted to allow the computation of the probabilities of different bounty lengths (lengths of adequate rainfall). However, more importantly, we compute the extreme order statistics of drought and bounty length. Also the standard deviation of this drought length is provided, allowing for the computation of the degree of variability of confidence intervals for the maximum drought length. Since this computation comes directly from the derived model, this computation is again a function of parameters that the hydrologist has access to, namely q and n, the number of years over which the prediction is estimated.

The generating function approach is not the only possible mode of computation available to stochastic hydrologists. Direct enumeration [3] has been employed in the past. However, this is most easily executed if the sample sizes are small. We prefer the generating function approach since it provides a direct solution for any value of $n > 0$. In addition, the recursive relationship developed here allows freedom in considering the run lengths of interest through the choice of K and L such that $0 < K \leq L$. Thus, one can choose to focus on minimum run lengths, and easily obtain the distribution for the minimum run length from the solution provided in the methods. Alternatively, one might focus on the probability that the run lengths are neither extremely small nor extremely large. In this setting, $1-\mathbf{T}_{,[K,L]}(n)$ is the probability of an extremely short ($< K$) or extremely long ($>L$) drought if q is the probability of inadequate water resources. Also, the closed form solution for $\mathbf{T}_{0,[K,L]}(n)$ does not confine the worker to first solve for all smaller values of n of lesser interest.

The users of run theory in the past have not translated their run theory computations from the probability arena to the hydrology arena successfully. This is a difficulty of the parameterization of the problem. This translation is essential if the develops from run theory are to be applied to hydrology. The work provided here offers the direct translation from run theory to probabilities for drought length order statistics, the expected length of the worst drought and the expected length of the smallest bounty.

The developed model here is extremely flexible. For example, one may consider the definition of a drought as a consecutive, uninterrupted string of successes as too confining. A drought could occur when there is one year of adequate water resources, preceded and followed by consecutive years of inadequate water resources. The current model could be modified to this scenario by allowing as a drought a success followed by consecutive years of failure $q^{k-m-1}pq^{m}$. The implications of a model defining a drought as such have yet to be explored.

A criticism of this model is the use of Bernoulli trials. However, a prospective run theory model based on independent Bernoulli trials has been demonstrated to work sufficiently well in predicting rainfall in Texas [2]. Thus, although some authors may believe there is correlation from year to year in rainfall amounts, the correlation is sufficiently small to allow the independence assumption as a reasonable approximation. Nevertheless, the theoretical underpinnings of this model would be substantially strengthened if the model could be developed allowing correlation between events on different trials. Although work continues in this area, the present development of stochastic hydrology continue to provide useful estimates of the occurrence or recurrent hydrologic phenomena.

The development presented in this chapter is a first step to extending a potentially useful application of difference equations to the area of water resource prediction.

Problems

1. Compute the probability that in a sequence of Bernoulli trials, failure run lengths are between two and five, by enumeration.

2. Show that $ps\sum_{n=0}^{K-1} s^n T_{0,[K,L]}(n) = ps$

3. Verify equation 7.29.

$$\sum_{n=0}^{\infty}\sum_{j=K}^{L} s^n q^j p T_{0,[K,L]}(n-j-1) = \sum_{j=K}^{L} q^j p s^{j+1}\left(G(s) - \sum_{n=0}^{K-j-1} s^n T_{0,[K,L]}(n)\right)$$

4. Write the family of difference equations for $T_{K,K}(n)$. Solve this family of difference equations using the generating function approach to find the probability that when failures occur in the string of Bernoulli trials, they occur only in runs of length K where $0 < K \leq n$.

5. Solve the run cluster problem. Assume that each trial is a Bernoulli trial. Let the probability of success be p and the probability of failure be $q = 1 - p$. Compute the probability that, when runs occur they are of length 1 or in the interval $[k - 1, k + 1]$. This model says that runs occur in clusters are as singletons.

6. Adapt the $T_{0,[K,L]}(n)$ model developed in this chapter for the circumstance that the experiment on any given trial may have three outcomes, success with probability p, failure with probability q, or un-resolvable with probability r, such that $p + q + r = 1$. Show specifically how the solution for this model is different than that for $T_{0,[K,L]}(n)$.

7. In a sequence of Bernoulli trials with known probability of success p, find the probability of "singleton failures", i.e., the probability that failures occur in isolation (i.e., they do not appear as consecutive sequence of failures, but always followed by a success or the end of the sequence).

8. Consider a cell receptor designed to respond to a stimulus in a short time interval. During this time interval, the receptor can be filled by a neurotransmitter with probability p and the receptor is considered "on" for that time interval. Conversely, the receptor is not filled by the neurotransmitter, and the receptor is "off" with probability $q = 1 - p$, where p is constant from cycle to cycle. After the time interval has elapsed, the receptor resets, ready to be "on" or "off" for the next time interval. Define a cycle as ten consecutive time intervals. Let the receptor activation threshold K be the number of consecutive time intervals in the cycle during which the receptor must be on for the receptor to be considered activated. Develop a family of difference equations that allows you to compute computes the probability that in this time cycle, the receptor is activated. Find the probability the receptor is activated as the threshold of activation is 4, 5, 6 or 7 as a function of p and q.

References

1. Yevjevich V. Stochastic Processes in Hydrology. Water resource. Publ. Fort Collins, Colorado. 1972.
2. Moyé L.A., Kapadia A.S., Cech I., Hardy R.J. The theory of runs with application to drought prediction. *Journal of Hydrology* 103:127-137. 1988.
3. Şen Z. The theory of runs with application to drought prediction – comment. *Journal of Hydrology* 110: 383-391. 1989.
4. Şen Z. Statistical analysis of hydrologic critical droughts. *Journal of Hydrology* Divs. ASCE, 106(H)99-115. 1980.
5. Moyé L.A., Kapadia A.S. Predictions of drought length extreme order statistics using run theory. *Journal of Hydrology* 169: 95-110. 1995.

8

Difference Equations: Rhythm Disturbances

8.1 Introduction

In this chapter, we change public health arenas, moving from the problem of predicting large hydrologic phenomena affecting communities and states to effects that operate inside the human body. Attention now focuses on the rhythm of the human heart. Alternations in its rhythm can have important clinical implications for the well being and survival of the patient, yet, the predictions of these rhythms can be extremely difficult to forecast. After an introduction to heart rhythm and its disturbances, we will examine potential applications of difference equations to predictions of rhythm disturbances.

8.2 Heart Rhythm

8.2.1 Normal heart rhythm

Heartbeats are taken for granted. If the reader spends three hours on this chapter, their heart will have beat almost 13,000 times. The heart is a pump pulling deoxygenated blood, high in carbon dioxide content, from the body, moving this blood quickly through the right side of the heart, to the lungs where the blood is enriched with oxygen and cleansed of CO_2. This replenished blood is then pumped into the left side of the heart, and from there forced out to the general circulation, where it nourishes muscles, kidneys, the digestive track, and the brain, and other organs. The movement of blood through the heart must be carefully synchronized if it is to pump effectively. This synchronization takes place through the meshing of valve operation and beating rhythm of the heart.

The heart has four chambers, two smaller upper ones called atria, and two huge lower chamber called ventricles. The ventricles are the main squeezing chambers. When the ventricles relax, they fill with blood from the two upper chambers. When they contract blood is ejected from the right side of the heart to the lungs, while simultaneously, blood is ejected from the left side of the heart to the systemic circulation. However, there is more to this operation than muscle strength.

This operation must be coordinated, and it is coordinated electrically. Located deep in the atria is a node, termed the sinoatrial node (SA node). It is the master control switch for the beating of the four chambers of the heart. It sends out a signal that first tells the atria to contract, pushing blood into the large ventricles. The same signal then progresses to the ventricles, causing them to contract. Timing is essential, and normally the signal for the ventricles reaches them when they are full and ready to contract.

This regular rhythm is easily assessed just by listening to one's own heart (a favorite pastime for children, and a worthwhile exercise for medical and nursing students). However, the flow of electricity through the heart is measured by the electrocardiogram. This tool is the cardiologist's objective measure of the electrical conductivity of the heart.

8.2.2 Heart rhythm disturbances

There are many heart rhythm disturbances, some affect the atria only, others affect the ventricles. The range of these electrical aberrations range from the simple, relatively infrequent, and escalate to the more severe. For the difference equation model, we will focus on arrhythmias that affect the ventricles.

8.2.3 Premature ventricular contractions

Most everyone remembers when their "heart skips a beat". Children (and adults) do this to each other fairly easily by jumping out from a closet or a darkened hallway. The effect on the frightened person is immediate. The fright causes a series of nervous system and neurohormonal responses and aberrations. One of the most common is that the ventricles contract prematurely, before they are completely full. It is a fascinating observation that the off rhythm beat is so noticeable, especially when the hundreds of thousands of regular beats go relatively unnoticed. Nevertheless, we all can recall the thud in our chest when so startled. For the majority of us, this one premature ventricular contraction (PVC) is at most a short-lived, unpleasant sensation. The SA node quickly reasserts its predominance as the source of electrical activity of the heart, and the heart quickly returns to its normal sinus rhythm.

8.2.4 Bigeminy, trigeminy, and quadregeminy

Occasionally, PVCs do not occur in isolation. They can occur more frequently. This is seen in patients who have heart disease. For example, patients who have heart attacks, when a portion of heart muscle dies, may have the electrical conductivity system of the heart affected. As the SA node functions less and less efficiently, other parts of the electrical conducting system try to break through, attempting to exert their own control over the flow of electricity through the heart. PVCs begin to occur more and more frequently, finally occurring as couplets or two consecutive ones. The occurrence of such a couple is termed bigeminy. Trigeminy is the occurrence of three consecutive PVCs and quadgeminy defines the consecutive occurrence of four such PVCs.

These short bursts of ectopic ventricular activity can occur in the normal heart as well, as one of the authors can attest. On his first day as an intern, he was learning about his new responsibilities as a physician, while smoking a pipe and drinking coffee. The combination of stress from the new job, the nicotine from the pipe (a heart stimulant) and the caffeine (another heart stimulant) from the coffee produced a very memorable burst of trigeminy.

8.2.5 Ventricular arrhythmias and sudden death

As the consecutive occurrence of PVCs becomes more frequent, even the runs of bigeminy, trigeminy, and quadgeminy begin to come together into a burst of ventricular tachycardia (rapid heart rate). These episodes of ventricular tachycardia (sometimes termed bouts of VT) are extremely dangerous. As we pointed out in the previous section, when a PVC occurs in isolation, normal sinus rhythm is quickly restored, and the heart returns to its efficient filling mechanism. However, when many of these abnormal ventricular beats occur in a row, the movement of blood through the heart is profoundly disturbed. The ventricles contract rapidly, but they contract far too rapidly, well before the atria have the opportunity to fill the ventricles with blood. Thus in VT, the ventricles contract to no good end, since they are not pumping blood at all. This condition, if not treated, can deteriorate to ventricular fibrillation (VF) where the complicated interrelated muscle system in the ventricles no longer contracts as a cohesive unit, and the ability to move blood out to the systemic circulation is effectively destroyed.

This destruction results in the sudden death of the individual. One moment the person appears fine–the next moment they are on the floor within seconds of death. In the case of drowning, even though there has been no breathing, the heart has continued to pump throughout the accident. Thus, even though the victim's brain is receiving only deoxygenated blood, they are still receiving blood, and survival is prolonged. However, in sudden death, the heart stops beating at once. The brain receives no blood, and brain cells die by the tens of millions in a matter of seconds. Certainly, sudden death syndrome must be avoided at all costs.

8.2.6 The arrhythmia suppression hypothesis

Cardiologists have long recognized the sudden death syndrome and have worked hard to identify both risk factors for the syndrome and a way to treat it. Over the past thirty years, many esteemed workers in cardiology have constructed a theory that has come to be known as the arrhythmia suppression hypothesis. This hypothesis states that the harbinger of sudden death was not the occurrence of "runs of VT" but the occurrence of bigeminy, trigeminy, and even PVCs. Since these relatively mild rhythm abnormalities sometimes deteriorate to runs of VT, and from VT to ventricular fibrillation and death, preventing the mild ventricular rhythm disturbances would prevent sudden death. This hypothesis was put to the test in a complicated experiment called CAST (Cardiac Arrhythmia Suppression Trial) by testing drugs known to reduce mild ventricular arrhythmias. However, the drugs that were tested produced even worse rhythms then the ones they were designed to reduce and the experiment had to be stopped. Many cardiologists still believe in the arrhythmia suppression hypothesis, and are waiting for the correct drug to be developed.

8.2.7 Difference equations and cardiac arrhythmias

One difficulty with the arrhythmia model is that it would seem to require a way to convert the occurrence of PVCs to a predicted frequency of occurrence of bigeminy, trigeminy, and longer runs of unusual ventricular beats. Our goal will be to tailor difference equations developed in earlier chapters to model the occurrence of runs of ventricular beats. Ultimately, we will want to examine the relationship between the probability of a PVC and the expected occurrence of short and long runs of ventricular tachycardia. If the correspondence between the level of PVC prevalence and occurrence of ventricular tachycardia of various lengths could be demonstrated, the cardiologist may be able to know how a decrease in the prevalence of PVC through medical intervention would effect the occurrence of ventricular tachycardia.

8.3 Application of the $T_{[K,L]}(n)$ Model

For all computations in the cardiac rhythm domain, it will be assumed that each beat represents a Bernoulli trial. There are only two possibilities for the beat; it is either a normal sinus beat, which occurs with probability p, or a PVC (abnormal beat) which occurs with probability q. Thus, a failure is the occurrence of a premature, ventricular contraction. We may therefore think of a burst of bigeminy as a failure run of length two, since bigeminy is two consecutive aberrant ventricular beats. Analogously, the occurrence of three successive irregular ventricular beats (trigeminy) will be considered as a failure run of length three and so on. We will explore several applications of the $T_{[K,L]}(n)$ to model these occurrences.

Recall from Chapter 7 that the $\mathbf{T}_{[K,L]}(n)$ model predicts the probability that all failure run lengths are between length K and length L where $0 < K \leq L \leq n$ in a sequence of n independent Bernoulli trials with fixed probability of success p. As before

$\mathbf{T}_{[K,L]}(n)$ = P[the minimum and maximum failure run lengths is in the interval K to L when $K \leq L$ in n trials]

For example, $\mathbf{T}_{[2,6]}(12)$ is the probability that in 12 Bernoulli trials all failure runs have lengths 2, 3, 4, 5, or 6, with all other failure run lengths excluded. Recall that

$\mathbf{T}_{[K,K]}(n)$ = P[all failure run lengths are exactly of length K]
$\mathbf{T}_{[0,L]}(n)$ = P[all failure runs are $\leq$ L in length]
1-$\mathbf{T}_{[0,L]}(n)$ = P[at least one failure run length is greater than L]

If $M_F(n)$ is the maximum failure run length in n trials, then

$$E\left[M_F(n)\right] = \sum_{L=0}^{n} L\left[T_{[0,L]}(n) - T_{[0,L-1]}(n)\right]$$

$$\mathrm{Var}\left[M_F(n)\right] = \sum_{L=0}^{n} L^2\left[T_{[0,L]}(n) - T_{[0,L-1]}(n)\right] - E^2\left[M_F(n)\right] \tag{8.1}$$

.

Recollect that $\mathbf{T}_{0,[K,L]}(n)$ is simply $\mathbf{T}_{[K,L]}(n)$ - p^n. The boundary conditions for $\mathbf{T}_{0,[K,L]}(n)$ are

$T_{0,[K,L]}(n) = 0$ for all $n < 0$:
$T_{0,[K,L]}(n) = 1$ for $n = 0$:
$T_{0,[K,L]}(n) = p^n$ for $0 < n < K$

Using the indicator function, we may write the recursive relationship for $T_{0,[K,L]}(n)$ for $0 < K \leq \min(L, n)$

$$\begin{aligned} T_{0,[K,L]}(n) = pT_{0,[K,L]}(n-1) + q^K pT_{0[K,L]}(n-K-1) + q^{K+1} pT_{0,[K,L]}(n-K-2) \\ + q^{K+2} pT_{0,[K,L]}(n-K-3) + \ldots + q^L pT_{0,[K,L]}(n-L-1) + q^n I_{K \leq n \leq L} \end{aligned} \quad (8.2)$$

$I_{x \in A} = 1$ if x is in the set A and zero otherwise. This difference equation may be rewritten as

$$T_{0,[K,L]}(n) = pT_{0,[K,L]}(n-1) + \sum_{j=K}^{L} q^j pT_{0,[K,L]}(n-j-1) + q^n I_{K \leq n \leq L} \quad (8.3)$$

As before, assume that the run length interval bounds K and L are known. Assume that the probability of failure, q is known, and that
$p = 1 - q$.

8.4 Difference Equations for Isolated PVCs

The plan will be to tailor equation (8.3) to a model to predict runs of irregular ventricular beats. Let n be the total number of heartbeats that are observed. Begin by computing the probability that in a consecutive sequence of heartbeats, the only ectopy (i.e., abnormal heart rhythm) that occurs is isolated PVCs. An isolated PVC is a failure run of length one. Converting this to model terminology, $K = L = 1$ and we compute $n > 0$.

$$T_{0,[1,1]}(n) = pT_{0,[1,1]}(n-1) + qpT_{0,[1,1]}(n-2) + qI_{n=1} \quad (8.4)$$

Equation (8.4) is very illuminating. It provides for failure run lengths of length zero or length one, and does not permit failure runs of length > 1 for $n \geq 1$. From the boundary conditions we have defined $T_{0,[1,1]}(0) = 1$. We compute from equation (8.4) $T_{0,[1,1]}(1) = p + q$ and $T_{[1,1]}(1) = q$. We can compute from this same equation that $T_{0,[1,1]}(2) = p(p+q) + qp = p^2 + 2pq$ and $T_{[1,1]}(2) = T_{0,[1,1]}(2) - p^2 = 2pq$.

The goal is to solve equation (8.4) in its entirety, finding a general solution for $T_{1,1}(n)$ using the generating function approach. Define

$$G(s) = \sum_{n=0}^{\infty} s^n T_{0,[1,1]}(n) \tag{8.5}$$

proceed by multiplying each side of equation (8.4) by s^n and begin the conversion process.

$$\begin{aligned} s^n T_{0,[1,1]}(n) &= s^n p T_{0,[1,1]}(n-1) + s^n qp T_{0,[1,1]}(n-2) + s^n q I_{n=1} \\ \sum_{n=1}^{\infty} s^n T_{0,[1,1]}(n) &= \sum_{n=1}^{\infty} s^n p T_{0,[1,1]}(n-1) + \sum_{n=1}^{\infty} s^n qp T_{0,[1,1]}(n-2) + \sum_{n=1}^{\infty} s^n q I_{n=1} \end{aligned} \tag{8.6}$$

Of course, the term on the left side of the equality of equation (8.6) is G(s)-1. The last term on the right side of this equation is qs. Evaluating the first two terms of equation (8.6) as follows: $\sum_{n=1}^{\infty} s^n p T_{0,[1,1]}(n-1)$ becomes

$$\sum_{n=1}^{\infty} s^n p T_{0,[1,1]}(n-1) = ps \sum_{n=1}^{\infty} s^{n-1} T_{0,[1,1]}(n-1) = ps \sum_{n=0}^{\infty} s^n T_{0,[1,1]}(n) = psG(s) \tag{8.7}$$

and the next term may be reworked

$$\begin{aligned} \sum_{n=1}^{\infty} s^n qp T_{0,[1,1]}(n-2) &= qps^2 \sum_{n=1}^{\infty} s^{n-2} T_{0,[1,1]}(n-2) \\ &= qps^2 \sum_{n=0}^{\infty} s^n T_{0,[1,1]}(n) = qps^2 G(s) \end{aligned} \tag{8.8}$$

We may rewrite equation (8.6) in terms of G(s) and simplify

$$
\begin{aligned}
&G(s)-1 = ps + psG(s) + qps^2G(s) + qs \\
&G(s)\left[1 - ps - qps^2\right] = s+1 \qquad (8.9)\\
&G(s) = \frac{s+1}{1-ps-qps^2}
\end{aligned}
$$

This is a polynomial generating function with no helpful roots for the denominator, so we proceed with the inversion by collecting coefficients

$$
G(s) = \frac{s+1}{1-ps-qps^2} = \frac{s+1}{1-(1+qs)ps} \qquad (8.10)
$$

and following the method of coefficient collection outlined in Chapter 3, work to expand each term of the infinite series $(1+qs)^n p^n s^n$. Applying the binomial theorem to the first factor

$$
(1+qs)^n = \sum_{j=0}^{n}\binom{n}{j}q^j s^j \qquad (8.11)
$$

and the nth term of the series $(1+qs)^n p^n s^n$ becomes $p^n\sum_{j=0}^{n}\binom{n}{j}q^j s^{n+j}$. Now collecting coefficients to see that

$$
G(s) = \frac{s+1}{1-ps-qps^2} \triangleright \left\{\sum_{m=0}^{n-1}p^m\sum_{j=0}^{m}\binom{m}{j}q^j I_{m+j=n-1} + \sum_{m=0}^{n}p^m\sum_{j=0}^{m}\binom{m}{j}q^j I_{m+j=n}\right\} \qquad (8.12)
$$

and

$$
T_{[1,1]}(n) = \sum_{m=0}^{n-1}p^m\sum_{j=0}^{m}\binom{m}{j}q^j I_{m+j=n-1} + \sum_{m=0}^{n}p^m\sum_{j=0}^{m}\binom{m}{j}q^j I_{m+j=n} - p^n \qquad (8.13)
$$

8.5 Difference Equation for Bigeminy Only

With the experience from the isolated PVC model, we can expand our application of the difference equation approach to arrhythmia occurrence. In this section, we will focus solely on failure run lengths of two alone, i.e., the $T_{0\,[2,2]}(n)$ model. Writing the family of difference equations for $n > 0$

$$T_{0,[2,2]}(n) = pT_{0,[2,2]}(n-1) + q^2pT_{0,[2,2]}(n-K-1) + q^2I_{n=2} \tag{8.14}$$

We will proceed as we did in the previous section. Thus $T_{0\,[2,2]}(0)$ is defined as 1, $T_{0,[2,2]}(1) = p$ (remember that $T_{0,[2,2]}(n)$ is the probability that either the failure runs are of length 2 or that there are no failure runs at all). $T_{0,[2,2]}(2) = p(p) + q^2 = p^2 + q^2$. We now proceed to find the general solution for the probability of bigeminy alone. Define

$$G(s) = \sum_{n=1}^{\infty} s^n T_{0,[2,2]}(n) \tag{8.15}$$

and proceed with conversion. Begin by multiplying each term in equation (8.14) by s^n to find

$$s^n T_{0,[2,2]}(n) = s^n pT_{0,[2,2]}(n-1) + s^n q^2 pT_{0,[2,2]}(n-3) + s^n q^2 I_{n=2} \tag{8.16}$$

The next step requires taking the sum from n equal one to infinity in each term of equation (8.16) revealing

$$\sum_{n=1}^{\infty} s^n T_{0,[2,2]}(n) = \sum_{n=1}^{\infty} s^n pT_{0,[2,2]}(n-1) + \sum_{n=1}^{\infty} s^n q^2 pT_{0,[2,2]}(n-3) + \sum_{n=1}^{\infty} s^n q^2 I_{n=2} \tag{8.17}$$

Note that the first term on the right side of equation (8.17) may be written as

$$\sum_{n=1}^{\infty} s^n pT_{0,[2,2]}(n-1) = psG(s) \tag{8.18}$$

Now rewrite equation (8.17) in terms of G(s).

$$G(s)-1=psG(s)+q^2ps^3G(s)+q^2s^2 \tag{8.19}$$

Simplification leads to

$$\begin{aligned} G(s)\left[1-ps-q^2ps^3\right] &= 1+q^2s^2 \\ G(s) &= \frac{1}{1-ps-q^2ps^3}+\frac{q^2s^2}{1-ps-q^2ps^3} \end{aligned} \tag{8.20}$$

and inversion here presents no difficulty. We invert

$$\frac{1}{1-ps-q^2ps^3} = \frac{1}{1-\left(1+q^2s^2\right)ps} \triangleright_s \left\{\left(1+q^2s^2\right)^n p^n\right\} \tag{8.21}$$

And now applying the binomial theorem to the first factor in the nth term of the sequence to find.

$$\left(1+q^2s^2\right)^n = \sum_{j=0}^{n}\binom{n}{j}q^{2j}s^{2j} \tag{8.22}$$

and we can proceed with the inversion

$$\frac{1}{1-ps-q^2ps^3} \triangleright \left\{\sum_{m=0}^{n}p^m\sum_{j=0}^{m}\binom{m}{j}q^{2j}I_{m+2j=n}\right\} \tag{8.23}$$

applying this intermediate finding to G(s) from equation (8.17) we find that

$$G(s) = \frac{1}{1 - ps - q^2ps^3} + \frac{q^2s^2}{1 - ps - q^2ps^3}$$
$$\triangleright \left\{ \sum_{m=0}^{n} p^m \sum_{j=0}^{m} \binom{m}{j} q^{2j} I_{m+2j=n} + q^2 \sum_{m=0}^{n-2} p^m \sum_{j=0}^{m} \binom{m}{j} q^{2j} I_{m+2j=n-2} \right\} \tag{8.24}$$

and we write

$$T_{[2,2]}(n) = \sum_{m=0}^{n} p^m \sum_{j=0}^{m} \binom{m}{j} q^{2j} I_{m+2j=n} + q^2 \sum_{m=0}^{n-2} p^m \sum_{j=0}^{m} \binom{m}{j} q^{2j} I_{m+2j=n-2} - p^n \tag{8.25}$$

8.6 Solution for the Ventricular Tachycardia Runs of Length k

With the experience of the isolated PVC and bigeminy problem, we are now in a position to provide a general solution for the probability of the occurrence of ventricular tachycardia of runs of length k by solving the $T_{0,[K,K]}(n)$ model that will be the purpose of this demonstration. The solution can be provided for bigeminy, trigeminy, quadgeminy, or for any run of ventricular tachycardia of length k. The solution derived is the probability for k-geminy only. In the next section, we will expand this solution to include the probability of ventricular tachycardia run of length K. As before begin with the $T_{0,[K,K]}(n)$ model, $n > 0$.

$$T_{0,[K,K]}(n) = pT_{0,[K,K]}(n-1) + q^K pT_{0,[K,K]}(n-K-1) + q^K I_{n=K} \tag{8.26}$$

Define G(s) as

$$G(s) = \sum_{n=1}^{\infty} s^n T_{0,[K,K]}(n) \tag{8.27}$$

and proceed with conversion and consolidation. Begin by multiplying each term in equation (8.26) by s^n to find

$$s^n T_{0,[K,K]}(n) = s^n pT_{0,[K,K]}(n-1) + s^n q^K pT_{0,[K,K]}(n-K-1) + s^n q^K I_{n=K} \tag{8.28}$$

Summing each term in equation (8.28) from $n = 1$ to ∞ reveals

$$\sum_{n=1}^{\infty} s^n T_{0,[K,K]}(n) = \sum_{n=1}^{\infty} s^n p T_{0,[K,K]}(n-1) + \sum_{n=1}^{\infty} s^n q^K p T_{0,[K,K]}(n-K-1) + \sum_{n=1}^{\infty} s^n q^K I_{n=K} \quad (8.29)$$

The goal is to write equation (8.29) in terms of G(s). Begin by writing

$$G(s) - 1 = \sum_{n=1}^{\infty} s^n p T_{0,[K,K]}(n-1) + \sum_{n=1}^{\infty} s^n q^K p T_{0,[K,K]}(n-K-1) + s^K q^K \quad (8.30)$$

and continue

$$G(s) - 1 = ps \sum_{n=1}^{\infty} s^{n-1} T_{0,[K,K]}(n-1) + q^K p s^{K+1} \sum_{n=1}^{\infty} s^{n-K-1} T_{0,[K,K]}(n-K-1) + s^k q^k \quad (8.31)$$

We need to examine two of these terms in some detail. Begin with the second term on the left side of equation (8.31).

$$ps \sum_{n=1}^{\infty} s^{n-1} T_{0,[K,K]}(n-1) = ps \sum_{n=0}^{\infty} s^n T_{0,[K,K]}(n) = psG(s) \quad (8.32)$$

The last summation of the right side of equation (8.32) reduces to 1 since $T_{0,[K,K]}(n) = p$ for n=1 and $T_{0,[K,K]}(n) = p^n$ for $1 \le n \le K - 2$.

For $q^K p s^{K+1} \sum_{n=1}^{\infty} s^{n-K-1} T_{0,[K,K]}(n-K-1)$, observe that

$$q^K p s^{K+1} \sum_{n=1}^{\infty} s^{n-K-1} T_{0[K,K]}(n-K-1) = q^K p s^{K+1} G(s) \quad (8.33)$$

With these results, can complete the consolidation of G(s) as

$$G(s) - 1 = psG(s) + q^K p s^{K+1} G(s) + s^K q^K \quad (8.34)$$

or

$$G(s)\left(1 - ps - q^K p s^{K+1}\right) = 1 + s^K q^K \quad (8.35)$$

leading to

$$G(s) = \frac{1}{1 - ps - q^K ps^{K+1}} + \frac{s^K q^K}{1 - ps - q^K ps^{K+1}} \tag{8.36}$$

Proceeding with the inversion

$$\frac{1}{1 - ps - q^K ps^{K+1}} = \frac{1}{1 - \left(1 + q^K s^K\right) ps} \triangleright \left\{ \sum_{m=0}^{n} p^m \sum_{j=0}^{m} \binom{m}{j} q^{Kj} I_{m+Kj=n} \right\} \tag{8.37}$$

and we can compute the inversion of G(s)

$$G(s) = \frac{1}{1 - ps - q^K ps^{K+1}} + \frac{s^K q^K}{1 - ps - q^K ps^{K+1}}$$
$$\triangleright \left\{ \sum_{m=0}^{n} p^m \sum_{j=0}^{m} \binom{m}{j} q^{Kj} I_{m+Kj=n} + q^K \sum_{m=0}^{n-K} p^m \sum_{j=0}^{m} \binom{m}{j} q^{Kj} I_{m+Kj=n-K} \right\} \tag{8.38}$$

8.7 Predicting Mild Arrhythmias

The previous demonstrations have provided the solutions for the occurrence of isolated arrhythmias. However, such findings of only bigeminy or only trigeminy are of limited concern to cardiologists. Often findings of isolated PVCs occur in concert with bigeminy, trigeminy, and others. Although the previous derivations are illuminating, it would be more useful to compute the probability that a run of PVCs of length L or less have occurred. If, for example L = 4, then this probability would include incidents of isolated PVCs and/or bigeminy, and/or trigeminy, and/or quadgeminy. Further evaluation reveals that one minus this probability is the probability that at least one run of five or more beats of consecutive VT have occurred. This would be more useful to cardiologists then the probabilities of isolated runs of VT restricted to one and only one length.

This probability is accessible to us through another adaptation of the $T_{0,[K,L]}(n)$, recognizing that $T_{[0,L]}(n) = T_{0,[0,L]}(n)$. To apply this to the scenario outlined at the beginning of this section, set K = 0, and for $n \geq 0$, write

$$T_{[0,L]}(n) = pT_{[0,L]}(n-1) + qpT_{[0,L]}(n-2) + q^2pT_{[0,L]}(n-3) + q^3pT_{[0,L]}(n-4) + \ldots + q^LpT_{[0,L]}(n-L-1) + q^nI_{0\le n\le L} = \sum_{j=0}^{L} q^jpT_{[0,L]}(n-j-1) + q^nI_{0\le n\le L} \tag{8.39}$$

To verify this equation, compute the probability that all failure runs have lengths less than or equal to L in n=L trials. We know that this answer must be one. Examination of equation (8.39) reveals

$$\begin{aligned} T_{[0,L]}(L) &= pT_{[0,L]}(L-1) + qpT_{[0,L]}(L-2) + q^2pT_{[0,L]}(L-3) + q^3pT_{[0,L]}(L-4) \\ &\quad + \ldots + q^LpT_{0,L}(L-L-1) + q^L \\ &= p + qp + q^2p + q^2p + \ldots + q^{L-1}p + q^L \\ &= p\left(1 + q + q^2 + q^3 + \ldots + q^{L-1}\right) + q^L \\ &= p\frac{1-q^L}{1-q} + q^L = 1 - q^L + q^L = 1 \end{aligned} \tag{8.40}$$

and the intuitive answer is reproduced from the $T_{[0,L]}(L)$ model since $T_{[0,L]}(n) = 1$ for $0 \le n \le L$.

We now pursue the solution for $T_{[0,L]}(n)$. Define

$$G(s) = \sum_{n=0}^{\infty} s^n T_{[0,L]}(n) \tag{8.41}$$

and proceeding with conversion, we find

$$\begin{aligned} T_{[0,L]}(n) &= \sum_{j=0}^{L} q^jpT_{[0,L]}(n-j-1) + q^nI_{0\le n\le L} \\ s^nT_{[0,L]}(n) &= s^n\sum_{j=0}^{L} q^jpT_{[0,L]}(n-j-1) + s^nq^nI_{0\le n\le L} \\ \sum_{n=0}^{\infty} s^nT_{[0,L]}(n) &= \sum_{n=0}^{\infty} s^n\sum_{j=0}^{L} q^jpT_{[0,L]}(n-j-1) + \sum_{n=0}^{\infty} s^nq^nI_{0\le n\le L} \\ G(s) &= \sum_{n=0}^{\infty} s^n\sum_{j=0}^{L} q^jpT_{[0,L]}(n-j-1) + \sum_{n=0}^{L} s^nq^n \end{aligned} \tag{8.42}$$

The task before us now is to simplify the first term on the right side of the final row of equation (8.42).

$$
\begin{aligned}
&\sum_{n=0}^{\infty} s^n \sum_{j=0}^{L} q^j p T_{[0,L]}(n-j-1) = \sum_{j=0}^{L} q^j p \sum_{n=0}^{\infty} s^n T_{[0,L]}(n-j-1) \\
&= \sum_{j=0}^{L} q^j p s^{j+1} \sum_{n=0}^{\infty} s^{n-j-1} T_{[0,L]}(n-j-1) = \sum_{j=0}^{L} q^j p s^{j+1} G(s)
\end{aligned}
\tag{8.43}
$$

We can now complete the consolidation process. Rewrite equation (8.43) as

$$
G(s) = \sum_{j=0}^{L} q^j p s^{j+1} G(s) + \sum_{n=0}^{L} s^n q^n \tag{8.44}
$$

$$
G(s) = \frac{\sum_{n=0}^{L} s^n q^n}{1 - \sum_{j=0}^{L} q^j p s^{j+1}} \tag{8.45}
$$

and proceed with the inversion by noting

$$
\frac{1}{1 - \sum_{j=0}^{L} q^j p s^{j+1}} = \frac{1}{1 - \left[\sum_{j=0}^{L} q^j s^j \right] ps} \tag{8.46}
$$

then utilize the multinomial theorem to write

$$
\begin{aligned}
&\left(\sum_{j=0}^{L} q^j s^j \right)^N \\
&= \sum_{j_1=0}^{N} \sum_{j_2=0}^{N-j_1} \sum_{j_3=0}^{N-j_1-j_2} \cdots \sum_{j_L=0}^{N-\sum_{i=1}^{L-1} j_i} \binom{N}{j_1 \quad j_2 \quad j_3 \quad \cdots \quad j_L} q^{\sum_{i=1}^{L} i j_i} s^{\sum_{i=1}^{L} i j_i}
\end{aligned}
\tag{8.47}
$$

and invert:

$$\frac{1}{1-\sum_{j=0}^{L} q^j p s^{j+1}}$$

$$\triangleright \left\{ \sum_{m=0}^{n} p^m \sum_{j_1=0}^{m} \sum_{j_2=0}^{m-j_1} \sum_{j_3=0}^{m-j_1-j_2} \cdots \sum_{j_L=0}^{m-\sum_{i=1}^{L-1} j_i} \binom{m}{j_1 \quad j_2 \quad j_3 \quad \cdots \quad j_L} q^{\sum_{i=1}^{L} i j_i} I_{m+\sum_{i=1}^{L} i j_i = n} \right\} \tag{8.48}$$

completing the inversion of G(s) from equation (8.45)

$$G(s) = \frac{\sum_{h=0}^{L} s^h q^h}{1-\sum_{j=0}^{L} q^j p s^{j+1}}$$

$$\triangleright \left\{ \sum_{h=0}^{\min(L,n)} q^h \sum_{m=0}^{n-h} p^m \sum_{j_1=0}^{m} \sum_{j_2=0}^{m-j_1} \sum_{j_3=0}^{m-j_1-j_2} \cdots \sum_{j_L=0}^{m-\sum_{i=1}^{L-1} j_i} \binom{m}{j_1 \quad j_2 \quad j_3 \quad \cdots \quad j_L} q^{\sum_{i=1}^{L} i j_i} I_{m+\sum_{i=1}^{L} i j_i = n-h} \right\} \tag{8.49}$$

8.8 Model Predictions

It is of interest to see how these models perform. An example would be to examine the expected maximum run length as a function of the probability of an ectopic beat. This maximum run length is the same as the maximum expected duration of VT. Bigeminy reflects VT of length 2. The expected maximum run length is computed from the equation. Just as an illustration, we set n =100, equivalent to examining the patient's heart rhythm for just under two minutes. We then compute the expected run length as a function of q (Figure 8.1). This figure portrays that increasing probability of ectopic beats is associated with an increased expected maximum length of ventricular tachycardia. The relationships are quite detailed. If the probability of an ectopic beat is 0.1, the expected maximum burst of QT is 1.69 (approximately bigeminy). If the probability of an ectopic beat is 0.4, the expected VT_{max} = 4.61 (approximately quadgeminy). That is, changes in the probability of an ectopic beat can be linked to changes in expected VT_{max}. For example, if a medication can reduce the probability of an ectopic beat from 0.5 to 0.1, the expected worse burst of

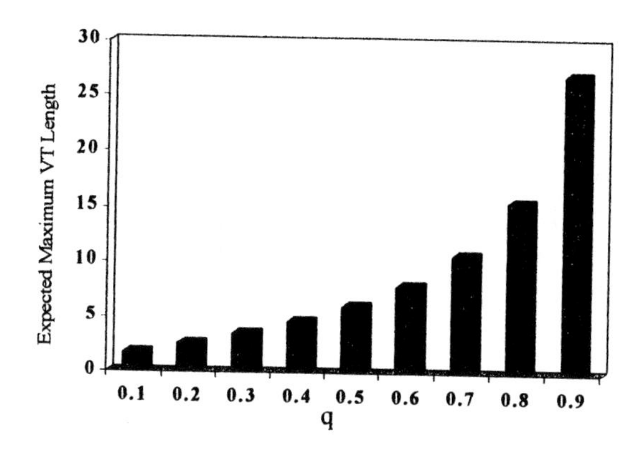

Figure 8.1 Expected Maximum VT length as a function of q, the probability of an ectopic beat.

ventricular tachycardia is reduced from a run of six ectoptic beats to a much milder run of length 2.

8.9 Future Directions

This application requires additional development before it can be fully applicable. The notion of looking at only 100 beats is a limitation. Holter monitors can store an individual's electrocardiogram for over 24 hours, over 105,000 heartbeats. Running the model with $n = 105{,}000$ is time consuming regardless of whether the computations are carried out iteratively or by the generating function approach.

In addition, one of the fundamental assumptions of the model must be critically examined. This model is predicated on the Bernoulli assumption. The notion of independence from heartbeat to heartbeat must be carefully considered, since many believe that the occurrence of single beat of ectopy increases the likelihood of closely following ectopic beats. The idea of dependence from beat to beat is not considered in this model.

Problems

1. Verify by enumeration that for a sequence of Bernoulli trials of lengths three, four, and five, that $T_{[1,1]}(n) = \sum_{m=0}^{n-1} p^m \sum_{j=0}^{m} \binom{m}{j} q^j I_{m+j=n} - p^n$.
2. Demonstrate using the method of cofficient collection and the binomial theorem that $\frac{1}{1-ps-q^2ps^3} \triangleright \left\{ \sum_{m=0}^{n} p^m \sum_{j=0}^{m} \binom{m}{j} q^{2j} I_{m+2j=n} \right\}$.
3. Compute the probability using the Bernoulli model of the probability that in five minutes, a patient has no burst of ventricular tachycardia of length greater than five.
4. How could you compute the expected maximum length of ventricular tachycardia from the $T_{[K,L]}(n)$ model?

9

Difference Equations: Follow-up Losses in Clinical Trials

9.1 Introduction

Clinical experiments have evolved for many hundreds of years. The current state-of-the-art clinical experiment is the randomized, controlled clinical trial. These advanced experiments have been used to demonstrate important relationships in public health. Two examples are 1) the association between reductions in elevated levels of blood pressure and the reduced incidence of strokes, and 2) the reduction of blood cholesterol levels and the reduced occurrence of heart attacks. However, many patients who agree to participate in this trial often choose not to return to the clinical trial physician for regularly scheduled followup visits. Although this behavior is understandable, and is in no way unethical, the absence of these patients from all future participation in the trial can complicate the interpretation of the trial. After a brief introduction to clinical trials, this chapter will apply difference equation methodology to predictions about the frequency of these required patient visits that would be useful to clinical trial workers.

9.2 Clinical Trials

Clinical health care research is the study of the effect of genetic and environment influences on the occurrence of disease. These efforts generally focus on the effect of a genetic factor or environmental factor on our likelihood of getting disease. Such research can be divided into observational research and experimental research. In observational research methods, the scientist does not control the factor they wish to study. For example, in working to determine the relationship between electrical fields and cancer, scientists do not choose who will be exposed to electrical fields and who will not. They simply observe those who by chance or by choice were exposed to electrical fields (the risk factor in this example) and measure the frequency of cancer, while simultaneously observing those who were not exposed to the electrical fields and measure their frequency of cancer. At the conclusion of the observation period the scientists attempt to draw some conclusions. The scientist in these studies did not control electrical field exposure; they chose instead to observe relationships that were already embedded in the sample of patients they studied.

Experiments are distinguished from observational studies by the control the scientist has over the experiment. In experiments, the scientist determines whether a patient is exposed to the risk factor (or the treatment). For example, a scientist can give some patients aspirin, while other patients receive placebo therapy, and observe which patients have heart attacks. Clinical trials are experiments performed on people, in which the scientists are very careful in determining which patients get therapy and which get placebo. If the scientist chooses the therapy randomly (i.e., makes sure that each patient has the same constant probability of receiving the intervention) the trial is called a randomized, controlled clinical trial. Randomization assures the clinical scientist that there are no other differences between those patients who received the therapy and those who received placebo, making it easy to decide to what the difference in patient outcomes can be attributed. In addition, if the patient does not know what medication they received, it is called a patient-blinded (or single blind) trial. If the physician does not know what medication their patients have received, it is called a physician-blinded trial. Blinding is important because it removes as a cause of the effect any inclination the patient or physician may report occurrences of disease based on their belief in the medication's effect. If both the physician and the patient are blinded in a trial that uses randomization, the trial is described as a randomized, double blind, controlled clinical trial.

9.3 Follow-up Losses

9.3.1 Definition of a patient "lost to follow-up"

The use of randomized controlled clinical trials reflect the accumulated experience and methodological advances in the evolution of scientific, experimental design in medicine. These experiments are very complicated, involving the complexities of choosing patients from the population at large, choosing the type and concentration of the intervention, deciding how long patients must stay in contact with the scientists who are controlling the trial (some trials take 1-2 days to complete, while others can take five years or more). One of the crucial features of clinical trials is the occurrence of endpoints, the clinical events that will be used to determine the effectiveness of the intervention. For example, consider a trial designed to reduce the occurrence of death. By this we mean, that the investigators believe at the trial's end, there will be more deaths that occurred in the placebo group than in the group receiving the active intervention (the active group). In this trial, death is the endpoint. A clinical trial that is designed to measure the effect of medication to reduce the total number of deaths from a disease during the next five years must follow every patient until the end of the trial to ensure that they know who died. Without following each individual patient, the scientists will be uncertain as to who died and who survived, and without a careful count of this number, these scientists will have difficulty in deciding the true reduction in the number of deaths to which the medication can be attributed.

This problem is sometimes complicated by the low frequency of the occurrence of endpoints in the study. In recent clinical trials [1,2,3], the rates at which endpoints are predicted to occur are relatively small. In these trials, the measure of therapy effectiveness, i.e., measuring the difference in death rates between the intervention and the control group, depends on the occurrence of relatively infrequently occurring endpoints. The incomplete ascertainment of these endpoints would weaken the trial by making the final endpoint death rate unclear.

An important reason for incomplete endpoint ascertainment is the patient who is "lost to follow-up." This means that patients choose not to return to their clinical trial doctors for scheduled (i.e, required) visits. They are seen for the first visit (in fact, in many circumstances, the first visit is the visit at which the patient often enters the clinical trial, or is "randomized." Such "follow-up losses" do not stay in contact with their doctors. These patients often will stop taking the medication they received in the study, and may refuse any attempt by the study to re-contact them. They completely "drop out" of the study. There are many reasons for dropping out of a study. Divorce, changing city, state (or country) of residence, a life of crime, joining reclusive cults are all reasons that are given for patients dropping out of a study. However, the occurrence of a drop out means that the study cannot determine if the patient is alive or dead at the trial's conclusion (i.e., cannot determine the patient's vital status). Tremendous effort is required on the part of clinical trial workers to insure that the vital status of all trial participants is obtained. In fact, a major constraint to the execution of long-term follow-up studies is often the time, money, ingenuity and perseverance required to successfully trace subjects [4-5].

9.3.2 The effect of follow-up losses

As follow-up losses make the effect of the therapy difficult to determine by precluding an accurate computation of the total number of deaths experienced by patients treated with placebo and those treated with the intervention. Clinical trial workers labor intently to find patients who are lost to follow-up, often resorting to private investigating agencies after the patient has stopped attending the required clinic visits. The strengths and weaknesses of mailings to participants to remind them of missed visits has been described [5,6] and the advantages and disadvantages of city directories, telephone contacts, and the use of postal services have been delineated [4]. A National Death Index was established to promote statistical research in health care [7], and search procedures using this index have been utilized to identify patients who have dropped out from clinical trials [8-12]. In addition, information from the Social Security Administration is also useful in ascertaining vital status [13].

9.4 Vital Status and Difference Equations

9.4.1 Vital status terminology and parameterization

The hallmark of patients who are eventually lost to follow-up is that they miss scheduled visits. Many different patient visit patterns are often observed in clinical trials. A patient's attendance for scheduled visits may first be perfect. Over time, the visit pattern becomes sporadic as the patient gradually misses scheduled visits at a greater frequency until they no longer attend visits and are lost from the study. Other patients may have perfect attendance and then suddenly, inexplicably, and completely drop out of the study. However, missed visits is the common theme among these disparate visit patterns. Thus, one way to identify patients who are at risk of loss to follow-up is to examine their visit pattern.

The pattern of missed visits for patients who will eventually be lost to follow-up is the occurrence of consecutive missed visits. In fact, this is a necessary condition for the patient to be lost to follow-up. Our purpose here is to develop and explore a monitoring rule applicable to clinical trials to identify potential follow-up losses, based on the number of consecutive visits that have been missed. Once a candidate rule for identifying patients who are potentially lost to follow-up is identified, it is useful to know how likely a patient is to violate this rule by chance alone. Another useful exploration would be to identify the sensitivity of this procedure to other known measurements in a clinical trial.

Over the course of the clinical trial, each patient is scheduled for a sequence of n consecutive visits. The total number of these visits is known in advance. We begin by denoting the probability that a patient will keep a scheduled visit by p and let this probability be a known constant, fixed over the entire sequence of n visits. Denote $q = 1 - p$ as the probability that the patient misses a particular visit. Consider the following monitoring rule V(L) for loss to follow-up patients:

$V(L) = 1$ if the patient has missed at least L consecutively scheduled visits (rule violator).
$V(L) = 0$ otherwise (nonviolator)

We need to compute $P(V(L) = 1)$, the probability that a patient has missed at least L consecutive visits as a function of q and n. Once this probability is estimated, its relationship with L can be used to identify the optimum value of L for the trial. Patients then meet the criteria can be targeted for special attention in an attempt to get them back into the mainstream of the clinical trial and a more standard visit pattern.

The monitoring rule V(L) is based on a string or run of consecutive failures. Thus, the monitoring rule is triggered when there is at least one run of failures of length L or greater. With this development, difference equations may be used to identify the crucial recursive relationships in the occurrence of failure runs of the required lengths.

Recall that $T_{[0,L-1]}(n)$ is the probability that in a collection of n consecutive Bernoulli trials, there are no failure runs of length L or greater. Thus, all failure runs must be of length L - 1 or less. In this development, $1 - T_{[0,L-1]}(n)$ is the probability that in n trials, there is at least one occurrence of a failure run of length greater than or equal to L, i.e., the monitoring rule is violated. Thus, $P(V(L) = 1) = 1 - T_{[0,L-1]}(n)$.

To find $T_{[0,L]}(n)$ (from which we can deduce the solution for $T_{[0,L-1]}(n)$, a recursive relationship for $n \geq 0$ is:

$$\begin{aligned} T_{[0,L]}(n) &= pT_{[0,L]}(n-1) + qpT_{[0,L]}(n-2) + q^2pT_{[0,L]}(n-3) + q^3pT_{[0,L]}(n-4) \\ &+ \ldots + q^LpT_{[0,L]}(n-L-1) + q^nI_{0 \leq n \leq L} = \sum_{j=0}^{L} q^jpT_{[0,L]}(n-j-1) + q^nI_{0 \leq n \leq L} \end{aligned} \tag{9.1}$$

The boundary conditions for the family of difference equations represented by equation (9.1) are $T_{[0,L]}(n) = 0$ for $n < 0$ and $T_{[0,L]}(n) = 1$ for $0 \leq n \leq L$.

9.4.2 Worst visit pattern

$1 - T_{[0,L-1]}(n)$ is the probability of at least one failure run of length L or greater in n trials. This is also the probability that the maximum failure run length is $\geq$ L. Similarly, $1 - T_{[0,L]}(n)$ is the probability that there is at least one failure run of length L + 1 or greater in n trials, or the probability that the maximum failure run length is $\geq$ L + 1 Thus the

$$\begin{aligned}&P[\text{maximum failure run length is} = L \text{ in n trials}]\\&= [1 - T_{[0,L-1]}(n)] - [1 - T_{[0,L]}(n)] = T_{[0,L]}(n) - T_{[0,L-1]}(n).\end{aligned}$$

Using this probability, we can compute in n trials the expected maximum failure run length E(n,q) and its standard deviation SD(n,q) in n trials as

$$E(n,q) = \sum_{L=0}^{n} L\left[T_{[0,L]}(n) - T_{[0,L-1]}(n)\right]$$
$$SD(n,q) = \sqrt{\sum_{L=0}^{n} L^2\left[T_{[0,L]}(n) - T_{[0,L-1]}(n)\right] - E^2(n,q)} \tag{9.2}$$

In the context of missed visits for a patient in a follow-up study, E(n,q) is the expected maximum number of visits to be missed consecutively by chance alone, i.e. the expected worst visit pattern, and SD(n,q) is its standard deviation.

9.5 $T_{[0,L]}(n)$ and Missed Visits

9.5.1 Exploration of the $T_{[0,L]}(n)$

This is a family of difference equations that are related to the equations we have solved in Chapter 8. We now pursue the solution for $T_{0,L}(n)$. Recall

$$G(s) = \sum_{n=0}^{\infty} s^n T_{[0,L]}(n) \tag{9.3}$$

Proceed with the conversion:

$$
\begin{aligned}
T_{[0,L]}(n) &= \sum_{j=0}^{L} q^j p T_{[0,L]}(n-j-1) + q^n I_{0 \le n \le L} \\
s^n T_{[0,L]}(n) &= s^n \sum_{j=0}^{L} q^j p T_{[0,L]}(n-j-1) + s^n q^n I_{0 \le n \le L} \\
\sum_{n=0}^{\infty} s^n T_{[0,L]}(n) &= \sum_{n=0}^{\infty} s^n \sum_{j=0}^{L} q^j p T_{[0,L]}(n-j-1) + \sum_{n=0}^{\infty} s^n q^n I_{0 \le n \le L} \\
G(s) &= \sum_{n=0}^{\infty} s^n \sum_{j=0}^{L} q^j p T_{[0,L]}(n-j-1) + \sum_{n=0}^{L} s^n q^n
\end{aligned}
\tag{9.4}
$$

and complete the consolidation process,

$$
\begin{aligned}
G(s) &= \sum_{j=0}^{L} q^j p s^{j+1} G(s) + \sum_{n=0}^{L} s^n q^n \\
G(s) &= \frac{\sum_{n=0}^{L} s^n q^n}{1 - \sum_{j=0}^{L} q^j p s^{j+1}}
\end{aligned}
\tag{9.5}
$$

Recall from Chapter 8,

$$
\frac{1}{1-\sum_{j=0}^{L} q^j p s^{j+1}}
\triangleright \left\{ \sum_{m=0}^{n} p^m \sum_{j_1=0}^{m} \sum_{j_2=0}^{m-j_1} \sum_{j_3=0}^{m-j_1-j_2} \cdots \sum_{j_L=0}^{m-\sum_{i=1}^{L-1} j_i} \binom{m}{j_1 \; j_2 \; j_3 \; \cdots \; j_L} q^{\sum_{i=1}^{L} i j_i} I_{m+\sum_{i=1}^{L} i j_i = n} \right\}
\tag{9.6}
$$

Complete the inversion of G(s) from equation (9.5):

$$G(s) = \frac{\sum_{h=0}^{L} s^h q^h}{1 - \sum_{j=0}^{L} q^j p s^{j+1}}$$

$$\triangleright \left\{ \sum_{h=0}^{\min(n,L)} q^h \sum_{m=0}^{n-h} p^m \sum_{j_1=0}^{m} \sum_{j_2=0}^{m-j_1} \sum_{j_3=0}^{m-j_1-j_2} \cdots \sum_{j_L=0}^{m-\sum_{i=1}^{L-1} j_i} \binom{m}{j_1 \quad j_2 \quad j_3 \quad \cdots \quad j_L} q^{\sum_{i=1}^{L} i j_i} I_{m+\sum_{i=1}^{L} i j_i = n-h} \right\} \tag{9.7}$$

Thus, $T_{[,0,L]}(n)$ is an explicit function of n, the total number of visits, q the probability of a missed visit, and L, the monitoring rule.

9.6 Results

Since the difference equations and their solutions are in terms of q, the probability of a missed clinic visit, n, the number of visits, and V_L, the vigilance rule, i.e., $T_{0,L-1}(n)$ can be examined as a function of these parameters. As a first example, the value of $1 - T_{0,L-1}(n)$, the probability of L or more consecutive missed visits was explored as a function of both q, the probability of a missed visit and L for n=10 visits. Table 9.1 displays the anticipated relationship between $1-T_{[0,L]}(n)$, and q for fixed L.

As the probability of a missed visit (q) increases, the probability of missing L or more consecutive visits also increases. For example, the probability of missing at least two consecutive visits increases from 0.021 for q=0.05 to 0.504 for q = 0.30. The probability of missing at least k consecutive visits as a function of L can also be observed from Table 9.1. For each value of q examined, the probability of L or more consecutive missed visits decreases rapidly. For example, when the probability of a missed visit is 0.10, the probability that, out of ten scheduled visits, a patient will miss at least one or more consecutive visits is 0.651. However, the probability that a patient will miss two or more consecutive visits decreases to 0.080 for the same value of q. This phenomena of rapid fall off in the probability of consecutive missed visits as L increases was observed for each value of q examined. This property of rapidly decreasing probability of greater run lengths suggests that a value of L may be readily identified that would trigger additional attention to the patient who misses this many consecutive visits.

Table 9.1 Probability of at Least L Consecutive Missed Visits out of 10 Scheduled Visits as a Function of the Probability of a Missed Visit (q)

	Probability of a Missed Visit					
	0.05	0.10	0.15	0.20	0.25	0.30
1	0.401	0.651	0.803	0.893	0.944	0.972
2	0.021	0.080	0.168	0.273	0.388	0.504
3	0.001	0.007	0.023	0.052	0.096	0.155
4	0.000	0.001	0.003	0.009	0.021	0.042
5	0.000	0.000	0.000	0.002	0.005	0.011
6	0.000	0.000	0.000	0.000	0.001	0.003
7	0.000	0.000	0.000	0.000	0.000	0.001
8	0.000	0.000	0.000	0.000	0.000	0.000
9	0.000	0.000	0.000	0.000	0.000	0.000
10	0.000	0.000	0.000	0.000	0.000	0.000

The development of $T_{[0,L]}(n)$ also provides a direct computation of the expected maximum number of consecutive missed visits E(n,q) a typical patient will experience, and its standard deviation sd(n,q). Table 9.2 demonstrates the relationship between the expected worst visit performance and the probability of a missed visit q and n. Note that E(n,q) increases as the total number of scheduled visits increases.

Table 9.2 Expected Worst Visit Performance E(n,q) as a Function of n and q

		q		
		0.10	0.20	0.30
n	10.00	0.7 (0.9)	1.2 (2.1)	1.7 (3.7)
	20.00	1.1 (1.5)	1.6 (3.2)	2.2 (5.9)
	30.00	1.2 (1.8)	1.8 (4.1)	3.0 (7.5)
	40.00	1.3 (2.1)	2.0 (4.8)	2.8 (8.8)

This satisfies intuition, since the more visits available, the greater the number of opportunities for missed consecutive visits. However the increase in expected worst visit performance increases only gradually as the total number of scheduled visits is quadrupled from 10 to 40. In addition, Table 9.2 and Figure 9.1 revealed the relationship between expected worst visit performance as a function of the probability of a missed visit, worst visit performance worsening as q increases.

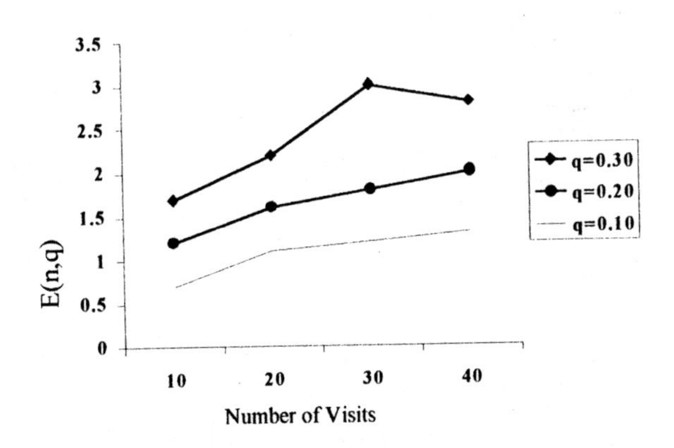

Figure 9.1 Expected worst visit performance by n and q.

The standard deviation of the maximum number of consecutive missed visits is more sensitive to changes in both the probability of a missed visit and the number of scheduled visits. This standard deviation increases from 0.9 to 3.7 as the probability of a missed visit increase from 0.10 to 0.30. In addition, this increased SD(n,q) is a function of n, the total number of scheduled visits.

Figure 9.2 displays the relationship between the upper 5% tail of the worst visit performance (mean + 1.96SD) and the probability of a missed visit. Even though the increase in expected maximum as a function of n is modest, the upper 5% of the distribution of the maximum is more sensitive to increases in the probability of a missed visit. Thus, although in a clinical trial with 20 scheduled visits, if it is anticipated that a typical patient will miss 1 in 5 scheduled visits, the expected maximum number of missed consecutive visits is 1.6. However, through the random mechanism of chance the distribution of the worst visit performance could be as large as 4 or 20% of scheduled visits.

Figure 9.3 depicts the relationship between the number of typical patients at risk for follow-up loss (false positives) in a 5000 patient trial as a function of both the probability of a missed visit and the vigilance rule V_L (i.e., they have missed at least L consecutive visits). The number of false positive patients at risk for a follow-up loss is the expected number of patients who violate the vigilance rule. We see that as the probability of a missed visit increases, the expected number of patients who violate this rule V_L (false positives) also increases. It is also clear that the expected number of false positives who violate V_L decrease as L increases. For each combination of q and V_L, the number of patients who require increased vigilance for follow-up loss can be determined.

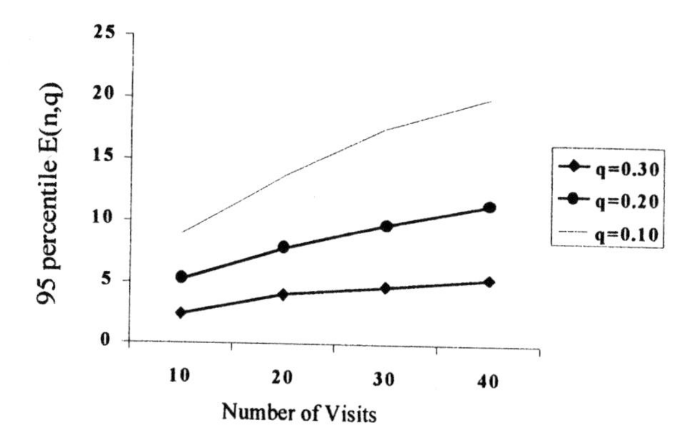

Figure 9.2. 95th Percentile of expected worst visit performance by n and q.

9.7 Expected Number of Violators

This work develops and explores the notion of missed clinic visits from the perspective of run theory. The salient features of the consecutive missed visits is drawn from the clinical arena, and the creation of a vigilance rule, V_L, that states that increased vigilance, and therefore increased trial resources, must be expended for a patient who misses at least k consecutive scheduled visits. The implications of such a rule are examined from the perspective of run theory first developing a recursive relationship and then using difference equations to solve for the probability of interest, $T_{[0,L]}(n)$. In this development the probability a patient violates the vigilance rule is $P[V(L)=1] = 1 - T_{[0,,L-1]}(n)$. The theoretical result for $T_{[0,L-1]}(n)$ is identified as a function of parameters that are available to clinical trial workers.

Clearly one obvious solution known to clinical trial workers is to keep attendance as close to perfect as possible in a follow-up study. The reality of clinical trial experience is that even patients with good visit compliance records will occasionally miss visits; we need to find a boundary rule that distinguishes between the predictable behavior of a patient who has had a short string of bad attendance from that patient whose bad attendance is a reliable indicator that they will eventually be lost to follow-up. There are two types of patients who will violate V_L; those who are false positives and those who are beginning the steps to become lost to follow-up. The predictor suggested here is the number of consecutive missed visits. Using the model $T_{0,L}(n)$ we can explore the implications of different choices of L for the allocation of resources to identify these patients. The expected number of violators V is computed from equation (9.8). We suggest keeping the number of false positives low. The rule for triggering additional attention and the assignment of potential lost to follow-up depends then on the probability of missing a visit (Table 9.3).

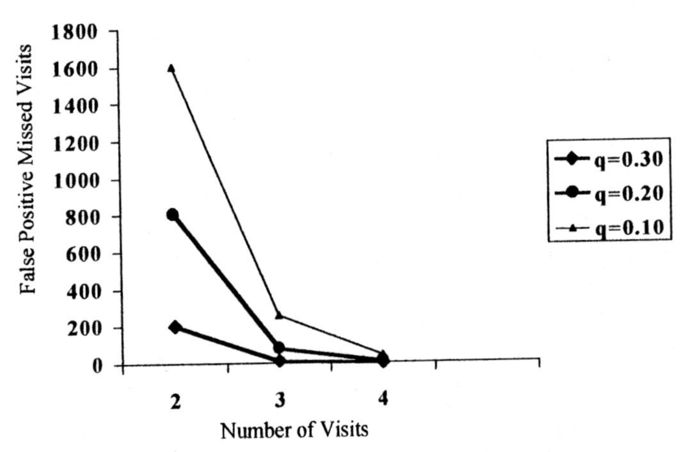

Figure 9.3 False positive missed visit frequency by q and violator rule L.

For example, in a trial randomizing 5000 patients with 20 scheduled visits, if the probability of a missed visit is 0.05, one may choose to wait until a patient has missed three consecutive visits before additional vigilance is required, V_L=3. V is then computed as

$$V = 5000\left[1 - T_{0,3}(20)\right] \tag{9.8}$$

This choice leads us to expect that from an examination of Table 9.4 that there would be V = 11 patients who would meet this criteria "by chance alone," that is, their behavior would not indicate poor performance but just a string of bad attendance that would be corrected. Thus, by applying this rule, any patient who misses at least three consecutive visits would receive greater attention with attempts to contact them. This would lead to 11 such patients who would be false positives and all other patients thus identified would be potential follow-up losses to be recaptured. On the other hand, in a clinical trial where attendance of visits may be low, and the probability of a missed visit being on the order of 15%, waiting until four consecutive missed visits before categorizing the patients as lost to follow-up may be in order. One could argue more

Table 9.3 Expected Number of Violators (V), Expected Worst Visit Performance (E(n,q)), with its 95th Percentile: n=2

		Probability of Failure (q)		
		0.05	0.1	0.15
	V	107	401	838
L=2	E(n,q)	0.4	0.7	1
	95% UB	1.4	1.9	2.4
	V	5	36	117
L=3	E(n,q)	0.4	0.7	1
	95% UB	1.4	1.9	2.4
	V	0	3	15
L=4	E(n,q)	0.4	0.7	1
	95% UB	1.4	1.9	2.4

conservatively for three consecutive visits, but with the likelihood of follow-up visits attendance being so low, instituting this procedure would start the search on 256 false positives, a number that may be prohibitive. The use of run theory to solve issues in clinical trials exclusive of endpoint analysis is a new application for this branch of statistics.

Table 9.4 Expected Number of Violators (V), Expected Worst Visit Performance (E(n,q)), with its 95th Percentile:n=20

L		Probability of Failure (q) 0.05	0.1	0.15
	V	222	806	96
L=2	E(n,q)	0.6	1.1	1.3
	95% UB	1.9	2.3	2.7
	V	11	81	256
L=3	E(n,q)	0.6	1.1	1.3
	95% UB	1.9	2.3	2.7
	V	1	8	37
L=4	E(n,q)	0.6	1.1	1.3
	95% UB	1.9	2.3	2.7

Table 9.5 Expected Number of Violators (V), Expected Worst Visit Performance (E(n,q)), with its 95th Percentile:n=30

L		Probability of Failure (q) 0.05	0.1	0.15
	V	222	806	1596
L=2	E(n,q)	0.6	1.1	1.3
	95% UB	1.9	2.3	2.7
	V	11	81	256
L=3	E(n,q)	0.6	1.1	1.3
	95% UB	1.9	2.3	2.7
	V	1	8	37
L=4	E(n,q)	0.6	1.1	1.3
	95% UB	1.9	2.3	2.7

Problems

1. Show using equation (9.1) that $T_{[0,K]}(K) = 1$.

2. Using equation (9.1), write the family of difference equations for $T_{[0,7]}(n)$ for $n \geq 0$.

3. Find the inverse of the generating function G(s) where

$$G(s) = \frac{\sum_{h=0}^{7} s^h q^h}{1 - \sum_{j=0}^{7} q^j p s^{j+1}}$$

4. Consider a clinical trial that will follow 40,000 patients for a total of seven years. Each patient should complete three visits a year for a total of twenty one visits by the conclusion of the trial. Let the probability of a missed visit be $q = 0.07$.
   a. Compute $P[V(2) = 1]$
   b. Compute $P[V(3) = 1]$
   c. Compute $P[V(4) = 1]$

References

1. The SHEP Cooperative Research Group. Prevention of stroke by antihypertensive drug therapy in older persons with isolated systolic hypertension: final results of the systolic hypertension in the elderly program (SHEP). *JAMA.* 265: 3255-3264.1991.
2. Pfeffer M.A., Brauwald E. Moyé L.A. et al. Effect of captopril on mortality and morbidity in patients with left ventricular dysfunction after myocardial infarction - results of the Survival and Ventricular Enlargement Trial. *N Eng J Med* 327: 669-677.1992.
3. Sacks F.M., Pfeffer M.A., Moyé L.A. et. al. The cholesterol and recurrent events trial (*CARE*): rationale, design and baseline characteristics of secondary prevention trial of lowering serum cholesterol and myocardial infarction. *Am J Card* 68(15):1436-1446.1991.
4. Boice J.D. Follow-up methods to trace women treated for pulmonary tuberculosis, 1930-1954. *Am J Epi.* 107. 127-138. 1978.

5. Cutter G., Siegfried H., Kastler J., Draus J.F., Less E.S., Shipley T., Stromer M. Mortality surveillance in collaborative trials. *Am J Public Health* 70:394-400. 1980.
6. Austin M.A., Berreysea E., Elliott J.L., Wallace R.B., Barrett-Connor E., Criqui M.H. Methods for determining long-term survival in a population based study. *Am J Epi.*110.747-752. 1979.
7. Patterson J.E. The establishment of a national death index in the United States. Banbury Report 4: Cancer Incidence in Defined Populations. Cold Spring Harbor Laboratory.443-447. 1980.
8. Edlavitch S.A., Feinleib M., Anello C. A potential use of the national death index for post marketing drug surveillance. *JAMA* 253.1292-1295.1985.
9. Williams B.C., Bemitrack L.B., Bries B.E. The accuracy of the national death index when personal identifiers other than social security number are used *Am J Public Health.* 82:1145-1146.1992.
10. Stampfer M.J., Willett W.C., Speizer F.E., Dysert D.C., Lipnick R., Rosner B., Hennekens C.H. Test of the national death index. *Am Jour Epi* 119:837-839.1984.
11. Davis K.B., Fisher L., Gillespie M.J., Pettinger M. A test of the national death index using the coronary artery surgery study (CASS). *Cont Clin Trials* 6:179-191.1985.
12. Curb J.D., Ford C.E., Pressel S., Palmer M., Babcock C., Hawkins C.M. Ascertainment of the vital status through the national death index and the social security administration. *Am J Epi.* 121:754-766.1985.
13. Wentworth D.N., Neaton J.D., Rasmusen W.L. An evaluation of the social security administration master beneficiary record file and the national death index in the ascertainment of vital status. *Am J Public Health* 73:1270-1274.1983.

10

Difference Equations: Business Predictions

10.1 Introduction

The last two decades have rocked the health care industry with political, economic, and organizational turmoil. The development and inexorable intrusion of managed care plans into the customary health care delivery systems routinely offered by private physicians working out of their wholly owned businesses has created continuing confusion and instability among health care providers. In this chapter, difference equations will be applied to predictions of short burst of business cycles in ambulatory health care clinics.

10.2 Ambulatory Health Care Clinics

Over the past several years the health care industry in the United States has seen a number of alternative forms of health care delivery, as administrators and physicians seek to find the balance between its cost and equity of distribution. One of the most innovative of these new approaches is the ambulatory "urgent care" clinic. These clinics offer patients who have minor emergencies and who do not have a family doctor the opportunity to see a physician, allowing them to avoid the expensive and time consuming trips to large emergency rooms[1-2].

These ambulatory care clinics are usually minimally staffed with a licensed physician, nurse, receptionist, and x-ray and/or laboratory technician. They are often open beyond the customary regular business hours both during the week and on weekends. By restricting the type of health problems the urgent care centers treat, they are able to keep their operation costs at a minimum. Thus, the urgent care centers offer strong competition to the conventional hospital emergency rooms that are traditionally more adept at taking care of life-threatening crises but not emergent day-to-day health care needs of otherwise healthy patients.

However, both the combination of changes in the health care needs of the community and economic turmoil have made the urgent care industry a volatile one. In particular, the declining economic fortunes of Houston in 1985-1990 have decreased the average number of patients seen at these centers. The introduction of HMOs has further eroded the urgent care center's patient base[3]. Many of these urgent care centers went out of business, and those that remained were precariously balanced for survival as they struggled to stabilize their declining patient numbers and revenues, rising malpractice premiums, increasing supply costs, and issues of quality control.

Since there facilities were often privately owned, there was usually a minimum of financial resources from which they could draw if the clinic's business declined. This lack of financial reserve demanded that business plans be exact, and, therefore, accurate prediction of future daily patient visits and generation of revenue was critical. In some instances, an unexpected three or four consecutive days when the number of patient visits was unexpectedly low could lead to inability to meet payroll, and/or to adequately replenish supplies. These dire circumstances could have disastrous results both for clinic morale and the quality of patient care.

However, no satisfactory method has yet evolved to accurately predict the number of patients to be treated in a given day in the future, or to predict cash flow. The output measures for hospitals are well enumerated [4] and some workers believe that many hospital cost studies have floundered on the appropriate defintion of the choice of output measures on which costs are to be assessed [5]. Furthermore, most of the work on ambulatory care clinics has been an examination of their characteristics from the patient's perspective [6]. Many of the underlying trends that administrators of these facilities used to predict patient counts and incoming revenue had been swept away by the tidal wave of economic depression leaving behind the unrecognizable economic and demographic debris.

Since traditional financial forecasting and manpower resource planning were unproductive in the post depression terra incognita of the current environment, health care managers were often at a loss as to how to plan. Administrators of these facilities typically computed the minimum daily revenue required to meet the clinic costs. Alternatively, some managers identify the minimum number of patients required to be treated by each clinic each day, that provided for the clinics an ethically sound, daily goal toward which to work. These managers can also identify how many consecutive days they can keep cash flow acceptable in the circumstances in which they did not generate the minimum revenue or see the minimum number of patients a day (termed a weak business day). Unfortunately, what they did not know was the likelihood the clinic will experience the predefined number of consecutive days of weak business. Knowledge of this probability would aid a clinic very likely to experience such a run of weak business to begin taking appropriate action in time to ensure its survival (e.g., to cut its fixed costs, reduce its hours, or engage in further attempts to expand its patient base). However, such data were not available to clinic managers. The goal of this chapter is to develop a model that would provide this information in the form of a system of difference equations.

10.3 Return of the $R_{0,k}(n)$ Model

Bernoulli trials play an important role in this text, and will be useful to here as we transform this prediction problem from a medical–economic to a probabilistic context. Consider the following formulation: On any given day, clinic revenue requirements are either met with the probability p or not met with probability q, where $p + q = 1$. Thus, if the knowledge of a previous day's result (i.e., whether enough revenue was generated or not) does not aid in determining a succeeding day's result, and the probability of achieving the minimum clinic revenue remains unchanged across days, the clinic's experience across consecutive days may be considered a sequence of Bernoulli trials.

If, in addition, on k consecutive days, the clinic's revenue requirements are not met, then it is observed that a *weak business run of length k* has occurred. Furthermore, one may specify a minimum run length of interest, k_d thereby focusing consideration on only weak business runs of at least length k_d days. Thus, any ability to predict the occurrence of runs in a sequence of Bernoulli trials yet to be observed may be applied to the prediction of "weak business runs." [7]

If the runs under consideration are weak business runs, and k_d is the length of minimum run of weak business of interest, then we can return to the $\mathbf{R}_{ik}(n)$ model to provide information about the relative frequency of occurrence of these weak business time periods. To review, the $\mathbf{R}_{ik}(n)$ model is the probability that there are exactly i failure runs of length k in n trials. It is a family of difference equations in two variables. For fixed i, the model represents a difference equation in n. However, as seen in Chapter 6, the model can also be approached as a difference equation in i.

Beginning with $\mathbf{R}_{0,k}(n)$, the boundary conditions may be written as

$$\begin{aligned} R_{0,k}(n) &= 1 \qquad \text{for } n < k \\ R_{0,k}(n) &= 0 \qquad \text{for } n < 0 \end{aligned} \tag{10.1}$$

The probability of no runs of length k when n < k must of course be 1. We will define $\mathbf{R}_{0,k}(1) = 1$. Using the boundary conditions provided in the relationships laid out in (10.1) the general difference equation for the probability of exactly no runs of length k in n trials when $n \geq 0$ may be written as

$$\begin{aligned} R_{0,k}(n) &= pR_{0,k}(n-1) + qpR_{0,k}(n-2) + q^2pR_{0,k}(n-3) + q^3pR_{0,k}(n-4) \\ &+ \ldots + q^{k-1}pR_{0,k}(n-k) + q^{k+1}pR_{0,k}(n-k-2) \\ &+ \ldots + q^{n-1}pR_{0,k}(n-k-2) + q^n I_{n \neq k} \end{aligned} \tag{10.2}$$

As in Chapter 6, each term on the right side of equation (10.2) permits a failure run of a specified length. Each term in (10.2) begins with the probability of a failure run of a specified length followed by the probability of no failures of length k in the remaining trials. Every possible failure run length is permitted except a run of length k, a requirement of $\mathbf{R}_{0,k}(n)$. Thus, $\mathbf{R}_{0,k}(n)$ is exactly the probability of permitting all possible failure run lengths except a failure run of length k.

A similar set of equations may be defined to compute $\mathbf{R}_{1k}(n)$ the probability of exactly one failure run of length k in n trials as follows

$$\begin{aligned} R_{1,k}(n) &= 0 \text{ for } n < k \\ R_{1,k}(n) &\equiv 0 \text{ for } n = 0 \\ R_{1,k}(n) &= pR_{1,k}(n-1) + qpR_{1,k}(n-2) + q^2pR_{1,k}(n-3) \\ &+ q^3pR_{1,k}(n-3) + \ldots + q^{k-1}pR_{1,k}(n-k) + q^{k+1}pR_{1,k}(n-k-2) \\ &+ \ldots + q^{n-1}pR_{1,k}(0) + q^k pR_{0,k}(n-k-1) \end{aligned} \tag{10.3}$$

For $i > 1$, there is a minimum number of observations n_i, such that for $n < n_i$ $\mathbf{R}_{ik}(n) = 0$. It is easily shown that $n_i = i(k+1) - 1$. Also, as demonstrated in Chapter 6, it is easily seen that

$$R_{i,k}(n_i) = q^k \left(pq^k\right)^{i-1} \tag{10.4}$$

Then

$$R_{i,k}(n) = 0 \text{ for } n < n_i$$

$$R_{i,k}(n) \equiv 0 \text{ for } n = 0$$

$$\begin{aligned} R_{i,k}(n) = {} & pR_{i,k}(n-1) + qpR_{i,k}(n-2) + q^2pR_{i,k}(n-3) + q^3pR_{i,k}(n-3) \\ & + \ldots + q^{k-1}pR_{i,k}(n-k) + q^{k+1}pR_{i,k}(n-k-2) \\ & + \ldots + q^{n-1}pR_{i,k}(0) + q^kpR_{i-1,k}(n-k-1) \end{aligned} \tag{10.5}$$

10.4 Applicability of $R_{IK}(n)$ to Ambulatory Business Cycle

10.4.1 Initial considerations

The derivation of $\mathbf{R}_{i,k}(n)$ permits, for a given q, k, and n, the computation of the exact probabilities of the occurrence of runs of length k under a variety of interesting and useful assumptions. The applicability of this distribution to predicting this phenomenon of weak business run lengths must be tested however. To do this, we examined the daily records of a major ambulatory health care center in Houston, Texas.

From this health care facility, daily weekday data were available for an eight month period from October 1986 to May 1987. These data consisted of a tabulation of the days in which each of the following three criteria were met:

1 - revenues generated less than $1,000.00
2 - revenues generated less than $1,500.00
3 - revenues generated less than $1,800.00

To provide assurances that the applicability of the $\mathbf{R}_{i,k}(n)$ model to this business prediction environment would not be a function of the weak business definition, the $\mathbf{R}_{i,k}(n)$ model was tested for each of the above definitions as in Moyé and Kapadia [1].

In order to calculate the expected run frequencies, one must first

1 - estimate q
2 - determine n, k
3 - compute probabilities for the distribution of interest
4- compute the expected number of runs for each value of k

For each definition of a weak business day provided above, a value for q, the probability of the occurrence of a weak business day must be computed. This value might be obtained by dividing the total number of weekdays in the past when the day's revenue was less than $1,000.00 by the number of weekdays of observation. For the three definitions, these estimates of q were 0.13 (revenues generated less than $1,000.00 – Table 10.1), 0.14 (revenues generated less than

$1,500.00), and 0.15 (revenues generated less than $1,800.00). Data from October 1986 through December 1986 were used to estimate q, then the model was used to predict the distribution of weak business runs over the next five months, from January, 1987 to May 1987. Thus, the period over which q was computed was distinct from that over which the model was fitted, thus testing the model under the conditions under which it could be used in the health care field.

Table 10.1 Distribution of Days When Total Patient Charges are Less Than $1000

Day	Charge	Day	Charge	Day	Charge	Day	Charge
1		28		55		82	990
2		29		56		83	
3		30		57		84	
4		31		58		85	
5		32		59		86	898
6		33		60		87	
7		34		61		88	
8		35		62		89	
9		36		63		90	
10		37		64		91	
11		38		65		92	
12		39		66		93	
13		40		67		94	
14	997	41		68		95	
15		42		69		96	
16		43		70		97	
17		44		71		98	
18		45		72	919	99	
19		46		73		100	
20		47		74		101	
21		48		75		102	
22		49		76		103	
23		50		77		104	
24		51		78		105	
25		52		79		106	
26		53		80		107	
27		54		81		108	

For the first definition of a weak business day as being < \$1000.00 in revenue there are only four runs of length 1. Table 10.2 provides a comparison of the observed results with those predicted by the model for revenue.

Alternatively, one could define a weak business day as one in which less than \$1500 in revenues was generated. Table 10.3 provides the distribution of run lengths observed for this definition, and Table 10.4 compares the distribution of runs with what the model predicted. The value of q for this definition of a weak business day was 0.14.

10.5 Comments on the Applicability of $R_{IK}(n)$

The health care administrator has access to q (the probability of a weak day estimated using past data), n (the number of days for which the predictions are

Table 10.2 Comparison of Observed Runs vs. Expected Weak Business Runs for Revenue < \$1000

Length of weak business runs	Observed	Expected
1 day	4	4.8
2 days	0	0.2
3 days	0	0.0
4 days	0	0.0
5 days	0	0.0
6 days	0	0.0
7 days	0	0.0
8 days	0	0.0
9 days	0	0.0
10 days	0	0.0

to be made), and k (the defined weak business run length). It is our contention that this information is sufficient to obtain accurate estimates of future weak business run lengths. The approach offered here demonstrates the attainment of the required distribution using only this information and is therefore of interest from a number of perspectives. First, it offers an exact solution for the prediction of the future occurrence of runs of an arbitrary length in a sequence of

Bernoulli trials, with no approximations required. Second, the solution is in terms of parameters with which the health care administrator is familiar and has direct access. Third, the outcome measures of this model (expected number of weak business runs of an arbitrary length and average duration) are measures of administrative importance. Thus, starting from estimates of parameters and

Table 10.3 Distribution of Days When Total Patient Charges are Less Than $1500

Day	Charge	Day	Charge	Day	Charge	Day	Charge
1		28		55		82	990
2		29		56		83	
3		30		57		84	
4	1136	31		58	1275	85	1146
5	1430	32	1143	59		86	898
6		33	1404	60		87	
7		34	1042	61		88	
8		35		62	1196	89	
9		36		63	1461	90	
10	1387	37		64	973	91	1049
11		38		65	1146	92	1407
12	1029	39	1242	66		93	
13		40	1488	67		94	
14	997	41		68		95	
15		42		69		96	
16	1060	43	1472	70	1405	97	
17		44		71	1272	98	1453
18	1400	45	1296	72	919	99	
19		46	1393	73		100	
20	1142	47		74		101	1333
21		48		75	1481	102	1311
22	1090	49	1240	76		103	659
23		50	1087	77		104	1470
24	1305	51	1148	78	1285	105	
25		52		79		106	
26		53	1213	80		107	
27		54	1018	81		108	

Table 10.4 Comparison of Observed Runs vs. Expected Weak Business Runs for Revenue < $1500.00

Length of weak business runs	Observed	Expected
1 day	14	16.0
2 days	7	5.4
3 days	3	1.8
4 days	2	0.6
5 days	0	0.2
6 days	0	0.0
7 days	0	0.0
8 days	0	0.0
9 days	0	0.0
10 days	0	0.0

using the presented model, the administrator will gain a pertinent, accurate assessment of the likelihood of runs of lengths of interest.

The Bernoulli model represented by $\mathbf{R}_{i,k}(n)$ works in predicting runs. Using alternative run definitions we observed that the model performed reasonably well in predicting both revenue and census runs. These results are encouraging, but it must be emphasized that more theoretical development is required along the following lines. First, a rigorous goodness of fit test must be developed. Such a test is not presently available. The traditional chi-square test will not be entirely appropriate since it assumes that all expectations are derived from the same probability distribution. In fact, the expected frequency of each run length has its own distribution. Secondly, it was assumed that the value of q is region-specific and remains constant over time. Initial observations suggest that the model is not overly sensitive to variations in q. Nevertheless, additional estimation techniques for q might need to be developed. Third, it is possible that random fluctuations in business occurring when the unit of time is a day may all but disappear when the unit of time is a week or a month. Hence, the unit of time to be studied is crucial and is a function of the size of the business and how long it has been in operation.

Table 10.5 Distribution of Days When Total Patient Charges are Less Than $1500

Day	Charge	Day	Charge	Day	Charge	Day	Charge
1	1552	28	1613	55		82	990
2		29	1691	56	1577	83	1700
3	1817	30	1705	57	1616	84	1721
4	1136	31		58	1275	85	1146
5	1430	32	1143	59	1539	86	898
6		33	1404	60	1740	87	1502
7	1133	34	1042	61		88	
8	1553	35		62	1196	89	1543
9		36	1672	63	1461	90	
10	1387	37	1764	64	973	91	1049
11		38		65	1146	92	1407
12	1029	39	1242	66	1695	93	
13	1515	40	1488	67	1598	94	1607
14	997	41	1600	68	1750	95	
15	1506	42	1572	69	1616	96	
16	1060	43	1472	70	1405	97	
17	1576	44	1549	71	1272	98	1453
18	1400	45	1296	72	919	99	1655
19	1616	46	1393	73		100	
20	1142	47		74	1600	101	1333
21		48	1538	75	1481	102	1311
22	1090	49	1240	76		103	659
23		50	1087	77		104	1470
24	1305	51	1148	78	1285	105	
25		52		79		106	
26		53	1213	80	1583	107	
27		54	1018	81		108	

Table 10.6 Comparison of Observed Runs vs. Expected Weak Business Runs for Revenue < $1800.00

Length of weak business runs	Observed	Expected
1 day	8	11.2
2 days	7	6.5
3 days	3	3.7
4 days	2	2.1
5 days	1	1.2
6 days	1	0.7
7 days	0	0.4
8 days	1	0.1
9 days	1	0.1
10 days	1	0.0

10.6 Worst Business Run

This model may be used to predict the performance of a clinic over a period of time. Defining the worst weak business run as the expected maximum run length of weak business. Figure 10.1 depicts the relationship between the likelihood of a weak business day (q) and the length of the worst weak business run. Also, the 95% upper confidence interval (i.e., Mean + 1.96SD) is provided

One of the authors's experience in modeling [8] suggests that getting the theoretical model to precisely predict actual data can be a difficult and time consuming process, and modeling assumptions must be scrupulously examined for applicability to real world situations. The Bernoulli model assumes independence in the occurrence of the event of interest over consecutive days. This assumption may or may not be valid, depending on the period of observation. Because the clinics are generally not as busy on the weekends as they have been during the week, it was felt that the inclusion of the weekend data was inappropriate in a Bernoulli model.

However, given this limitation, the ability of this model to accurately predict weak business runs for this clinic system is remarkable. It is possibly that, although predictions of revenue and census magnitude vary from day to

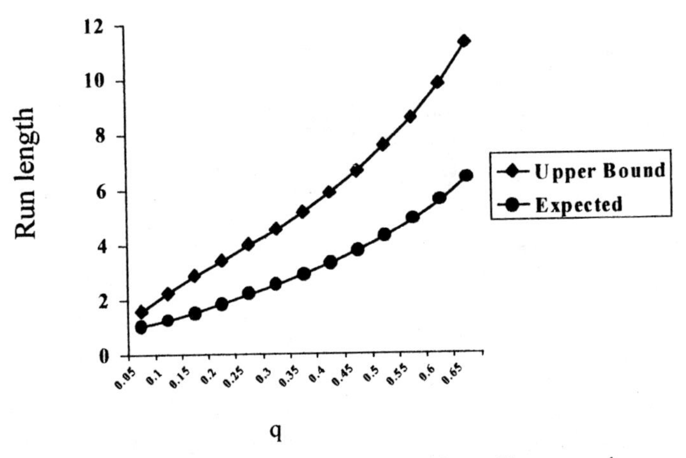

Figure 10.1 Expected worst weak business run by q. The expected number of consecutive days of weak business increases as the probability of a weak business day increases.

day and might be serially correlated, this is less crucial an assumption in predicting the actual event of run. Alternatively, an extended model that would take into account possible Markov-type memory among annual events would be expected to perform even better. An examination of the tail of the fit in Table 10.4 suggests that the occurrence of longer run lengths than predicted may be due to serial correlation of revenue generation. Efforts to develop this option suggest that this more sophisticated alternative would use as the probability of a run length of length k whose individual failures have values $q_1, q_2, \ldots, q_k$, and not q. In this more complicated circumstance, q_i would be the probability that the ith Bernoulli trial is a q given the preceding $i - 1$ trials are each a failure. A more complex model would incorporate these second- or higher-order Markovian dependencies into the statistical run theory.

The applicability of the solutions to these difference equations presented here are not restricted to daily data. Weekly, monthly or yearly data easily could be applied. In the same vein, this model may be of value to hospital administrators who have at their disposal information on the likelihood of consecutive weeks or months of hospital admissions. The model presented here does not supplement time-tested and proven techniques of stock control manpower/resource planning and financial forecasting. The development presented in this paper is offered as a first step to extend a potentially useful theory to health care management needs in a turbulent economic environment. We continue to work on the development of this model along the avenues outlined above. The limitations notwithstanding, the present developments offer an encouraging picture for the prospect of run prediction.

Problems

1. Enumerate the probability that, in a ten day period, there are exactly 3 runs of weak business when a weak business run is defined as two consecutive days of poor business revenues.
2. Compute the probabality that, in a ten day period, there is at least one run of weak business when a weak business run is defined as at least two consecutive days of poor business revenue.
3. Using equation (10.5), when $q = 0.5$ and $k = 2$ find the following probabilities:
   -There are no runs of weak business in nine days
   -There are between three and four runs of weak business in nine days
   -There are exactly three runs of weak business in nine days
4. Enumerate the events that comprise $R_{0,3}(5)$. Compare your derivation with equation (10.2) for $k = 3$, $n = 5$.
5. Using the formula, Prob(all failure runs are of length k in n trials) = $R_{0,k-1}(n) - R_{0,k}(n)$, obtain the expected failure run length for $n = 5$, $k = 3$.
6. Derive the difference equation for $R_{i,k}(n)$ for $1 < i < k < n$ for case of $p = q = ½$.
7. Enumerate the events that comprise $R_{2,5}(6)$ if a run of four consecutive failures is not possible.
8. Simplify the solution to problem seven for $p = 1 = ½$.
9. If successes and failures are governed by the transition probability matrix A below, then obtain the general form for $R_{0,k}(n)$ (a) assuming the first trial was a success and (b) assuming the first trial was a failure.

$$A = \begin{array}{c} \\ S \\ F \end{array}\!\!\begin{array}{c} \begin{array}{cc} S & F \end{array} \\ \begin{bmatrix} p & q \\ q & p \end{bmatrix} \end{array}$$

10. Obtain the difference equations when successes and failures are governed by the transition probability matrix B when (a) the first trial was a success and (b) the first trial was a failure. In this case, $p_1+q_1 = 1$, and $p_2+ q_2 = 1$.
11. Derive the difference equation in problem 10 for the case of $p_1 = 0.4$, and $p_2 = 0.6$.

$$A = \begin{array}{c} \\ S \\ F \end{array}\!\!\begin{array}{c} \begin{array}{cc} S & F \end{array} \\ \begin{bmatrix} p_1 & q_1 \\ q_2 & p_2 \end{bmatrix} \end{array}$$

References

1. Erman D. and Gabel J. The changing face of American health care – multiple hospital systems, emergency centers, and surgery centers. *Medical Care* 23.1980.
2. Eisenberg H. Convenience clinics: your newest rival for patients. *Medical Economics* 24:71-84. 1980.
3. Aday L.A. et al. Health Care in the US – Equitable for whom? Sage Publications. Beverly Hills, CA.
4. Lave J. A review of methods uses to study hospital costs. *Inquiry* 57-81. May 1966
5. Berke, S.E. Hopital Economics Lexington. Lexington Books. 1972.
6. Miller K.E., Lairson D.R., Kapadia A.S., Kennedy V.C. Patient characteristics and the demand for care in freestanding emergency centers. *Inquiry* 22(4):418-425. 1985.
7. Craven H.L. and Leadbetter M.R. Stationary and related stochastic processes. New York. Wiley and Sons. 1967.
8. Moyé L.A., Kapadia A.S., Cech I., Hardy R.J. A prediction model for outpatient visits and cash flow: development in prospective run theory. *Statistics* 39.(4):399-414.1990.
9. Moyé L.A. and Roberts S.D. Modeling the pharmacologic treatment of hypertension. *Management Science* 28:781-797.1982.

11

Difference Equations in Epidemiology: Basic Models

11.1 Introduction

11.1.1 The role of this chapter

The purpose of this chapter is to provide an introduction to the difference equation perspective often found useful by epidemiologists as they work to understand the spread of a disease in a population. After a brief discussion of the purposes of epidemiology, we will develop both the nomenclature and the difference equations used to predict the number of patients with a disease under a mix of interesting and useful assumptions. These equations will be well recognized by the advanced worker as immigration models, emigration models, birth models, and death models, and will be so labeled in this chapter. The next chapter will be a continuation of this work from the difference equation perspective for more complex models that perhaps more realistically portray changes in the numbers of patients with an illness. In each of these developments, the focus here will be to elaborate the role of difference equations in providing solutions to these dynamic systems.

During this development, we will confront for the first time in this book, not just difference equations, but differential equations as well. The reader with no background in differential equations need not be alarmed, however. Only the concepts of differential equations relevant to our immediate needs for these chapters' discussions and absorption will be presented. Furthermore, this narrow need will be fully developed with solutions provided in complete detail.

11.1.2 The role of epidemiologists

Epidemiology is the study of the determinants (who, why, where, when and how) of disease. One can think of epidemiologists as "disease detectives." It is a science that can trace its roots to writings that are over one thousand years old and has assumed a role of increasing importance in health care research. The identification of many health care risks that are now taken for granted were demonstrated by epidemiologists. For example, the casual observation that we all avoid drinking soiled, dirty water is credited to careful epidemiologic work in 19th century London, England, by John Snow who revealed the patterns of gastrointestinal illnesses and matched these to the degree of water contamination through human use. The observation that cigarette smoking increases the risk of atherosclerotic heart disease, peripheral vascular disease, and cancer is credited to epidemiology. In addition the role of thalidomide in producing birth defects was determined by the use of epidemiologic methods in the 1960s. Epidemiologists and biostatisticians are more alike than different, but do have some distinguishing features. Biostatistics is focused on the development of probabilitistic and statistical tools in studying natural phenomena. Epidemiologists use these tools to augment both careful observations and deductive reasoning to determine whether a risk factor is a cause of a disease.

11.2 Immigration Model–Poisson Process

11.2.1 Introduction to the immigration process

For this first model, the goal is to compute the number of patients who have a disease. The assumption that the disease is not contagious, and that no one in the population has the illness at the beginning of the process will be made. Since the disease does not spread from individual to individual, the only people who will have the illness are those that arrive with the disease.

If one knew how many were arriving with the disease, or even the arrival rate, in this circumstance of no person-to-person disease spread, it would be easy to project the total number of subjects with the disease (or cases). With this knowledge, one would merely multiply the arrival rate by the length of time over which the arrivals took place to compute the total number of patients with the disease. With no deaths, and no emigration, the computation is easy. However, this computation does not consider the randomness inherent in the entry of disease. In this process, it is impossible to know whether any one individual diseased patient will enter – what is known instead is the probability of that introduction. It is the replacement of certainty by probability that makes the process stochastic rather than deterministic. We will then give up on the idea of computing exactly how many cases there are, focusing instead on computing the probability that there will be n cases of patients with the disease at time t, for any value of $n \geq 0$ and for all $t > 0$.

Assume that patients with the disease arrive at the rate of λ arrivals per unit of time. The rate of these arrivals is independent of the number of patients in the diseased population. If ten diseased patients arrive per week, then, on average in two weeks we would expect twenty diseased patients, in three weeks, thirty diseased patients would be expected etc. In general, in the short time interval Δt, we would expect $\lambda\Delta t$ arrivals. With neither deaths, cures, nor exits, cases can only accumulate in the system (Figure 11.1).

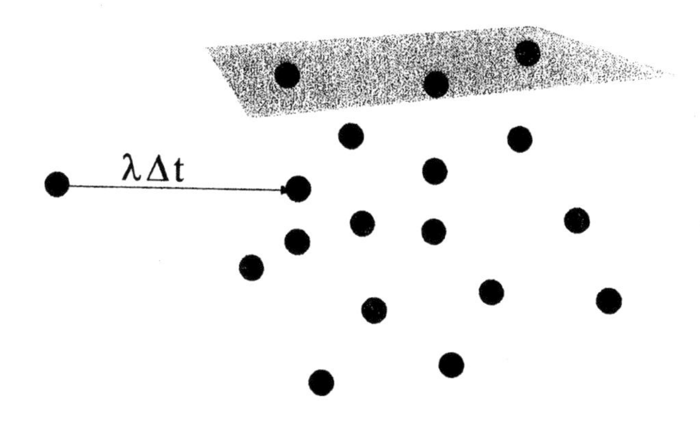

Figure 11.1 Immigration Model: The only source of disease is patients arriving into the population at a constant rate.

11.2.2 Parameterization of the immigration process

The task at hand is to compute the probability that at time t there are n subjects with the disease, a probability that we denote by $P_n(t)$. To express this, a family of difference equations is first developed. Assume that the movement of time can be slowed down, so that we can observe the enry of patients with the disease in slow motion, one event at a time. This is equivalent to making Δt so small that in the time interval (t, t + Δt), there will be time for only one event to take place. From this admittedly unique observation point, note that either 1) a single immigration may take place (a diseased individual joins the ranks of the diseased individuals or 2) no immigration takes place. It is not known with certainty whether there will be an arrival in the time interval Δt. However, Δt is so small that the probability of a single immigration or arrival may be written as $\lambda\Delta t$.

To develop the difference equation, consider how there may be n cases (diseased individuals) at time t + Δt. Following our previous discussion and, given that Δt is so small that only an event involving one subject can occur, there are two ways that there will be n diseased patients at time t. One possibility is that there are n – 1 diseased individuals at time t, and there is exactly one arrival that occurs with probability $\lambda\Delta t$. The other possibility is that there are n patients with the disease at time t, and there is no arrival in the time interval Δt, an event that occurs with probability 1 - $\lambda\Delta t$.

11.2.3 Chapman–Kolmogorov forward equations

We are now in a position to write a difference equation that reflects how there will be n patients with the disease at time t + Δt, when only one event can occur from time t to time t + Δt.

$$P_n(t+\Delta t) = \lambda\Delta t P_{n-1}(t) + P_n(t)(1-\lambda\Delta t) \tag{11.1}$$

This is known as a Chapman–Kolmogorov forward equation. Equation (11.1) looks "forward" from the current time of t to what the number of diseased patients with the disease will be at the future time t + Δt. Define $P_n(t) = 0$ for n < 0. For n = 0, equation (11.1) gives

$$P_0(t+\Delta t) = P_0(t)(1-\lambda\Delta t) \tag{11.2}$$

It can be seen that equation (11.1) holds for $n \geq 0$. We can now proceed with the solution of this difference equation in n, but first the term Δt that is contained in equation (11.1) requires attention.

11.2.4 Difference–differential equations

The first task in this effort will be somewhat unusual. Recall that equation (11.1) holds for $n \geq 0$. In order to remove the Δt terms, we proceed as follows.

$$
\begin{aligned}
P_n(t+\Delta t) &= \lambda\Delta t P_{n-1}(t) + P_n(t)(1-\lambda\Delta t) \\
P_n(t+\Delta t) &= \lambda\Delta t P_{n-1}(t) + P_n(t) - \lambda\Delta t P_n(t) \\
P_n(t+\Delta t) - P_n(t) &= \lambda\Delta t P_{n-1}(t) - \lambda\Delta t P_n(t) \\
\frac{P_n(t+\Delta t) - P_n(t)}{\Delta t} &= \lambda P_{n-1}(t) - \lambda P_n(t)
\end{aligned}
\tag{11.3}
$$

Next, examine the behavior of this equation as Δt approaches zero. The right side of the equation is unaffected since it includes no term involving Δt. However, assuming $P_n(t)$ is a smooth, continuous function of t,

$$\lim_{\Delta t \to 0} \frac{P_n(t+\Delta t) - P_n(t)}{\Delta t} = \frac{dP_n(t)}{dt} \tag{11.4}$$

equation (11.3) may be re written as

$$\frac{dP_n(t)}{dt} = \lambda P_{n-1}(t) - \lambda P_n(t) \tag{11.5}$$

for $n = 0$ to ∞. Note that this is still a difference equation in n. Although terms involving Δt have been removed, we are "stuck" with a derivative. Thus, the family of equations (11.3) may be described not as a family of differential equations, but as a family of difference-differential equations.

11.2.5 The generating function argument

For the time being we will not treat the $\frac{dP_n(t)}{dt}$ term with any special attention, and will proceed in our accustomed manner. Begin by defining

$$G_t(s) = \sum_{n=0}^{\infty} s^n P_n(t) \tag{11.6}$$

The subscript "t" is added to the generating function reflecting the realization that $P_n(t)$ is a function not just of n, but of t. With this comment, we are now ready to initiate our familiar conversion and consolidation process to the family of equations (11.5)

$$s^n \frac{dP_n(t)}{dt} = s^n \lambda P_{n-1}(t) - s^n \lambda P_n(t)$$
$$\sum_{n=0}^{\infty} s^n \frac{dP_n(t)}{dt} = \lambda \sum_{n=0}^{\infty} s^n P_{n-1}(t) - \lambda \sum_{n=0}^{\infty} s^n P_n(t) \tag{11.7}$$

Each of these terms is simplified individually, beginning with the last term on the right side of equation (11.7)

$$\lambda \sum_{n=0}^{\infty} s^n P_n(t) = \lambda G_t(s) \tag{11.8}$$

$$\lambda \sum_{n=0}^{\infty} s^n P_{n-1}(t) = \lambda s \sum_{n=0}^{\infty} s^{n-1} P_{n-1}(t) = \lambda s \sum_{n=0}^{\infty} s^n P_n(t) = \lambda s G_t(s) \tag{11.9}$$

$$\sum_{n=0}^{\infty} s^n \frac{dP_n(t)}{dt} = \frac{d \sum_{n=0}^{\infty} s^n P_n(t)}{dt} = \frac{dG_t(s)}{dt} \tag{11.10}$$

The last line of equation (11.10) is worthy of additional comment. While it is true in general that the derivative of a finite sum is the sum of the derivatives of the summands, this convenient relationship breaks down when the sum is over an infinite series of terms. The reversal of the summand and the derivative over an infinite sum is only true when certain conditions are met.* However, these conditions are met here, and we may continue to write

$$\frac{dG_t(s)}{dt} = \lambda s G_t(s) - \lambda G_t(s)$$
$$\frac{dG_t(s)}{dt} = \lambda G_t(s)(s-1) \tag{11.11}$$

Thus an infinite number of equations in n have been consolidated into one differential equation involving $G_t(s)$.

---

* G(s) must be uniformly convergent, which poses no problem here. See Chiang [1] for details.

11.2.6 Solving the simple differential equation

Although many differential equations are difficult to solve, there is fortunately no difficulty in identifying the solution to equation (11.11). Since the right side of this equation is not a function of t, rearrange terms in equation (11.11) and proceed as follows:

$$\frac{dG_t(s)}{dt} = \lambda G_t(s)(s-1)$$
$$\frac{dG_t(s)}{G_t(s)} = \lambda(s-1)dt \tag{11.12}$$

Integrating each side of this equation, the left side with respect to $G_t(s)$, the right side with respect to t.

$$\int \frac{dG_t(s)}{G_t(s)} = \int \lambda(s-1)dt$$
$$\ln G_t(s) = \lambda t(s-1) + C \tag{11.13}$$

where C is a constant of integration. In order to evaluate this constant of integration, the boundary condition for equation (11.11) are utilized. Let us assume that when the process starts, there are no cases. Therefore at t = 0

$$G_0(s) = \sum_{n=0}^{\infty} s^n P_n(0) = s^0 = 1 \tag{11.14}$$

Using the observation that ln(1) = 0, when t = 0 we see from equation (11.13) that C = 0. Thus

$$\ln G_t(s) = \lambda t(s-1)$$
$$G_t(s) = e^{\lambda t(s-1)} \tag{11.15}$$

Recall from section 2.8.5 that

$$G_t(s) = e^{\lambda t(s-1)} \triangleright \left\{ \frac{(\lambda t)^n}{n!} e^{-\lambda t} \right\} \tag{11.16}$$

revealing that, if X(t) is the number of diseased patients at time t, then

$$P[X(t)=n]=P_n(t)=\frac{(\lambda t)^n}{n!}e^{-\lambda t} \tag{11.17}$$

11.2.7 Final comments on the immigration process

The solution of the difference equation (11.1) reveals that the probability of having n patients with the disease at time t follows a Poisson probability distribution with parameter λ. An arrival process with the characteristics of the immigration process described in detail here is termed a Poisson process. The mean number of diseased patients at time t is λt, that is exactly the number one would have estimated by multiplying the arrival rate by the time elapsed since there were no cases at time t = 0.

An interesting question is, What would have the result been if the assumption that at time t = 0 the system already had a_0 diseased individuals? Before returning to the system of equations previously developed in this chapter, we can first think this process through, to arrive at a solution. If there are a_0 cases in the system at time 0, then by the assumptions made for the immigration process (i.e., there are neither deaths nor exits of diseased patients), these patients are still available at time t. In addition, since it is assumed that the disease is not contagious, these a_0 patients have not contributed to the number of other cases in the system. All other cases must have arrived with arrival rate λ. Thus, if there are to be n diseased individuals at time t, there can only have been $n - a_0$ arrivals, resulting in

$$P_n(t)=\frac{(\lambda t)^{n-a_0}}{(n-a_0)!}e^{-\lambda t}I_{n>a_0} \tag{11.18}$$

where $I_{n>a_0}$ represents a form of the indicator function introduced in Chapter 3. This result can be reproduced for the analytical approach, beginning with equation (11.13).

$$\ln G_t(s)=\lambda t(s-1)+C \tag{11.19}$$

However, in this circumstance, evaluating $G_t(0)$ reveals $G_0(s)=s^{a_0}$. Thus $C=\ln\left(s^{a_0}\right)$. Inserting this result into equation (11.19) note that

$$\begin{aligned}\ln G_t(s)&=\lambda t(s-1)+\ln s^{a_0}\\ G_t(s)&=s^{a_0}e^{\lambda t(s-1)}\end{aligned} \tag{11.20}$$

Inversion of $G_t(s)$ from equation (11.20) requires only the sliding tool from Chapter 2 to identify

$$P_n(t) = \frac{(\lambda t)^{n-a_0}}{(n-a_0)!} e^{-\lambda t} I_{n>a_0} \tag{11.21}$$

11.3 Emigration Model

11.3.1 Initial comments

In this section, the opposite of the immigration model considered previously is studied. In the immigration model, patients with disease enter a population. Once these patients arrive, they stay–they neither die nor exit. In this section, patients with disease can only leave. At the conclusion of this section $P_n(t)$ will be identified for the new circumstance where diseased patients can only leave. It is assumed that diseased patients leave at the rate of $\mu\Delta t$ per unit time (Figure 11.2). The rate of these departures or emigrations is independent of the number of patients in the diseased population. This is an assumption which distinquishes the emigration process from the death process, which will be developed later in this chapter.

11.3.2 The intuitive approach

Some initial thought might suggest the form of the answer. Assume that, at time 0 there are a_1 patients with the disease in the system. Then, in the emigration

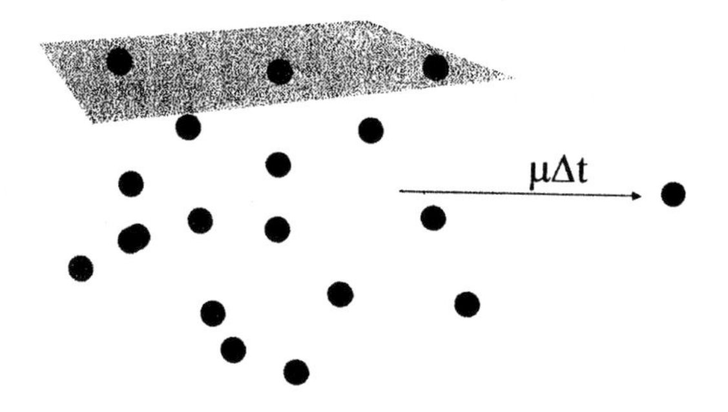

Figure 11.2 Emigration Model: Patients with disease leave the population at a constant rate.

model, the number of cases, n, must be between 0 and a_1. If $0 \le n \le a_1$, then there can only be n patients with disease in the system when $a_1 - n$ patients have left. For example, if there were 30 patients in the system at time $t = 0$, then the probability that there are 18 patients in the system at time t is the probability that 12 patients have exited the system.

These 12 patients leave the system by entering another system outside the original system (i.e. they exit by entering another system). This sounds like an immigration process – patients who leave the original system leave by entering (or immigrating to) another. Thus our thought process for this system reveals that for $n \ge 0$

$$\begin{aligned} P_{18}(t) &= P\,[18 \text{ patients in the system}] \\ &= P\,[12 \text{ patients leave the system}] \\ &= \frac{(\mu t)^{12}}{12!} e^{-\mu t} \end{aligned} \tag{11.22}$$

Therefore, our experience from the immigration process in section 11.2 reveals that for the emigation model:

$$P_n(t) = \frac{(\mu t)^{a_1 - n}}{(a_1 - n)!} e^{-\mu t} \tag{11.23}$$

We will now provide this solution using a generating function approach.

11.3.3 The Chapman–Kolmogorov equation for the emigration process

We will produce the Chapman–Kolmogorov forward equation for the emigration process paralleling the model used for the immigration process. As before, consider the dynamics of patients with the disease as time moves from time t to time t + Δt when Δt is very small, allowing only a change in the number of diseased patients by one. In this framework, the number of cases at time t can only be 1) reduced by one or 2) remain the same. The number of patients with disease is reduced by one with probability $\mu\Delta t$, or the number can stay the same with probability $1 - \mu\Delta t$. Therefore the Chapman–Kolmogorov forward equation may be written for $n > 0$ as

$$P_n(t + \Delta t) = \mu \Delta t P_{n+1}(t) + P_n(t)(1 - \mu\Delta t) \tag{11.24}$$

Proceed with this solution as was executed for the immigration process; first removing the Δt term and identifying a derivative, and then applying a generating function argument.

$$\begin{aligned}
&P_n(t+\Delta t) = \mu\Delta t P_{n+1}(t) + P_n(t)(1-\mu\Delta t) \\
&P_n(t+\Delta t) - P_n(t) = \mu\Delta t P_{n+1}(t) - \mu\Delta t P_n(t) \\
&\frac{P_n(t+\Delta t) - P_n(t)}{\Delta t} = \mu P_{n+1}(t) - \mu P_n(t) \\
&\lim_{\Delta t \to 0} \frac{P_n(t+\Delta t) - P_n(t)}{\Delta t} = P_{n+1}(t)\mu - P_n(t)\mu \\
&\frac{dP_n(t)}{dt} = P_{n+1}(t)\mu - P_n(t)\mu
\end{aligned} \tag{11.25}$$

11.3.4 Application of the generating function–conversion and consolidation

Continuing with the development of the emigration process, define $G_t(s)$ as

$$G_t(s) = \sum_{n=0}^{\infty} s^n P_n(t) \tag{11.26}$$

and proceed with conversion and consolidation of equation (11.25)

$$\begin{aligned}
&\frac{dP_n(t)}{dt} = \mu P_{n+1}(t) - \mu P_n(t) \\
&s^n \frac{dP_n(t)}{dt} = s^n \mu P_{n+1}(t) - s^n \mu P_n(t) \\
&\sum_{n=0}^{\infty} s^n \frac{dP_n(t)}{dt} = \sum_{n=0}^{\infty} s^n \mu P_{n+1}(t) - \sum_{n=0}^{\infty} s^n \mu P_n(t)
\end{aligned} \tag{11.27}$$

Simplifying each of these terms on the right side of equation (11.27) in the familiar fashion. The term $\sum_{n=0}^{\infty} s^n \frac{dP_n(t)}{dt}$ is converted to $\frac{dG_t(s)}{dt}$ by switching the derivative and summation sign as in solving the solution for the immigration process. The resulting simplification leads to

$$\frac{dG_t(s)}{dt} = \mu s^{-1}\left[G_t(s) - P_0(t)\right] - \mu G_t(s) \tag{11.28}$$

At this point the simplifying assumption that $P_0(t) = 0$ is made. This is the assertion that at no time are the number of patients with disease reduced to zero. This might occur, for example, when μ, the emigration rate, is very much smaller than the number of patients with disease initially. With this simplifying assumption we can proceed with the conversion.

$$\begin{aligned}
\frac{dG_t(s)}{dt} &= \mu s^{-1}G_t(s) - \mu G_t(s)\\
\frac{dG_t(s)}{dt} &= \mu\left(s^{-1} - 1\right)G_t(s)\\
\frac{dG_t(s)}{G_t(s)} &= \mu\left(s^{-1} - 1\right)dt\\
\int\frac{dG_t(s)}{G_t(s)} &= \int\mu\left(s^{-1} - 1\right)dt\\
\ln G_t(s) &= \mu t\left(s^{-1} - 1\right) + C
\end{aligned} \tag{11.29}$$

The boundary condition for this model is that at $t = 0$, $n = a_1$. Thus $G_0(s) = s^{a_1}$, and solving for C

$$\ln G_t(s) = \mu t\left(s^{-1} - 1\right) + \ln s^{a_0} \tag{11.30}$$

or

$$G_t(s) = s^{a_0} e^{\mu t\left(s^{-1} - 1\right)} \tag{11.31}$$

The generating function in equation (11.31) appears to be related to that of the Poisson process. However, the s^{-1} term is new, and requires attention.

11.3.5 Inversion process

To begin with the inversion of $G_t(s)$, focus on the second term in the equation (11.31). Rewrite this as

$$e^{\mu t\left(s^{-1} - 1\right)} = e^{-\mu t}e^{\mu t s^{-1}} \tag{11.32}$$

We may write $e^{\mu t s^{-1}}$ using a Taylor series expansion.

$$e^{\mu t s^{-1}} = \frac{\mu t s^{-1}}{1!} + \frac{(\mu t)^2 s^{-2}}{2!} + \frac{(\mu t)^3 s^{-3}}{3!} + \ldots + = \sum_{k=0}^{\infty} \frac{(\mu t)^k s^{-k}}{k!} \tag{11.33}$$

Thus, the coefficient of s^{-n} is $\frac{(\mu t)^n}{n!}$. Applying this result to equation (11.31) find that

$$G_t(s) = s^{a_0} e^{\mu t (s^{-1} - 1)} \quad \triangleright \left\{ \frac{(\mu t)^{a_0 - n}}{(a_0 - n)!} e^{-\mu t} \right\} \tag{11.34}$$

through the application of the sliding tool. Thus, the probability that there are n subjects with the disease at time t is the probability that $a_0 - n$ diseased subjects have departed by time t.

11.4 Immigration–Emigration Model

11.4.1 Introduction

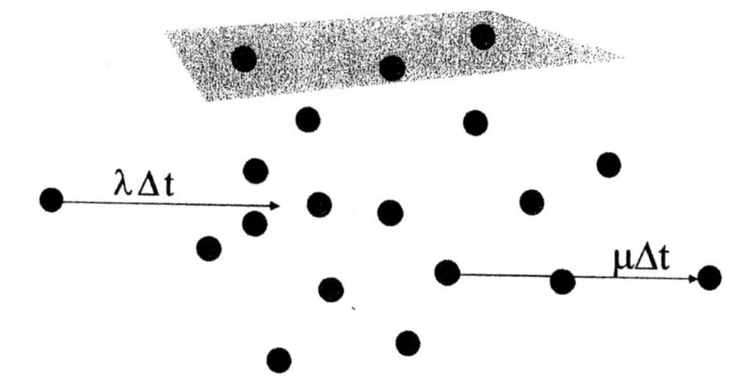

Figure 11.3 Immigration–Emigration Model: Patients with disease enter the system at a constant rate while others leave at a different rate.

This section will increase the complexity of the modeled system by combining the modeling features of the immigration model with those of the emigration model. The goal will be to identify the probability distribution of the number of patients with the disease at time t, given that patients can both enter and leave. It is assumed that there are a_0 patients at time t=0.

11.4.2 Intuitive approach

Figure 11.3 demonstrates the dynamics of the immigration-emigration model. Begin by assuming that, at time zero, there are a_0 patients with the disease. New diseased patients can arrive into the system. Assume that they arrive with an arrival rate of λ. That is, in a small period of time Δt, the probability that a single new diseased patient arrives is $\lambda\Delta t$. Furthermore, assume that diseased patients leave the system at the rate of μ, and that in a small period of time Δt, the probability a diseased patient leaves is $\mu\Delta t$. Thus the number of diseased individuals may either increase or decrease depending on the relative values of λ and μ. The goal is to compute $P_n(t)$, the probability that there are n patients with the disease at time t.

We may have enough insight into the dynamics of these models from our previous considerations of the immigration model and the emigration model. As a first attempt to identify the probability distribution of the total number of patients with the disease at time t, consider the following simpler problem. Under the above assumptions of an arrival rate $\lambda\Delta t$ and a departure rate $\mu\Delta t$, and assuming that we start with a_0 diseased patients in the system at time t = 0, then what is the probability that at time t, we again have a_0 diseased patients? That is, what is the probability that the total number of diseased patients has not changed? We observe that this will only happen if an equal number of diseased patients arrive and leave by time t. Thus computing

$$\begin{aligned}
P_{a_0}(t) &= \sum_{m=0}^{\infty} P\left[\text{m arrivals and m departures}\right] \\
&= \sum_{m=0}^{\infty}\left[\frac{(\lambda t)^m}{m!}e^{-\lambda t}\right]\cdot\left[\frac{(\mu t)^m}{m!}e^{-\mu t}\right] \\
&= \sum_{m=0}^{\infty}\left[\frac{(\lambda t)^m(\mu t)^m}{m!m!}e^{-(\lambda+\mu)t}\right]
\end{aligned} \tag{11.35}$$

that can be rewritten as

$$P_{a_0}(t) = \sum_{m=0}^{\infty}\left[\frac{(2m)!}{m!\,m!}\left(\frac{\lambda}{\lambda+\mu}\right)^m\left(\frac{\mu}{\lambda+\mu}\right)^m\right]\left[\frac{[(\lambda+\mu)t]^{2m}}{(2m)!}e^{-(\lambda+\mu)t}\right] \tag{11.36}$$

Equation (11.36) contains the product of two probabilities. The second factor is the probability that there are 2m events that follow a Poisson distribution with parameter $\lambda + \mu$. The first factor we recognize as a term from the binomial probability distribution function. It is the probability that given 2m events have occurred, half of them are arrivals and the other half are departures.

Now expand this consideration to the more general case where $P_{a_0}(0) = 1$; we are interested in identifying $P_n(t)$. Assume here that $0 < n < a_0$, i.e. their have been more departures then arrivals. Write

$$\begin{aligned} P_n(t) &= \sum_{m=0}^{\infty} P[\text{m arrivals and } m + a_0 - n \text{ departures}] \\ &= \sum_{m=0}^{\infty}\left[\frac{(\lambda t)^m}{m!}e^{-\lambda t}\right]\cdot\left[\frac{(\mu t)^{m+(a_0-n)}}{(m+a_0-n)!}e^{-\mu t}\right] \end{aligned} \tag{11.37}$$

This can be rewritten as

$$P_n(t) =$$

$$\sum_{m=0}^{\infty}\left[\frac{(2m+a_0-n)!}{m!(m+a_0-n)!}\left(\frac{\lambda}{\lambda+\mu}\right)^m\left(\frac{\mu}{\lambda+\mu}\right)^{m+a_0-n}\right]\left[\frac{[(\lambda+\mu)t]^{2m+a_0-n}}{(2m+a_0-n)!}e^{-(\lambda+\mu)t}\right] \tag{11.38}$$

As in the previous case, it can be seen that the last expression of equation (11.37) is the sum of the product of binomial and Poisson probabilites. The Poisson term is the probability that there are exactly $2m + a_0 - n$ Poisson $(\lambda t + \mu t)$ events. The binomial probability is the probability that, given that there are $2m + a_0 - n$ events, exactly m of them are arrivals and the remaining $m + a_0 - n$ are departures.

It is easy to see that if we allowed $n > a_0$, there would be $n - a_0$ more arrivals than departures, and $P_n(t)$ would be

$$P_n(t) = \sum_{m=0}^{\infty} \left[\frac{(2m+n-a_0)!}{m!(m+n-a_0)!} \left(\frac{\lambda}{\lambda+\mu} \right)^{m+n-a_0} \left(\frac{\mu}{\lambda+\mu} \right)^m \right] \left[\frac{[(\lambda+\mu)t]^{2m+n-a_0}}{(2m+n-a_0)!} e^{-(\lambda+\mu)t} \right] \tag{11.39}$$

This is the solution of the immigration-emigration model from an intuitive approach. Its formal derivation through the use of difference equations and generating functions follows.

11.4.3 Derivation of the difference–differential equation

In this section the difference equation reflecting the combined immigration–emigration process is developed. The reasoning will be analogous to that of the immigration process. We are concerned about the dynamics of this population as time moves from t to t + Δt. Let Δt become so small that one and only one event can occur. In this very short period of time, the number of patients with the diesease or cases can increase by one, an event that occurs with probability $\lambda\Delta t$. Also, the number of diseased patients could decrease by one, an event that occurs with probability $\mu\Delta t$. Finally, the size of the diseased population could remain constant from time t to time t + Δt, an event that would occur with probability 1 - $\lambda\Delta t$ - $\mu\Delta t$. The Chapman–Kolmogorov forward equation for n $\geq$ 0 may now be written as

$$P_n(t+\Delta t) = \lambda\Delta t P_{n-1}(t) + \mu\Delta t P_{n+1} + P_n(t)(1-\lambda\Delta t-\mu\Delta t) \tag{11.40}$$

Combining terms and taking the limit as Δt approaches 0 as in the previous two sections

$$\frac{dP_n(t)}{dt} = \lambda P_{n-1}(t) + \mu P_{n+1}(t) - \lambda P_n(t) - \mu P_n(t) \tag{11.41}$$

for n = 0 to ∞.

11.4.4 Application of the generating function argument

Define

$$G_t(s) = \sum_{n=0}^{\infty} s^n P_n(t) \tag{11.42}$$

Proceed with the generating function conversion and consolidation.

$$
\begin{aligned}
s^n \frac{dP_n(t)}{dt} &= \lambda s^n P_{n-1}(t) + \mu s^n P_{n+1}(t) - \lambda s^n P_n(t) - \mu s^n P_n(t) \\
\sum_{n=0}^{\infty} s^n \frac{dP_n(t)}{dt} &= \sum_{n=0}^{\infty} \lambda s^n P_{n-1}(t) + \sum_{n=0}^{\infty} \mu s^n P_{n+1}(t) - \sum_{n=0}^{\infty} \lambda s^n P_n(t) - \sum_{n=0}^{\infty} \mu s^n P_n(t) \\
\frac{dG_t(s)}{dt} &= \lambda s G_t(s) + \mu s^{-1}\left[G_t(s) - P_0(t)\right] - \lambda G_t(s) - \mu G_t(s)
\end{aligned}
\tag{11.43}
$$

Further development shows

$$
\frac{dG_t(s)}{dt} = \lambda G_t(s)[s-1] + \mu G_t(s)\left[s^{-1}-1\right] \tag{11.44}
$$

after making the simplifying assumption that $P_0(t) = 0$. Equation (11.44) is a first-order differential equation. Fortunately, this differential equation is easily solved since the terms involving $G_t(s)$ and the terms involving t can be separated. We can directly solve for $G_t(s)$ by the following development:

$$
\begin{aligned}
\frac{dG_t(s)}{dt} &= \lambda G_t(s)[s-1] + \mu G_t(s)\left[s^{-1}-1\right] \\
\frac{dG_t(s)}{G_t(s)} &= \left(\lambda[s-1] + \mu\left[s^{-1}-1\right]\right)dt \\
\int \frac{dG_t(s)}{G_t(s)} &= \int\left(\lambda[s-1] + \mu\left[s^{-1}-1\right]\right)dt \\
\ln G_t(s) &= \lambda t[s-1] + \mu t\left[s^{-1}-1\right] + C
\end{aligned}
\tag{11.45}
$$

This is the general solution to the differential equation (11.43). To make further progress, we use the initial condition that $G_0(s) = s^{a_0}$, coming from the fact $P_{a_0}(0) = 1$, and observe that $C = \ln s^{a_0}$. Incorporating this result into the last equation in (11.45), observe that

$$
\begin{aligned}
\ln G_t(s) &= \lambda t[s-1] + \mu t\left[s^{-1}-1\right] + \ln s^{a_0} \\
G_t(s) &= s^{a_0} e^{\lambda t[s-1]} e^{\mu t\left[s^{-1}-1\right]}
\end{aligned}
\tag{11.46}
$$

This may be recognized as a convolution of the immigration and emigration process. However, the s^{-1} term requires us to examine this explicitly rather than to simply invoke the convolution rule of Chapter 2. Let

$$a_n = \frac{(\lambda t)^n}{n!} e^{-\lambda t} : b_n = \frac{(\mu t)^n}{n!} e^{-ut} \tag{11.47}$$

Write

$$e^{\lambda t[s-1]} e^{\mu t[s^{-1}-1]} = (a_0 + a_1 s + a_2 s^2 + a_3 s^3 + \ldots)(b_0 + b_1 s^{-1} + b_2 s^{-2} + b_3 s^{-3} + \ldots) \tag{11.48}$$

and observe that

$$s^{a_0} e^{\lambda t[s-1]} e^{\mu t[s^{-1}-1]} \triangleright \left\{ \sum_{n=0}^{k} a_n b_{a_0 - k + n} \right\} \tag{11.49}$$

This is similar to a single–server queuing model with Poisson input and with an exponential service time distribution. An interesting rearrangement of the terms in expression (11.49) is left to the problems.

11.5 Birth Model

11.5.1 Introduction to the process of births

The immigration, emigration, and combined immigration-emigration models served as useful tools for the introduction to the use of difference-differential equations in modeling the movement of patients with disease through a population. However, helpful as they have been, these models have not included an important naturally observed phenomenon of disease dynamics–that is the spread of the disease from one infected individual to another susceptible person. This section will examine this important concept. Unfortunately, the inclusion of this component adds a new complication to the difference-differential equation required, namely, the equation becomes a partial difference-differential equation. However, the solution to these systems of equations will be straightforward, and will provide all of the structure for the solution to guide the reader whose background is relatively weak in the solution of differential equations. However, our focus continues to be on the use of difference equations to derive a solution to this problem in population dynamics.

11.5.2 Motivation for the birth process

Figure 11.4 demonstrates the fundamental dynamics of the birth process. In this figure, new cases (i.e., patients with disease) are produced from established cases. Even without an immigration process, the disease spreads on its own. However, its ability to spread depends on the number of patients with the disease. The larger the number of cases, the greater the spread of the disease will be. A small number of cases will diminish the rate at which the disease will spread throughout population.

Let the probability that a patient will contract the disease from time t to time $t + \Delta t$ be $\nu\Delta t$. Then if n patients are at risk of developing the disease, the probability that k of them actually contract the disease follows a binomial probability distribution, i.e.:

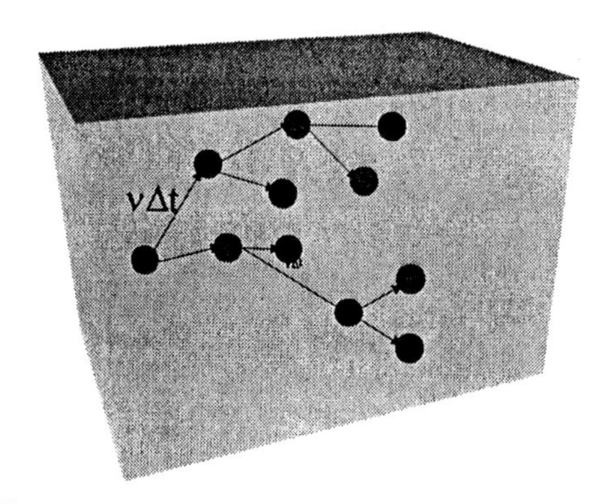

Figure 11.4 Birth Process: Patients embedded in the population with the disease spread the disease to others in the population.

$$P[\text{k of n patients contract the disease in } \Delta t] = \binom{n}{k} (\nu\Delta t)^k (1-\nu\Delta t)^{n-k} \qquad (11.50)$$

These are Bernoulli trials where the probability of spreading the disease in time Δt is $\nu\Delta t$ Essentially, each case has a probability of spreading the disease. For Δt very small, the probability that one of n patients spreads the disease in time Δt is approximately $n\nu\Delta t$.* This rate is a function of the number of patients with the disease. This relationship between the number of patients with the disease

* Other terms in this expression have in them expression $(\Delta t)^k$ which are neglible in this derivation

and the birth rate distinquishes the birth process from the immigration process, in which the arrival rate is independent of the number of patients with the disease.

11.5.3 Chapman–Kolmogorov forward equations

To construct these equations, we will need to determine in what way the number of cases can change from time t to time $t + \Delta t$. Assume that there are a_0 diseased patients in the system at time $t = 0$. Begin by considering how many different ways there can be n patients with disease at time $t + \Delta t$. Also, as before, assume Δt is so small that the number of cases can only change by one during this short time span. In these circumstances, there can either be a new case, an event that occurs with probability $(n - 1)\nu\Delta t$, or there is no change, an event that occurs with probability $1 - n\nu\Delta t$. The Chapman–Kolmogorov forward equations for the birth process are

$$P_n(t+\Delta t) = P_{n-1}(t)(n-1)\nu\Delta t + P_n(t)(1-n\nu\Delta t) \tag{11.51}$$

This equation is true for $n \geq a_0$, where a_0 is the number of patients in the population at time $t = 0$. We assume that $P_n(t)$ is zero for negative $n < a_0$. We proceed with converting this family of difference equations into a family of difference differential equations in n by collecting the terms that involve Δt and taking the limit as Δt approaches 0.

$$\begin{aligned} P_n(t+\Delta t) &= (n-1)\nu\Delta t P_{n-1}(t) + (1-n\nu\Delta t)P_n(t) \\ P_n(t+\Delta t) - P_n(t) &= (n-1)\nu\Delta t P_{n-1}(t) - n\nu\Delta t P_n(t) \end{aligned} \tag{11.52}$$

Continuing this process reveals

$$\begin{aligned} \frac{P_n(t+\Delta t) - P_n(t)}{\Delta t} &= (n-1)\nu P_{n-1}(t) - n\nu P_n(t) \\ \lim_{\Delta t \to 0} \frac{P_n(t+\Delta t) - P_n(t)}{\Delta t} &= (n-1)\nu P_{n-1}(t) - n\nu P_n(t) \end{aligned} \tag{11.53}$$

Recognizing the last line of equation (11.53) as a derivative, write

$$\frac{dP_n(t)}{dt} = \nu P_{n-1}(t)(n-1) - \nu P_n(t)\,n \tag{11.54}$$

11.5.4 Application of the generating function

Next apply the generating function to the difference-differential equation family in equation (11.54). Define $G_t(s)$ as

$$G_t(s) = \sum_{n=a_0}^{\infty} s^n P_n(t) \tag{11.55}$$

and proceed by multiplying each side of equation (11.55) by the term s^n

$$\begin{aligned}
s^n \frac{dP_n(t)}{dt} &= \nu(n-1)s^n P_{n-1}(t) - \nu n s^n P_n(t) \\
\sum_{n=a_0}^{\infty} s^n \frac{dP_n(t)}{dt} &= \sum_{n=a_0}^{\infty} \nu(n-1)s^n P_{n-1}(t) - \sum_{n=a_0}^{\infty} \nu n s^n P_n(t) \\
\frac{dG_t(s)}{dt} &= \sum_{n=a_0}^{\infty} \nu(n-1)s^n P_{n-1}(t) - \sum_{n=a_0}^{\infty} \nu n s^n P_n(t)
\end{aligned} \tag{11.56}$$

Each of the terms on the right side of equation (11.56) will be evaluated separately, since they include not just a power of s (with which we are familiar) but also a term involving n. The first term on the right side of equation (11.56) is evaluated as

$$\begin{aligned}
\sum_{n=a_0}^{\infty} \nu(n-1)s^n P_{n-1}(t) &= \nu s^2 \sum_{n=a_0}^{\infty} (n-1)s^{n-2} P_{n-1}(t) = \nu s^2 \sum_{n=a_0}^{\infty} \frac{ds^{n-1}}{ds} P_{n-1}(t) \\
&= \nu s^2 \sum_{n=a_0}^{\infty} \frac{d\left[s^{n-1}P_{n-1}(t)\right]}{ds} = \nu s^2 \frac{d\sum_{n=a_0}^{\infty}\left[s^{n-1}P_{n-1}(t)\right]}{ds} \\
&= \nu s^2 \frac{d\sum_{n=a_0}^{\infty} s^n P_n(t)}{ds} = \nu s^2 \frac{dG_t(s)}{ds}
\end{aligned} \tag{11.57}$$

This term now includes a derivative of $G_t(s)$, this time with respect to s. Use an analogous procedure to evaluate the next term

$$\sum_{n=a_0}^{\infty} \nu n s^n P_n(t) = \nu \sum_{n=a_0}^{\infty} n s^n P_n(t) = \nu s \sum_{n=a_0}^{\infty} n s^{n-1} P_n(t) = \nu s \sum_{n=a_0}^{\infty} \frac{ds^n}{ds} P_n(t)$$
$$= \nu s \sum_{n=a_0}^{\infty} \frac{d\left[s^n P_n(t)\right]}{ds} = \nu s \frac{d \sum_{n=a_0}^{\infty} \left[s^n P_n(t)\right]}{ds} = \nu s \frac{dG_t(s)}{ds} \tag{11.58}$$

Rewrite the last expression of equation (11.56) as a single partial differential equation.

$$\frac{\partial G_t(s)}{\partial t} = \nu s^2 \frac{\partial G_t(s)}{\partial s} - \nu s \frac{\partial G_t(s)}{\partial s}$$
$$\frac{\partial G_t(s)}{\partial t} = \nu s(s-1) \frac{\partial G_t(s)}{\partial s} \tag{11.59}$$

11.5.5 General solutions to birth model–partial differential equations

The increased complexity of the birth process has complicated the solution for $P_n(t)$ by resulting in not just an ordinary differential equation (i.e., an equation in that the derivative is with respect to one variable), but a partial differential equation. Fortunately, partial differential equations such as equation (11.59) can be solved in a very straightforward manner. What is required is the solution of several equations, and the timely incorporation of a boundary condition. The general scheme is as follows: Given the partial differential equation

$$P \frac{\partial F(x,y)}{\partial x} + Q \frac{\partial F(x,y)}{\partial y} = R \tag{11.60}$$

where P, Q and R are meant to represents constants, i.e., they are not functions of F(x,y), then the general solution is found by considering the solution to the collection of auxiliary or subsidiary equations

$$\frac{dx}{P} = \frac{dy}{Q} = \frac{dF}{R} \tag{11.61}$$

These equations are used to find a general solution for F(x,y) [1]. The specific solution is obtained from the boundary conditions.

11.5.6 Specific solution to the partial differential equation for the birth process

Return now to the partial differential equation previously given in equation (11.59)

$$\frac{\partial G_t(s)}{\partial t} = vs(s-1)\frac{\partial G_t(s)}{\partial s} \tag{11.62}$$

and rewrite as

$$\frac{\partial G_t(s)}{\partial t} - vs(s-1)\frac{\partial G_t(s)}{\partial s} = 0 \tag{11.63}$$

This results in the auxiliary (or subsidiary) equations

$$\frac{dt}{1} = \frac{ds}{-vs(s-1)} = \frac{dG_t(s)}{0} \tag{11.64}$$

Note that the equations (11.64) are notational devices only, providing how the solution to the partial differential equation should proceed. The individual terms in these equations should not be viewed as quotients with permissible denominators of zeros. Equating the first and third expressions in (11.64)

$$\frac{dt}{1} = \frac{dG_t(s)}{0} \tag{11.65}$$

resulting in $dG_t(s) = 0$ or $G_t(s)$ is a constant that we will write as

$$G_t(s) = C_1 = \Phi(C_2) \tag{11.66}$$

Now evaluating the first and second terms of equation (11.64)

$$\frac{dt}{1} = \frac{ds}{-\nu s(s-1)}$$

$$-\nu dt = \frac{ds}{s(s-1)}$$

$$\int -\nu dt = \int \frac{ds}{s(s-1)} \tag{11.67}$$

$$-\nu t = \int \frac{ds}{s(s-1)}$$

The term on the right side of equation (11.67) is

$$\int \frac{ds}{s(s-1)} = \ln(s-1) - \ln s + C$$
$$= \ln\left(\frac{s-1}{s}\right) + C \tag{11.68}$$

Combining equations (11.66) and the last line of expression (11.68), observe

$$-\nu t = \ln\left(\frac{s-1}{s}\right) + C$$

$$C = \nu t + \ln\left(\frac{s-1}{s}\right) = -\nu t + \ln\left(\frac{s}{s-1}\right) \tag{11.69}$$

$$C_2 = \frac{s}{s-1} e^{-\nu t}$$

Combining equation (11.65) and equation (11.69) note that

$$G_t(s) = \Phi(C_2) = \Phi\left[\frac{s}{s-1} e^{-\nu t}\right] \tag{11.70}$$

This is the general solution to the partial differential equation expressed in (11.63). For its specific solution, we must incorporate the boundary conditions for this process. Recall that $P_0(a_0) = 1$, that is, there are a_0 cases at time $t = 0$. Using this information in equation (11.70)

$$s^{a_0} = G_0(s) = \Phi\left[\frac{s}{s-1}\right] \tag{11.71}$$

Thus

$$s^{a_0} = \Phi\left[\frac{s}{s-1}\right] \tag{11.72}$$

The task is now to solve for the function Φ. Let $z = \frac{s}{s-1}$ then $s = \frac{z}{z-1}$. So

$\Phi(z) = \left(\frac{z}{z-1}\right)^{a_0}$ and we can re-examine equation (11.70) and find

$$G_t(s) = \Phi\left[e^{-vt}\frac{s}{s-1}\right] = \left[\frac{e^{-vt}\frac{s}{s-1}}{e^{-vt}\frac{s}{s-1} - 1}\right]^{a_0} = \left[\frac{se^{-vt}}{se^{-vt} - s + 1}\right]^{a_0} = \left[\frac{se^{-vt}}{1 - s\left(1 - e^{-vt}\right)}\right]^{a_0}$$

(11.73)

To invert, return to Chapter 2 to see that this is the generating function of the negative binomial distribution.

$$P_n(t) = \binom{n-1}{a_0 - 1} e^{-a_0 vt}\left(1 - e^{-vt}\right)^{n-a_0} \tag{11.74}$$

This distribution can be used to obtain the expected number of cases of disease at time t as $a_0 e^{vt}$.

11.6 The Death Process

11.6.1 Introduction

Just as the birth model led to the propagation of new patients with disease in a way where the force of growth of the disease cases was related to the number of cases available, it stands to reason that there are circumstances where the number of cases will decline, the force of decline being related to the number of cases. This is called the death process. The plan here is to identify the probability distribution of the number of cases at time t when the number of cases cannot increase, but only decrease due to death. As with the birth model, we will need to develop and solve a family of partial differential equations. As with the birth model, however, the solution to these equations will be straightforward.

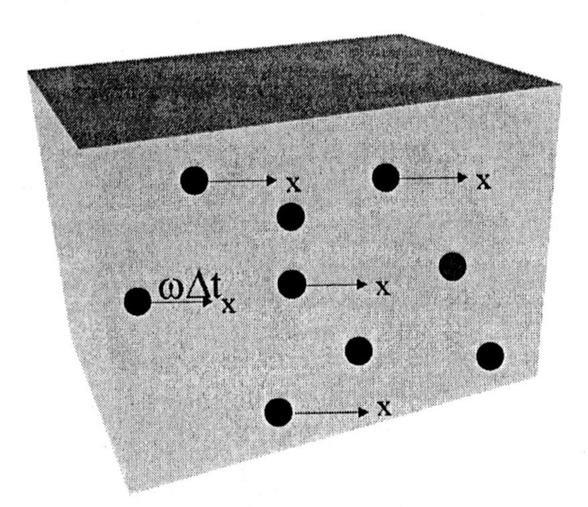

Figure 11.5 Death Model: Deaths occur only among the initial patients in the population (x signifies death).

11.6.2 Motivation for the death process

Figure 11.5 demonstrates the fundamental dynamics of the death process. This figure demonstrates that the unique feature of the death process is the removal of disease. This extinquishment could mean the patients with disease die (the death of the patient) or a complete cure (the death of the disease with no possibility of re-infection). As with the birth process, the force of death is related to the number of patients with disease. The larger the number of cases, the greater the likelihood of death. This relationship between the number of patients with the disease and the death rate is what distinquishes the death process from the emigration process, in which the departure rate is independent of the number of patients with the disease. This distinction is contrary to conventional wisdom in which the dynamics of the death and emigration processes are the same.

We begin here as we did in the birth process. Let the probability that a patient with disease is alive at time t will die in time t + Δt be $\omega\Delta t$. If n patients with the disease are at risk of dying then, the probability that k of them actually die from the disease in Δt follows a binomial probability distribution, i.e.,

$$P[\text{k of n patients die}] = \binom{n}{k}(\omega\Delta t)^k (1-\omega\Delta t)^{n-k} \tag{11.75}$$

These are a sequence of Bernoulli trials where the probability of dying from the disease is $\omega\Delta t$. As before, assume that Δt is so small that the probability of a death of one patient given n patients have the disease is $n\omega\Delta t$.

11.6.3 Chapman–Kolmogorov forward equations

To construct these equations, we will need to determine in what way the number of cases can change from time t to time t + Δt if there are to be n cases at time t + Δt. Also, assume Δt is so small that the number of cases can only change by one. In this short span of time only one of two events can occur. There can be a death with probability $(n+1)\omega\Delta t$ or there is no change, that occurs with probability $1 - n\omega\Delta t$. We can now write the Chapman–Kolmogorov forward equation for the death process.

$$P_n(t+\Delta t) = (n+1)\omega\Delta t P_{n+1}(t) + P_n(t)(1-n\omega\Delta t) \tag{11.76}$$

The above equation is true for $n = 0$ to ∞. Proceed with converting this family of difference equations in n to a family of difference differential equations in n.

$$\begin{aligned}
P_n(t+\Delta t) &= (n+1)\omega\Delta t P_{n+1}(t) + P_n(t)(1-n\omega\Delta t) \\
P_n(t+\Delta t) - P_n(t) &= (n+1)\omega\Delta t P_{n+1}(t) - n\omega\Delta t P_n(t) \\
\frac{P_n(t+\Delta t) - P_n(t)}{\Delta t} &= (n+1)\omega P_{n+1}(t) - n\omega P_n(t) \\
\lim_{\Delta t\to 0}\frac{P_n(t+\Delta t) - P_n(t)}{\Delta t} &= (n+1)\omega P_{n+1}(t) - n\omega P_n(t)
\end{aligned} \tag{11.77}$$

As in our treatment of the birth process in the previous section, the last relationship in expression (11.77) may be rewritten as

$$\frac{dP_n(t)}{dt} = \omega P_{n+1}(t)(n+1) - \omega P_n(t)\, n \tag{11.78}$$

11.6.4 Application of the generating function

We can now apply the generating function argument to the difference-differential equation family in equation (11.53). Let a_0 be the number of cases at time $t = 0$. Define $G_t(s)$ as

$$G_t(s) = \sum_{n=0}^{a_0} s^n P_n(t) \tag{11.79}$$

and proceed as before.

$$\begin{aligned}
s^n \frac{dP_n(t)}{dt} &= \omega(n+1)s^n P_{n+1}(t) - \omega n s^n P_n(t) \\
\sum_{n=0}^{a_0} s^n \frac{dP_n(t)}{dt} &= \sum_{n=0}^{a_0}\omega(n+1)s^n P_{n+1}(t) - \sum_{n=0}^{a_0}\omega n s^n P_n(t) \\
\frac{dG_t(s)}{dt} &= \sum_{n=0}^{a_0}\omega(n+1)s^n P_{n+1}(t) - \sum_{n=0}^{a_0}\omega n s^n P_n(t)
\end{aligned} \tag{11.80}$$

Each of the terms on the right side of equation (11.80) will be evaluated separately, since they have not just a power of s but a factor involving n. The process here is analogous to the procedure we followed for the evaluation of the birth process. The first term on the right side of equation (11.80) is evaluated as

$$\sum_{n=0}^{a_0}\omega(n+1)s^nP_{n+1}(t)=\omega\sum_{n=0}^{a_0}(n+1)s^nP_{n+1}(t)=\omega\sum_{n=0}^{a_0}\frac{ds^{n+1}}{ds}P_{n+1}(t)$$

$$=\omega\sum_{n=0}^{\infty}\frac{ds^{n+1}P_{n+1}(t)}{ds}=\omega\frac{d\sum_{n=0}^{a_0}s^{n+1}P_{n+1}(t)}{ds}=\omega\frac{d\left[\sum_{n=0}^{a_0}s^nP_n(t)-P_0(t)\right]}{ds} \quad (11.81)$$

$$=\omega\frac{dG_t(s)}{ds}$$

This computation is correct since (1) $P_n(t)$ is zero for $n \geq a_0$ and (2) the derivative of $P_0(t)$ with respect to s is equal to zero. Following our observation from the birth process, equation (11.81) now includes a derivative of $G_t(s)$ with respect to s. Using an analogous procedure to evaluate the next term

$$\sum_{n=0}^{a_0}\omega ns^nP_n(t)=\omega\sum_{n=0}^{a_0}ns^nP_n(t)=\omega s\sum_{n=0}^{a_0}ns^{n-1}P_n(t)=\omega s\sum_{n=0}^{a_0}\frac{ds^n}{ds}P_n(t)$$

$$=\omega s\sum_{n=0}^{a_0}\frac{ds^nP_n(t)}{ds}=\omega s\frac{d\sum_{n=0}^{a_0}s^nP_n(t)}{ds}=\omega s\frac{dG_t(s)}{ds} \quad (11.82)$$

Rewrite the last line of equation (11.80) as a single partial differential equation as follows:

$$\frac{\partial G_t(s)}{\partial t} = \omega\frac{\partial G_t(s)}{\partial s} - \omega s\frac{\partial G_t(s)}{\partial s}$$

$$\frac{\partial G_t(s)}{\partial t} = \omega(1-s)\frac{\partial G_t(s)}{\partial s} \quad (11.83)$$

11.6.5 General solutions to the death process partial differential equation

Rewrite equation (11.83) as

$$\frac{\partial G_t(s)}{\partial t} - \omega(1-s)\frac{\partial G_t(s)}{\partial s} = 0 \quad (11.84)$$

Being guided by the approach to the solution of partial differential equation which arose from the birth process, note that the subsidiary equations for the death process are

$$\frac{dt}{1} = \frac{ds}{-\omega(1-s)} = \frac{dG_t(s)}{0} \tag{11.85}$$

Examining the first and third terms in equation (11.85)

$$\frac{dt}{1} = \frac{dG_t(s)}{0} \tag{11.86}$$

resulting in $dG_t(s)=0$ or $G_t(s)$ is a constant C_1. This was the case with the birth process as well. Writing this equation as

$$G_t(s) = C_1 = \Phi(C_2) \tag{11.87}$$

will evaluate the first and second terms of equation (11.85)

$$\begin{aligned} \frac{dt}{1} &= \frac{ds}{-\omega(1-s)} \\ -\omega dt &= \frac{ds}{1-s} \\ \int -\omega dt &= \int \frac{ds}{1-s} \\ -\omega t &= \int \frac{ds}{1-s} \end{aligned} \tag{11.88}$$

The term on the right side of equation (11.88) is

$$\int \frac{ds}{1-s} = -\ln(1-s) + C \tag{11.89}$$

Combining equations (11.88) and (11.89) find

$$
\begin{aligned}
-\omega t &= -\ln(1-s)+C \\
C &= \ln(1-s)-\omega t \\
C_2 &= (1-s)e^{-\omega t}
\end{aligned}
\tag{11.90}
$$

Combining equation (11.87) and the last relationship in expression (11.90), observe

$$
G_t(s) = C_1 = \Phi(C_2) = \Phi\left[(1-s)e^{-\omega t}\right] \tag{11.91}
$$

We are now ready to incorporate the boundary conditions to find the specific solution of the partial differential equation for the generating function of the death process. When $t = 0$, there are a_0 cases. Using this information in equation (11.91) we have

$$
s^{a_0} = \Phi[1-s] \tag{11.92}
$$

and letting $z = 1-s$, observe that $\Phi(z) = (1-z)^{a_0}$. Apply this to equation (11.91):

$$
\begin{aligned}
&\Phi\left[e^{-\omega t}(1-s)\right] = \left[1-e^{-\omega t}(1-s)\right]^{a_0} \\
&= \left[1-e^{-\omega t}+se^{-\omega t}\right]^{a_0} = \left[\left(1-e^{-\omega t}\right)+se^{-\omega t}\right]^{a_0}
\end{aligned}
\tag{11.93}
$$

and we have the generating function. To invert, we return to Chapter 2 to see that this is the generating function of the binomial distribution. Thus,

$$P_n(t) = \binom{a_0}{n} e^{-n\omega t}\left(1 - e^{-\omega t}\right)^{a_0 - n} \tag{11.94}$$

for $0 < n \leq a_0$. Note that the average number of patients with disease in the system at time t is $E[X_t]$

$$E[X_t] = a_0 e^{-n\omega t} \tag{11.95}$$

and the varance of X_t is

$$Var[X_t] = a_0 e^{-n\omega t}\left(1 - e^{-\omega t}\right) \tag{11.96}$$

Problems

1. By modifying the Chapman–Kolmogorov forward equations for the immigration process, find the distribution of the number of patients with disease given that there are three sources of immigrants, one source arrives with arrival rate λ_1, the second source submits arrivals with arrival rate λ_2, and the third arrives with arrival rate λ_3.
2. Obtain equation (11.16) for the case $a_0 = 10$, $n = 15$, and $\lambda = 2$.
3. Obtain $P_n(t)$ using equation (11.18) for $a_0 = 10$ and $\lambda = 2$.
4. Using the relationship in equation (11.29), obtain $P_{12}(t)$ for $a_0 = 10$, and μ – 2.
5. Write the difference equation for $P_n(t)$ for the immigration-emigration model when $\lambda = \mu$.
6. Solve the difference equation derived from problem 5 above for $P_n(t)$.
7. Compare $P_n(t)$ derived from problem 6 above with that in equation (11.34) for $\lambda = \mu$.
8. Simplify equation (11.47) for the case when $a_0 = \lambda = \mu = 1$.
9. Solve equation (11.56) for the case of $v = 1.5$.
10. Derive equation (11.49) for the case when there are three types of births with rates v_1, v_2, and v_3, respectively.
11. Using equation (11.88), obtain the expected number in the system at time t for the death model.
12. Write down the Chapman–Kolmogorov forward equation for the death model when there are two causes of death with rates ω_1 and ω_2, respectively.

13. Show that from the immigration–emigration process:

$$\sum_{m=n-a_0}^{\infty} \frac{(\lambda t)^m}{m!} e^{-\lambda t} \frac{(\mu t)^{m-(n-a_0)}}{(m-(n-a_0))!} e^{-\mu t}$$

$$= \sum_{m=n-a_0}^{\infty} \frac{2m-(n-a_0)!}{m!(m-(n-a_0))!} \left(\frac{\lambda t}{\lambda t+\mu t}\right)^m \left(\frac{\mu t}{\lambda t+\mu t}\right)^{m-(n-a_0)} \frac{\left[(\lambda+\mu)t\right]^{2m-(n-a_0)}}{\left[2m-(n-a_0)\right]!} e^{-(\lambda+\mu)t}$$

$$= \sum_{m=n-a_0}^{\infty} \left[\binom{2m-(n-a_0)}{m} \left(\frac{\lambda}{\lambda+\mu}\right)^m \left(\frac{\mu}{\lambda+\mu}\right)^{m-(n-a_0)}\right] \left[\frac{\left[(\lambda+\mu)t\right]^{2m-(n-a_0)}}{\left[2m-(n-a_0)\right]!} e^{-(\lambda+\mu)t}\right]$$

(11.97)

14. Show that if

$$G_t(s) = \left[\frac{se^{-vt}}{1-s\left(1-e^{-vt}\right)}\right]^{a_0} \quad (11.98)$$

then

$$P_n(t) = \binom{n-1}{a_0-1} e^{-a_0 vt} \left(1-e^{-vt}\right)^{n-a_0} \quad (11.99)$$

References

1. Chiang C.L. An Introduction to Stochastic Processes and Their Applications. Huntington, New York. Robert E. Krieger Publishing Company. 1980.

12

Difference Equations in Epidemiology: Advanced Models

12.1 Introduction

In the previous chapter, the concept of modeling the spread of a disease process through a population was presented using difference equations (as the basis of the difference–differential equation), a concept that has been useful for epidemiologists and practitioners in identifying risk factors of diseases. Each of the four models reviewed in Chapter 11 isolated one component of the spread of the disease and provided a focus only on that component. The immigration process focused on the arrival of cases, i.e., the influx of patients with disease. The birth process and the death process concentrated on the spread of the disease and the removal of cases either by death or cure when the force of this movement was proportional to the number of patients. The emigration process allowed for cases to move out of the population. Both the immigration and emigration processes were passive, with arrivals or departures occurring independently of the total number of cases. Using these models as building blocks, this chapter will construct more complicated models, assembling realistic paradigms of the spread of diseases through populations.

12.2 Building from Component Parts

In Chapter 11 the distribution of the number of cases based on one of four component parts, immigration, birth, death, or emigration, was developed. In reality, these four forces act in combination. New cases arrive, then spread the disease to susceptible people, who spread the disease even further. Some of these cases die or are cured and immune from future attacks. Others with the disease remove themselves. This chapter focuses on building models from combinations of the four component processes studied thus far. In Chapter 12, the immigration, emigration, birth, and death processes are used as building blocks, from which we will construct epidemiologic models to more accurately depict the dynamics of disease.

12.3 The Immigration–Birth Model

12.3.1 Motivation for the contagion process

Consider a population that is disease free. Over time, patients with a disease arrive–they immigrate. The disease spreads among people who are susceptible. The number of cases increases slowly at first, then accelerates as the disease spreads. This is the classic model of the spread of an infectious disease, e.g., poliomyelitis, with entry of new patients continuing passively, followed by the active spread of the disease (Figure 12.1). The potentially rapid spread of the disease is the hallmark of the immigration-birth process.

The derivation of $P_n(t)$, the probability that there are n patients in the system at time t, will follow the development implemented in the previous chapter. In the contagion process, the simultaneous consideration of two rates is required. The first is the immigration or the arrival rate. We will characterize this rate as λ arrivals per unit time. The birth rate will denoted by ν.

12.3.2 Chapman–Kolmogorov equation for the immigration-birth process

The development of the Chapman–Kolmogorov equations will parallel the derivation provided in the previous chapter. We will allow Δt to be so small that the only change from time t to time $t + \Delta t$ will involve a single individual. In this framework, if there are to be n cases at time $t + \Delta t$, there are three possible events that may have occurred by time t. The first possibility is that there were $n - 1$ patients in the system at time t, followed by a new case arriving from outside the population in the interval Δt, an event that occurs with probability

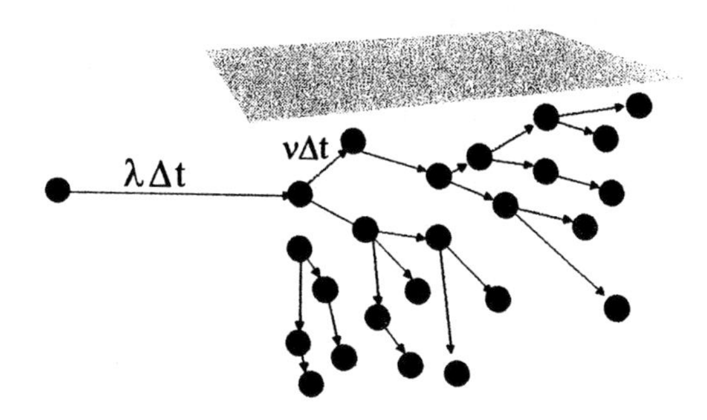

Figure 12.1 Contagion: Immigration-birth model. Patients with disease enter the system at a constant rate and spread the disease to others in the population.

$\lambda\Delta t$. A second event is that there may have been n – 1 cases at time t, followed by a new disease "birth", with probability $(n-1)\nu\Delta t$. Finally, there may have been n cases at time t, with neither a new birth nor immigration ocurring during the time interval, t, t + Δt. This event occurs with probability $1 - \lambda\Delta t - n\nu\Delta t$. We may now write the Chapman–Kolmogorov forward equation for the contagion process:

$$P_n(t+\Delta t) = \lambda\Delta t P_{n-1}(t) + (n-1)\nu\Delta t P_{n-1}(t) + (1-\lambda\Delta t - n\nu\Delta t)P_n(t) \qquad (12.1)$$

These equations are true for n = 0 to infinity if $P_n(t) = 0$ for $n < 0$.

12.3.3 Development of the partial difference–differential equation family

Proceeding with equation (12.1), begin its conversion into a difference–differential equation in t.

$$P_n(t+\Delta t) = \lambda\Delta t P_{n-1}(t) + (n-1)\nu\Delta t P_{n-1}(t) + (1-\lambda\Delta t - n\nu\Delta t)P_n(t)$$
$$P_n(t+\Delta t) - P_n(t) = \lambda\Delta t P_{n-1}(t) + (n-1)\nu\Delta t P_{n-1}(t) - \lambda\Delta t P_n(t) - n\nu\Delta t P_n(t)$$
$$\frac{P_n(t+\Delta t) - P_n(t)}{\Delta t} = \lambda P_{n-1}(t) + (n-1)\nu P_{n-1}(t) - \lambda P_n(t) - n\nu P_n(t)$$

(12.2)

This evaluation will follow the process developed in Chapter 11. First, take a limit

$$\lim_{\Delta t \to 0} \frac{P_n(t+\Delta t) - P_n(t)}{\Delta t} = \lambda P_{n-1}(t) + (n-1)\nu P_{n-1}(t) - \lambda P_n(t) - n\nu P_n(t) \quad (12.3)$$

and then recognize that $\lim_{\Delta t \to 0} \frac{P_n(t+\Delta t) - P_n(t)}{\Delta t} = \frac{dP_n(t)}{dt}$. Thus, equation (12.3) can be written as

$$\frac{dP_n(t)}{dt} = \lambda P_{n-1}(t) - \lambda P_n(t) + (n-1)\nu P_{n-1}(t) - n\nu P_n(t) \quad (12.4)$$

Already, some portions of the above equation are becoming recognizable. The first two terms on the right side of equation (12.4) are similar to the difference-differential family of equations from the immigration process. The last two terms are from the "pure" birth process of Chapter 11. To proceed, define $G_t(s)$ as in Chapter 11.

$$G_t(s) = \sum_{n=0}^{\infty} s^n P_n(t) \quad (12.5)$$

Follow through with the conversion and consolidation of equation (12.4):

$$s^n \frac{dP_n(t)}{dt} = \lambda s^n P_{n-1}(t) - \lambda s^n P_n(t) + \nu(n-1)s^n P_{n-1}(t) - \nu n s^n P_n(t)$$

$$\sum_{n=0}^{\infty} s^n \frac{dP_n(t)}{dt} = \lambda \sum_{n=0}^{\infty} s^n P_{n-1}(t) - \lambda \sum_{n=0}^{\infty} s^n P_n(t) + \nu \sum_{n=0}^{\infty} (n-1)s^n P_{n-1}(t) - \nu \sum_{n=0}^{\infty} n s^n P_n(t) \quad (12.6)$$

There are no terms in the last line of equation (12.6) which are unfamiliar. Each of these expressions may be recognized from either the immigration process or the pure birth process introduced in Chapter 11, and may be now written as

$$\sum_{n=0}^{\infty} s^n \frac{dP_n(t)}{dt} = \lambda \sum_{n=0}^{\infty} s^n P_{n-1}(t) - \lambda \sum_{n=0}^{\infty} s^n P_n(t) + \nu \sum_{n=0}^{\infty} (n-1)s^n P_{n-1}(t) - \nu \sum_{n=0}^{\infty} n s^n P_n(t) \quad (12.7)$$

That is,

$$\frac{\partial G_t(s)}{\partial t} = \lambda s G_t(s) - \lambda G_t(s) + \nu s^2 \frac{\partial G_t(s)}{\partial s} - \nu s \frac{\partial G_t(s)}{\partial s} \tag{12.8}$$

that may be rewritten as

$$\frac{\partial G_t(s)}{\partial t} = \lambda(s-1)G_t(s) + \nu s(s-1)\frac{\partial G_t(s)}{\partial s} \tag{12.9}$$

The partial differential equation as defined in equation (12.9) must now be solved to continue the exploration of the contagion process.

12.3.4 Recall partial differential equation solution

Recall from Chapter 11 that for the partial differential equation

$$P\frac{\partial F(x,y)}{\partial x} + Q\frac{\partial F(x,y)}{\partial y} = R \tag{12.10}$$

where P, Q, and R are constants. The general solution is found by considering the solution to the collection of auxiliary or subsidiary equations.

$$\frac{dx}{P} = \frac{dy}{Q} = \frac{dF}{R} \tag{12.11}$$

These equations are utilized to find a general solution for F(x,y). The specific solution is obtained from the boundary conditions. This is exactly the situation in equation (12.9) which may be rewritten as

$$\frac{\partial G_t(s)}{\partial t} - \nu s(s-1)\frac{\partial G_t(s)}{\partial s} = \lambda(s-1)G_t(s) \tag{12.12}$$

leading to the auxiliary equations

$$\frac{dt}{1} = \frac{ds}{-\nu s(s-1)} = \frac{dG_t(s)}{\lambda(s-1)G_t(s)} \tag{12.13}$$

We now do a sequence of two comparisons to identify the general form of $G_t(s)$. Compare the second and third terms of equation (12.13).

$$\frac{ds}{-\nu s(s-1)} = \frac{dG_t(s)}{\lambda(s-1)G_t(s)}$$
$$-\frac{\lambda}{\nu s}ds = \frac{dG_t(s)}{G_t(s)} \tag{12.14}$$
$$\int -\frac{\lambda}{\nu s}ds = \int \frac{dG_t(s)}{G_t(s)}$$
$$\ln G_t(s) = -\frac{\lambda}{\nu}\ln s + C_1$$

Define $e^{C_1} = \Phi(C_2)$, and proceed with this development to show

$$G_t(s) = s^{\frac{-\lambda}{\nu}}\Phi(C_2) \tag{12.15}$$

where $\Phi(C_2)$ is the constant of integration. Return to equation (12.13) and focus on the first two terms $\frac{dt}{1} = \frac{ds}{-\nu s(s-1)}$. The development of the constant term C_2 proceeds as follows:

$$\frac{dt}{1} = \frac{ds}{-\nu s(s-1)}$$
$$-\nu dt = \frac{ds}{s(s-1)} \tag{12.16}$$
$$\int -\nu dt = \int \frac{ds}{s(s-1)}$$

Integration of the left side reveals

$$-\nu t = \int \frac{ds}{s(s-1)} \tag{12.17}$$

The term on the right side of equation (12.17) is

$$\begin{aligned}\int \frac{ds}{s(s-1)} &= \ln(s-1) - \ln s + C \\ &= \ln\left(\frac{s-1}{s}\right) + C\end{aligned} \tag{12.18}$$

Combining equations (12.17) and (12.18)

$$\begin{aligned}-vt &= \ln\left(\frac{s-1}{s}\right) + C \\ C &= vt + \ln\left(\frac{s-1}{s}\right) = -vt + \ln\left(\frac{s}{s-1}\right) \\ C_2 &= \frac{s}{s-1} e^{-vt}\end{aligned} \tag{12.19}$$

This segment of the solution of the partial differential equation concludes with $C_2 = \frac{s}{s-1} e^{-vt}$. Combining this term for C_2 with the expression from $G_t(s)$ in equation (12.15), $G_t(s)$ can be expressed as

$$G_t(s) = s^{-\frac{\lambda}{v}} \Phi(C_2) = s^{-\frac{\lambda}{v}} \Phi\left(\frac{s}{s-1} e^{-vt}\right) \tag{12.20}$$

Use the boundary condition to tailor this solution to the specific circumstance of the contagion process by assuming that there are a_0 cases at time $t = 0$. Then

$$G_0(s) = s^{a_0} = s^{-\frac{\lambda}{v}} \Phi\left(\frac{s}{s-1} e^{-v0}\right) = s^{-\frac{\lambda}{v}} \Phi\left(\frac{s}{s-1}\right) \tag{12.21}$$

Proceed by letting $z = \frac{s}{s-1}$ or $s = \frac{z}{z-1}$. Applying these results to equation (12.21) demonstrates

$$s^{a_0} = \left(\frac{z}{z-1}\right)^{a_0} = \left(\frac{z}{z-1}\right)^{-\frac{\lambda}{v}} \Phi(z)$$
$$\Phi(z) = \left(\frac{z}{z-1}\right)^{a_0+\frac{\lambda}{v}} \tag{12.22}$$

and

$$G_t(s) = s^{-\frac{\lambda}{v}} \Phi(C_2) = s^{-\frac{\lambda}{v}} \Phi\left(\frac{s}{s-1} e^{-vt}\right) \tag{12.23}$$

With the function Φ defined, write $G_t(s)$ as

$$G_t(s) = s^{-\frac{\lambda}{v}} \left(\frac{\frac{s}{s-1} e^{-vt}}{\frac{s}{s-1} e^{-vt} - 1}\right)^{a_0+\frac{\lambda}{v}} = s^{-\frac{\lambda}{v}} \left(\frac{se^{-vt}}{1-\left(1-e^{-vt}\right)s}\right)^{a_0+\frac{\lambda}{v}} \tag{12.24}$$

Assume that $\lambda, v, \text{and} \frac{\lambda}{v}$ are each integers greater than zero. The last term of equation (12.24) is related to the probability generating function from a negative binomial random variable. Chapter 11 demonstrated that

$$\left(\frac{se^{-vt}}{1-s\left(1-e^{-vt}\right)}\right)^{a_0+\frac{\lambda}{v}} \triangleright \left\{\binom{n-1}{a_0+\frac{\lambda}{v}-1} e^{-v\left(a_0+\frac{\lambda}{v}\right)t} \left(1-e^{-vt}\right)^{n-\left(a_0+\frac{\lambda}{v}\right)}\right\} \tag{12.25}$$

Use of the sliding tool leads to

$$P_n(t) = \binom{n+\frac{\lambda}{v}-1}{a_0+\frac{\lambda}{v}-1} e^{-v\left(a_0+\frac{\lambda}{v}\right)t} \left(1-e^{-vt}\right)^{n-a_0} \tag{12.26}$$

12.4 The Immigration-Death Process

12.4.1 Introduction

This process is useful in considering the spread of a noncontagious disease that may lead to death. The only source of new cases is the immigration of cases. There are no "births" (i.e., the disease cannot spread from one individual to another), and cases cannot leave the system. Thus, here there are two competing processes. As for the immigration-birth process, we will develop a system of difference-differential equations using generating functions, collapsing this system down to a single partial differential equation that will be solved in detail.

12.4.2 Description of the immigration–death process

In this model, patients arrive with the disease. The disease does not spread, but the patient can die from the disease. This is depicted in Figure 12.2. Here, the Chapman–Kolmogorov equations will be developed around the competing effects of immigration that will increase the number of cases, and the independent contribution of deaths that will produce a decrease in the number of patients with disease.

12.4.3 The Chapman–Kolmogorov equations for the immigration-death process

The goal is to compute $P_n(t)$, the probability that there are n cases in the system at time t. This is achieved by enumerating how their can be n cases at time t + Δt, given that Δt is so small that one and only one event can take place during its passage. As with the contagion process, one of three outcomes may occur. First, there may be n - 1 cases at time t, and a new arrival occurs in time Δt, with probability $\lambda\Delta t$. Another way to have n cases in the time interval (t, t + Δt) is if there are n+1 cases at time t and a death occurs with probability $(n+1)\omega\Delta t$. Finally, it could be that there is no event occurring in the time interval (t, t + Δt), when there are n patients with the disease in the system at time t. Considering the occurrences of each of these events, the Chapman–Kolmogorov forward equation may be written as

$$P_n(t+\Delta t) = \lambda\Delta t P_{n-1}(t) + (n+1)\omega\Delta t P_{n+1}(t) + (1 - \lambda\Delta t - n\omega\Delta t) P_n(t) \qquad (12.27)$$

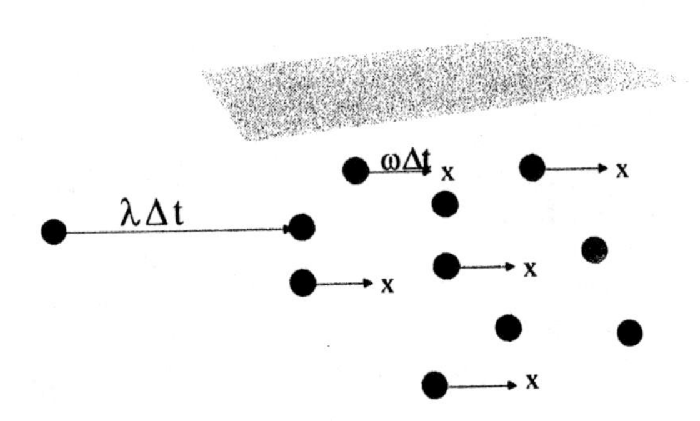

Figure 12.2 Immigration-death model: Patients with disease enter the system at a constant rate. Some of these patients die while in the population.

If $P_n(t)=0$ for negative n, equation (12.27) is true for n = 0 to infinity. We may now move forward with the manipulation required to remove Δt from this consideration, and in the process develop difference-differential equations.

$$
\begin{aligned}
P_n(t+\Delta t) &= \lambda\Delta t P_{n-1}(t)+(n+1)\omega\Delta t P_{n+1}(t)+(1-\lambda\Delta t-n\omega\Delta t)P_n(t)\\
P_n(t+\Delta t)-P_n(t) &= \lambda\Delta t P_{n-1}(t)+(n+1)\omega\Delta t P_{n+1}(t)-\lambda\Delta t P_n(t)-n\omega\Delta t P_n(t)\\
\frac{P_n(t+\Delta t)-P_n(t)}{\Delta t} &= \lambda P_{n-1}(t)+(n+1)\omega P_{n+1}(t)-\lambda P_n(t)-n\omega P_n(t)\\
\lim_{\Delta t\to 0}\frac{P_n(t+\Delta t)-P_n(t)}{\Delta t} &= \lambda P_{n-1}(t)-\lambda P_n(t)+(n+1)\omega P_{n+1}(t)-n\omega P_n(t)
\end{aligned}
\tag{12.28}
$$

and taking limits leads to the difference–differential equations

$$
\frac{dP_n(t)}{dt} = \lambda P_{n-1}(t)-\lambda P_n(t)+(n+1)\omega P_{n+1}(t)-n\omega P_n(t) \tag{12.29}
$$

12.4.4 Application of the generating function

Now applying the familiar generating function argument, define

$$G_t(s) = \sum_{n=0}^{\infty} s^n P_n(t) \tag{12.30}$$

and multiply each term in equation (12.29) by s^n to find

$$s^n \frac{dP_n(t)}{dt} = \lambda s^n P_{n-1}(t) - \lambda s^n P_n(t) + \omega(n+1)s^n P_{n+1}(t) - \omega n s^n P_n(t)$$

$$\sum_{n=0}^{\infty} s^n \frac{dP_n(t)}{dt} = \lambda \sum_{n=0}^{\infty} s^n P_{n-1}(t) - \lambda \sum_{n=0}^{\infty} s^n P_n(t) + \omega \sum_{n=0}^{\infty} (n+1)s^n P_{n+1}(t) - \omega \sum_{n=0}^{\infty} n s^n P_n(t) \tag{12.31}$$

Assuming $P_0(t) = 0$ allows simplification of equation (12.31) to

$$\frac{\partial G_t(s)}{\partial t} = \lambda s G_t(s) - \lambda G_t(s) + \omega \frac{\partial G_t(s)}{\partial s} - \omega s \frac{\partial G_t(s)}{\partial s} \tag{12.32}$$

this may be written as

$$\frac{\partial G_t(s)}{\partial t} = \lambda(s-1)G_t(s) - \omega(s-1)\frac{\partial G_t(s)}{\partial s} \tag{12.33}$$

Equation (12.33) is a partial differential equation in $G_t(s)$, requiring a solution.

12.4.5 Solution of the partial differential equation in $G_t(s)$

The general solution of equation (12.33) requires the solution of the auxiliary equations associated with the following partial differential equation:

$$\frac{\partial G_t(s)}{\partial t} - \omega(1-s)\frac{\partial G_t(s)}{\partial s} = \lambda(s-1)G_t(s) \tag{12.34}$$

To obtain subsidiary equations

$$\frac{dt}{1} = \frac{ds}{-\omega(1-s)} = \frac{dG_t(s)}{\lambda(s-1)G_t(s)} \tag{12.35}$$

Evaluation of the second and third term from equation (12.35) reveals

$$\begin{aligned}
&\frac{ds}{-\omega(1-s)} = \frac{dG_t(s)}{\lambda(s-1)G_t(s)} \\
&\frac{\lambda}{\omega}ds = \frac{dG_t(s)}{G_t(s)} \\
&\int\frac{\lambda}{\omega}ds = \int\frac{dG_t(s)}{G_t(s)} \\
&\frac{\lambda}{\omega}s + C_1 = \ln G_t(s) \\
&G_t(s) = e^{\frac{\lambda}{\omega}s}e^{C_1} = e^{\frac{\lambda}{\omega}s}\Phi(C_2)
\end{aligned} \tag{12.36}$$

from which we observe that $G_t(s) = e^{\frac{\lambda}{\omega}s}\Phi(C_2)$. From equation (12.35), observe that

$$\frac{dt}{1} = \frac{-ds}{\omega(1-s)} \tag{12.37}$$

Leading to

$$\begin{aligned}
&\int -\omega dt = \int\frac{ds}{(1-s)} \\
&-\omega t = -\ln(1-s) + C \\
&C = -\omega t + \ln(1-s) \\
&C_2 = e^{-\omega t}(1-s)
\end{aligned} \tag{12.38}$$

that leads to $\Phi[C_2] = \Phi\left[e^{-\omega t}(1-s)\right]$. Thus, $G_t(s)$ can be written as

$$G_t(s) = e^{\frac{\lambda}{\omega}s}\Phi\left[e^{-\omega t}(1-s)\right] \tag{12.39}$$

We now pursue the particular solution from equation (12.39) by using the boundary conditions. Specifically

$$G_0(s) = s^{a_0} = e^{\frac{\lambda}{\omega}s}\Phi\left[e^{-\omega(0)}(1-s)\right] = e^{\frac{\lambda}{\omega}s}\Phi\left[(1-s)\right] \tag{12.40}$$

Thus $\Phi\left[(1-s)\right] = e^{-\frac{\lambda}{\omega}s}s^{a_0}$ or, by letting $z = 1-s$

$$\Phi[z] = e^{\frac{\lambda}{\omega}(1-z)}(1-z)^{a_0} \tag{12.41}$$

To complete this examination, write

$$G_t(s) = e^{\frac{\lambda}{\omega}s}\Phi\left[e^{-\omega t}(1-s)\right] = e^{\frac{\lambda}{\omega}s}e^{-\frac{\lambda}{\omega}\left[1-e^{-\omega t}(1-s)\right]}\left[1-e^{-\omega t}(1-s)\right]^{a_0} \tag{12.42}$$

Observe that this is a convolution of two processes. The last component in expression (12.42) can be rearranged as

$$\left[1-e^{-\omega t}(1-s)\right]^{a_0} = \left[\left(1-e^{-\omega t}\right)+se^{-\omega t}\right]^{a_0} \tag{12.43}$$

combining results from Chapter 2 on probability generating functions and results from Chapter 11 on the death process reveals that

$$\left[\left(1-e^{-\omega t}\right)+se^{-\omega t}\right]^{a_0} \triangleright \left\{\binom{a_0}{k}e^{-k\omega t}\left(1-e^{-\omega t}\right)^{a_0-k}\right\} \tag{12.44}$$

The expression on the right side of equation (12.42) appears to be related to the Poisson distribution, but requires some simplification.

$$e^{\frac{\lambda}{\omega}s} e^{-\frac{\lambda}{\omega}\left[1-e^{-\omega t}(1-s)\right]} = e^{\frac{\lambda}{\omega}s\left(1-e^{-\omega t}\right)-\frac{\lambda}{\omega}\left(1-e^{-\omega t}\right)} = e^{\frac{\lambda}{\omega}\left[1-e^{-\omega t}\right][s-1]} \tag{12.45}$$

Write

$$G_t(s) = e^{\frac{\lambda}{\omega}\left(1-e^{\omega t}\right)(s-1)}\left[se^{-\omega t}+1-e^{-\omega t}\right]^{a_0} = G_1(s)G_2(s) \tag{12.46}$$

where $G_1(s)$ represents an infinite series of powers of s and $G_2(s)$ represents a finite sum of powers of s. To invert this, we invoke the findings from problem 47 in Chapter 2, that states if $G_1(s) = \sum_{j=0}^{J} a_j s^j$ and $G_2(s) = \sum_{j=0}^{\infty} b_j s^j$ then $G_1(s)G_2(s) \triangleright \left\{ \sum_{j=0}^{\min(J,k)} a_j b_{k-j} \right\}$. Applying this result

$$G_t(s) = e^{\frac{\lambda}{\omega}\left(1-e^{-\omega t}\right)(s-1)}\left[se^{-\omega t}+1-e^{-\omega t}\right]^{a_0}$$

$$\triangleright \left\{ \sum_{j=0}^{\min(a_0,k)} \left[\binom{a_0}{j} e^{-j\omega t}\left(1-e^{-\mu t}\right)^{a_0-j}\right]\left[\frac{\left(\frac{\lambda}{\omega}\left[1-e^{-\omega t}\right]\right)^{k-j}}{(k-j)!} e^{-\frac{\lambda}{\omega}\left[1-e^{-\omega t}\right]}\right]\right\} \tag{12.47}$$

12.5 The Emigration–Death Model

12.5.1 Introduction

The model in this section considers only the possibility of a decrease in the number of individuals with the disease. These diseased patients can either die from (or be cured of) the disease or they can leave the system. As before, we expect in this derivation of G(s) a partial differential equation in both t and s. Its solution will be straightforward.

12.5.2 Elaboration

In this circumstance, there are two processes working simultaneously. The first is the death force, that extinquishes the disease, either by death or cure. The second force is the emigration force, that removes patients with the disease. Recall that the distinction between a death process and the emigration process is that the death process produces deaths with a force that is proportional to the number of patients with the disease, while the emigration process is a force that is independent of the number of diseased patients. Of course, each of these two forces acts independently of the other. This is in contradistinction to the work of some [1] in which the force of emigration is the same as the force of death, each of which being proportional to the number of diseased patients.

12.5.3 Chapman–Kolmogorov forward equations

We begin the development of the family of difference equations that govern this process using the above parameterization. If time is slowed down sufficiently, how can we arrive at n cases at time $t + \Delta t$ allowing only one event and one event only to occur in the time interval $(t, t + \Delta t)$? This can happen one of two ways. The first way this may happen is that there are $n + 1$ patients at time t, and a death occurs, an event that occurs with probability $(n + 1)\omega\Delta t$. In addition, again with $n + 1$ patients at time t, a patient could leave the system with probability $\mu\Delta t$. Finally, there could be n patients in the system with neither a death, nor an emigration occurring in time Δt. Assume that $P_0(t) = 0$, and $P_{a_0}(0)=1$. The Chapman–Kolmogorov equations for this system are

$$P_n(t+\Delta t) = P_{n+1}(t)(n+1)\omega\Delta t + P_{n+1}(t)\mu\Delta t + P_n(t)\left(1-n\omega\Delta t-\mu\Delta t\right) \quad (12.48)$$

Collect all of the terms involving Δt on the left side of the equation and take a limit as Δt decreases in size to zero:

$$\begin{aligned}\frac{P_n(t+\Delta t)-P_n(t)}{\Delta t} &= (n+1)\omega P_{n+1}(t)+\mu P_{n+1}(t) \\ &\quad - n\omega t P_n(t)-\mu P_n(t) \\ \lim_{\Delta t\to 0}\frac{P_n(t+\Delta t)-P_n(t)}{\Delta t} &= (n+1)\omega P_{n+1}(t)+\mu P_{n+1}(t) \\ &\quad - n\omega P_n(t)-\mu P_n(t)\end{aligned} \quad (12.49)$$

leading to

$$\frac{dP_n(t)}{dt} = (n+1)\omega P_{n+1}(t)+\mu P_{n+1}(t)-n\omega P_n(t)-\mu P_n(t) \quad (12.50)$$

12.5.4 Application of the generating function argument

Begin as always by first defining the generating function

$$G_t(s) = \sum_{n=0}^{a_0} s^n P_n(t) \tag{12.51}$$

and move forward with the conversion and consolidation of equation (12.50).

$$\begin{aligned}
\frac{dP_n(t)}{dt} &= \omega(n+1)P_{n+1}(t) + \mu P_{n+1}(t) - \omega n P_n(t) - \mu P_n(t) \\
s^n \frac{dP_n(t)}{dt} &= \omega(n+1)\, s^n P_{n+1}(t) + \mu s^n P_{n+1}(t) - \omega n s^n P_n(t) - \mu s^n P_n(t)
\end{aligned} \tag{12.52}$$

The next step is to recognize that the last line of equation (12.52) represents a system of equations for $0 \le n \le a_0$, and write

$$\begin{aligned}
\sum_{n=0}^{a_0} s^n \frac{dP_n(t)}{dt} &= \omega \sum_{n=0}^{a_0} (n+1) s^n P_{n+1}(t) + \mu \sum_{n=0}^{a_0} s^n P_{n+1}(t) \\
&\quad - \omega \sum_{n=0}^{a_0} n s^n P_n(t) - \mu \sum_{n=0}^{a_0} s^n P_n(t) \\
\frac{d\sum_{n=0}^{a_0} s^n P_n(t)}{dt} &= \omega \sum_{n=0}^{a_0} (n+1) s^n P_{n+1}(t) + \mu \sum_{n=0}^{a_0} s^n P_{n+1}(t) \\
&\quad - \omega s \sum_{n=0}^{a_0} n s^{n-1} P_n(t) - \mu \sum_{n=0}^{a_0} s^n P_n(t)
\end{aligned} \tag{12.53}$$

Recognize these summands as functions of $G_t(s)$ and write

$$\begin{aligned}
\frac{\partial G_t(s)}{\partial t} &= \omega \frac{\partial G_t(s)}{\partial s} + \mu s^{-1} G_t(s) - \omega s \frac{\partial G_t(s)}{\partial s} - \mu G_t(s) \\
\frac{\partial G_t(s)}{\partial t} &= \omega(1-s) \frac{\partial G_t(s)}{\partial s} + \mu\left(s^{-1} - 1\right) G_t(s)
\end{aligned} \tag{12.54}$$

that becomes

$$\frac{\partial G_t(s)}{\partial t} - \omega(1-s)\frac{\partial G_t(s)}{\partial s} = \mu\left(s^{-1}-1\right)G_t(s) \tag{12.55}$$

12.5.5 Solution of the partial differential equation

Recognize equation (12.55) as a partial differential equation in t and in s, but of the form that will allow us to find a general solution using a subsidiary set of equations. Write these equations as

$$\frac{dt}{1} = \frac{ds}{-\omega(1-s)} = \frac{dG_t(s)}{\mu\left(s^{-1}-1\right)G_t(s)} \tag{12.56}$$

Continuing with the first two terms of equation (12.56), one can evaluate these equalities in two combinations to provide information on the form of the generating function $G_t(s)$. From the second and third terms of equation (12.56) note

$$\begin{aligned}
\frac{ds}{-\omega(1-s)} &= \frac{dG_t(s)}{\mu\left(s^{-1}-1\right)G_t(s)} \\
\frac{-\mu\left(s^{-1}-1\right)ds}{\omega(1-s)} &= \frac{dG_t(s)}{G_t(s)} \\
\frac{-\mu ds}{\omega s} &= \frac{dG_t(s)}{G_t(s)}
\end{aligned} \tag{12.57}$$

Integrating both sides of the last equation in (12.57) reveals

$$\begin{aligned}
\int\frac{-\mu ds}{\omega s} &= \int\frac{dG_t(s)}{G_t(s)} \\
-\frac{\mu}{\omega}\ln s + C &= \ln G_t(s) \\
\ln s^{\frac{-\mu}{\omega}} + C &= \ln G_t(s)
\end{aligned} \tag{12.58}$$

Exponentiating both sides gives

$$s^{\frac{\mu}{\omega}} C_1 = G_t(s) \tag{12.59}$$

that can be written in the form $G_t(s) = s^{-\frac{\mu}{\omega}} \Phi(C_2)$. The remaining requirement is to identify the form $\Phi(C_2)$. Using the first and second terms from equation (12.56) write

$$\frac{dt}{1} = \frac{ds}{-\omega(1-s)} \tag{12.60}$$

The above equation may be solved as follows:

$$\begin{aligned}
\frac{dt}{1} &= \frac{ds}{-\omega(1-s)} \\
\omega dt &= \frac{-ds}{(1-s)} \\
\int \omega dt &= \int \frac{-ds}{(1-s)} \\
\omega t + C &= \ln(1-s) \\
C &= -\omega t + \ln(1-s)
\end{aligned} \tag{12.61}$$

From which find

$$C_2 = e^{-\omega t}(1-s) \tag{12.62}$$

Combining $G_t(s) = s^{-\frac{\mu}{\omega}} \Phi(C_2)$ with $C_2 = e^{-\omega t}(1-s)$ gives

$$G_t(s) = s^{-\frac{\mu}{\omega}} \Phi\left(e^{-\omega t}(1-s)\right) \tag{12.63}$$

and we are now ready to identify the specific solution to the immigration-death-emigration model.

12.5.6 Specific solution for partial differential equation

Having identified $G_t(s) = s^{-\frac{\mu}{\omega}}\Phi\left(e^{-\omega t}(1-s)\right)$ now pursue a specific solution, beginning with the boundary conditions. At t = 0, if there are a_0 patients with the disease

$$G_0(s) = s^{a_0} = s^{-\frac{\mu}{\omega}}\Phi(1-s) \tag{12.64}$$

Now, let $z = 1-s$, then s = 1 – z, and substitute this result into equation (12.64)

$$(1-z)^{a_0} = (1-z)^{-\frac{\mu}{\omega}}\Phi(z) \tag{12.65}$$

or

$$\Phi(z) = (1-z)^{a_0+\frac{\mu}{\omega}} \tag{12.66}$$

and write

$$G_t(s) = s^{-\frac{\mu}{\omega}}\Phi\left(e^{-\omega t}(1-s)\right) = s^{-\frac{\mu}{\omega}}\left[1-e^{-\omega t}(1-s)\right]^{a_0+\frac{\mu}{\omega}} \tag{12.67}$$

12.5.7 The Inversion of $G_t(s)$

The inversion of $G_t(s)$ can be carried out in a straightforward manner. The first term on the right side of equation (12.67) will require use of only the sliding tool. Then the inversion of the second term of this equation we can recognize as

$$\left(1-e^{-\omega t}(1-s)\right)^{a_0+\frac{\mu}{\omega}} = \left[\left(1-e^{-\omega t}\right)+se^{-\omega t}\right]^{a_0+\frac{\mu}{\omega}} \tag{12.68}$$

that will be recalled from Chapter 2 is the probability generating function of the binomial distribution with $p = e^{-\omega t}$. Thus

$$\left[\left(1-e^{-\omega t}\right)+se^{-\omega t}\right]^{a_0+\frac{\mu}{\omega}} \triangleright \left\{\binom{a_0+\frac{\mu}{\omega}}{k}e^{-\omega tk}\left(1-e^{-\omega t}\right)^{a_0+\frac{\mu}{\omega}-k}\right\} \tag{12.69}$$

and the sliding tool's implementation leads to

$$P_k(t) \triangleright \left\{ \begin{pmatrix} a_0 + \frac{\mu}{\omega} \\ k + \frac{\mu}{\omega} \end{pmatrix} e^{-\omega t\left(k+\frac{\mu}{\omega}\right)} \left(1-e^{-\omega t}\right)^{a_0-k} \right\} \tag{12.70}$$

12.6 The Birth–Death Model

12.6.1 *Introduction*

In this section, the concept of death to the birth model is added. Each of these two processes has a force proportional to the number of cases.

12.6.2 *Motivation*

Figure 12.3 depicts the birth-death model in operation. The only new cases are due to births and the only exits are due to death. The disease spreads according to the birth process. However, the death process leads to a moderation in its effect since its force decreases the number of cases.

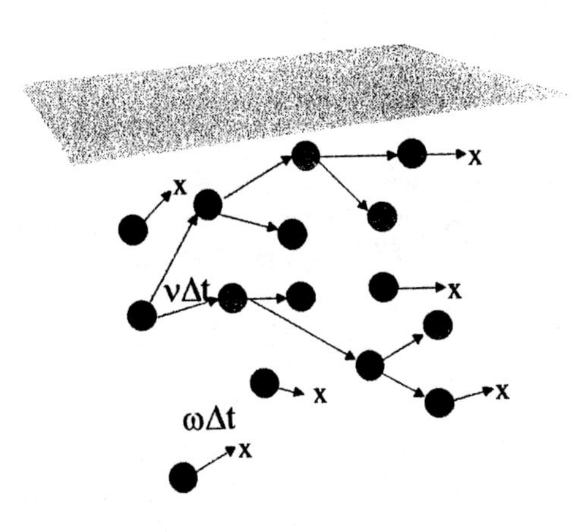

Figure 12.3 Birth-death model: Patients already in the population with disease can spread the disease within the population, die, or do neither.

12.6.3 The Chapman–Kolmogorov forward equations for the birth-death process

The process for developing the Chapman–Kolmogorov forward equations will be exactly analogous to their development thus far in this chapter. Begin with an enumeration of the three ways there can be n diseased patients at time $t + \Delta t$ if we allow Δt to be so small that only one event can occur in this short period of time. The first of these is that the number is (n - 1) at time t and there is a "birth" or spread of the disease from time t to time $t + \Delta t$, an event occurs with probability $(n-1)\nu\Delta t$. Alternatively, there may be (n + 1) individuals at time t and a death occurs, an event that occurs with probability $(n+1)\omega\Delta t$. Finally, there may be n individuals at time t and neither a birth nor a death occurs. The probability of the absence of these events is $1 - n\nu\Delta t - n\omega\Delta t$. We assume that ν is much greater than ω such that at no time in the system are there no individuals ($P_0(t) = 0$ for $t \geq 0$). Therefore, for $n \geq 0$, the Chapman–Kolmogorov forward equations for $n \geq 0$ may be written as:

$$\begin{aligned} &P_n(t+\Delta t) \\ &= (n-1)\nu\Delta t P_{n-1}(t) + (n+1)\omega\Delta t P_{n+1}(t) + (1 - n\nu\Delta t - n\omega\Delta t)P_n(t) \end{aligned} \tag{12.71}$$

proceeding to convert equations represented by (12.71) to the anticipated difference-differential equation.

$$\begin{aligned} P_n(t+\Delta t) &= (n-1)\nu\Delta t P_{n-1}(t) + (n+1)\omega\Delta t P_{n+1}(t) \\ &\quad + (1 - n\nu\Delta t - n\omega\Delta t)P_n(t) \\ P_n(t+\Delta t) - P_n(t) &= (n-1)\nu\Delta t P_{n-1}(t) + (n+1)\omega\Delta t P_{n+1}(t) \\ &\quad - n\nu\Delta t P_n(t) - n\omega\Delta t P_n(t) \end{aligned} \tag{12.72}$$

Gathering terms involving Δt to the right side gives

$$\begin{aligned} \frac{P_n(t+\Delta t) - P_n(t)}{\Delta t} &= (n-1)\nu P_{n-1}(t) + (n+1)\omega P_{n+1}(t) - n\nu P_n(t) - n\omega P_n(t) \\ \lim_{\Delta t \to 0} \frac{P_n(t+\Delta t) - P_n(t)}{\Delta t} &= (n-1)\nu P_{n-1}(t) - n\nu P_n(t) + (n+1)\omega P_{n+1}(t) - n\omega P_n(t) \end{aligned} \tag{12.73}$$

Evaluation of the limit obtained in equation (12.73) demonstrates

$$\frac{dP_n(t)}{dt} = (n-1)\nu P_{n-1}(t) - n\nu P_n(t) + (n+1)\omega P_{n+1}(t) - n\omega P_n(t) \quad (12.74)$$

We can now proceed with the application of the generating function.

12.6.4 The generating function in the birth-death model

Defining the generating function as

$$G_t(s) = \sum_{n=0}^{\infty} s^n P_n(t) \quad (12.75)$$

Multiplying both sides of equation (12.74) by s^n

$$s^n \frac{dP_n(t)}{dt} = (n-1)s^n \nu P_{n-1}(t) - ns^n \nu P_n(t) + (n+1)s^n \omega P_{n+1}(t) - ns^n \omega P_n(t) \quad (12.76)$$

Taking summations over the range of n demonstrates

$$\sum_{n=0}^{\infty} s^n \frac{dP_n(t)}{dt} = \sum_{n=0}^{\infty}(n-1)s^n \nu P_{n-1}(t) - \sum_{n=0}^{\infty} ns^n \nu P_n(t) + \sum_{n=0}^{\infty}(n+1)s^n \omega P_{n+1}(t) - \sum_{n=0}^{\infty} ns^n \omega P_n(t) \quad (12.77)$$

Continue the development

$$\frac{d\sum_{n=0}^{\infty} s^n P_n(t)}{dt} = \nu\sum_{n=0}^{\infty}(n-1)s^n P_{n-1}(t) - \nu\sum_{n=0}^{\infty} ns^n P_n(t) + \omega\sum_{n=0}^{\infty}(n+1)s^n P_{n+1}(t) - \omega\sum_{n=0}^{\infty} ns^n P_n(t) \quad (12.78)$$

assuming $P_0(t) = 0$. Converting these sums following the development in Chapter 11 can be written as

$$\begin{aligned}
\frac{\partial G_t(s)}{\partial t} &= \nu s^2 \frac{\partial G_t(s)}{\partial s} - \nu s \frac{\partial G_t(s)}{\partial s} + \omega \frac{\partial G_t(s)}{\partial s} - \omega s \frac{\partial G_t(s)}{\partial s} \\
\frac{\partial G_t(s)}{\partial t} &= \nu s(s-1) \frac{\partial G_t(s)}{\partial s} + \omega(1-s) \frac{\partial G_t(s)}{\partial s} \\
\frac{\partial G_t(s)}{\partial t} &= \left(\nu s(s-1) + \omega(1-s)\right) \frac{\partial G_t(s)}{\partial s} \\
\frac{\partial G_t(s)}{\partial t} &- \left(\nu s(s-1) + \omega(1-s)\right) \frac{\partial G_t(s)}{\partial s} = 0
\end{aligned} \tag{12.79}$$

that, when simplified, reveals

$$\frac{\partial G_t(s)}{\partial t} - \left((\nu s - \omega)(s-1)\right) \frac{\partial G_t(s)}{\partial s} = 0 \tag{12.80}$$

This conforms to the form of the partial differential equation to which we are accustomed, and can now proceed with outlining first its general solution, then its particular solution.

12.6.5 Solutions to the birth–death partial differential equation

Proceed to identify the subsidiary equations from equation (12.79) as

$$\frac{dt}{1} = \frac{ds}{-(\nu s - \omega)(s-1)} = \frac{dG_t(s)}{0} \tag{12.81}$$

From the first and third terms in equation (12.81), find that $dG_t(s)=0$ or $G_t(s)$ is a constant that will be denoted as $G_t(s) = C_1 = \Phi(C_2)$. Identification of the form of $\Phi(C_2)$ requires the use of the first two terms in equation (12.81).

$$\frac{dt}{1} = \frac{ds}{-(\nu s - \omega)(s-1)} \tag{12.82}$$

Using the technique of partial fractions introduced into Chapter 3, we may write

$$\frac{1}{(\nu s-\omega)(s-1)} = \frac{\nu/(\omega-\nu)}{\nu s-\omega} + \frac{1/(\nu-\omega)}{s-1} \tag{12.83}$$

So equation (12.82) may be written as:

$$-dt = \frac{\nu/(\omega-\nu)}{\nu s-\omega} + \frac{1/(\nu-\omega)}{s-1} \tag{12.84}$$

$$dt = \frac{\nu/(\nu-\omega)}{\nu s-\omega} - \frac{1/(\nu-\omega)}{s-1} \tag{12.85}$$

Integrating both sides with respect to t reveals

$$\begin{gathered}\int dt = \int\left[\frac{\nu/(\nu-\omega)}{\nu s-\omega} - \frac{1/(\nu-\omega)}{s-1}\right]ds \\ t+C = \frac{1}{\nu-\omega}\ln(\nu s-\omega) - \frac{1}{\nu-\omega}\ln(s-1) = \frac{1}{\nu-\omega}\ln\left(\frac{\nu s-\omega}{s-1}\right)\end{gathered} \tag{12.86}$$

Continue

$$\begin{gathered}(\nu-\omega)t + C = \ln\left(\frac{\nu s-\omega}{s-1}\right) \\ C = -(\nu-\omega)t + \ln\left(\frac{\nu s-\omega}{s-1}\right) \\ C_2 = e^{-(\nu-\omega)t}\left(\frac{\nu s-\omega}{s-1}\right)\end{gathered} \tag{12.87}$$

from which

$$C_2 = e^{-(v-\omega)t} \frac{vs-\omega}{s-1} \tag{12.88}$$

So we compute

$$G_t(s) = \Phi[C_2] = \Phi\left[e^{-(v-\omega)t} \frac{vs-\omega}{s-1}\right] \tag{12.89}$$

as the general solution to the partial differential equation (12.76), and we use the boundary solution to find the particular solution. Beginning with a_0 cases at time 0, we find

$$G_0(s) = s^{a_0} = \Phi\left[e^{-(v-\omega)0} \frac{vs-\omega}{s-1}\right] = \Phi\left[\frac{vs-\omega}{s-1}\right] \tag{12.90}$$

Letting $z = \frac{vs-\omega}{s-1}$ gives $s = \frac{z-\omega}{z-v}$ and allows us to write

$$\Phi(z) = \left[\frac{z-\omega}{z-v}\right]^{a_0} \tag{12.91}$$

and

$$\begin{aligned} G_t(s) &= \Phi\left[e^{-(v-\omega)t} \frac{vs-\omega}{s-1}\right]^{a_0} = \left[\frac{e^{-(v-\omega)t}\dfrac{vs-\omega}{s-1} - \omega}{e^{-(v-\omega)t}\dfrac{vs-\omega}{s-1} - v}\right]^{a_0} \\ &= \left[\frac{e^{-(v-\omega)t}(vs-\omega) - \omega(s-1)}{e^{-(v-\omega)t}(vs-\omega) - v(s-1)}\right]^{a_0} \\ &= \left[\frac{\left(ve^{-(v-\omega)t} - \omega\right)s + \omega\left(1 - e^{-(v-\omega)t}\right)}{\left(v - \omega e^{-(v-\omega)t}\right) - v\left(1 - e^{-(v-\omega)t}\right)s}\right]^{a_0} \end{aligned} \tag{12.92}$$

12.6.6 Inversion of $G_t(s)$

To begin the inversion of $G_t(s)$ from equation (12.92) by first simplifying, let $a_0 = 1$, and group terms containing s. Thus, equation (12.92) becomes

$$G_t(s) = \frac{\left(ve^{-(v-\omega)t} - \omega\right)s + \omega\left(1 - e^{-(v-\omega)t}\right)}{\left(v - \omega e^{-(v-\omega)t}\right) - v\left(1 - e^{-(v-\omega)t}\right)s} \quad (12.93)$$

The inversion of this quantity is straightforward since both the numerator and the denominator are merely linear functions of s. Recognize the denominator as being of the form a – bs; and since

$$\frac{1}{a - bs} \triangleright \left\{\left[\frac{1}{a}\right]\left[\frac{b}{a}\right]^k\right\} \quad (12.94)$$

then

$$\frac{1}{\left(v - \omega e^{-(v-\omega)t}\right) - v\left(1 - e^{-(v-\omega)t}\right)s} \triangleright \left\{\left[\frac{1}{v - \omega e^{-(v-\omega)t}}\right]\left[\frac{v\left(1 - e^{(v-\omega)t}\right)}{v - \omega e^{-(v-\omega)t}}\right]^k\right\} \quad (12.95)$$

and

$$G_t(s) = \frac{\left(ve^{-(v-\omega)t} - \omega\right)s + \omega\left(1 - e^{-(v-\omega)t}\right)}{\left(v - \omega e^{-(v-\omega)t}\right) - v\left(1 - e^{-(v-\omega)t}\right)s}$$

$$\triangleright \left\{\left[\frac{ve^{-(v-\omega)t} - \omega}{v - \omega e^{-(v-\omega)t}}\right]\left[\frac{v\left(1 - e^{(v-\omega)t}\right)}{v - \omega e^{-(v-\omega)t}}\right]^{k-1} + \left[\frac{\omega\left(1 - e^{-(v-\omega)t}\right)}{v - \omega e^{-(v-\omega)t}}\right]\left[\frac{v\left(1 - e^{(v-\omega)t}\right)}{v - \omega e^{-(v-\omega)t}}\right]^k\right\}$$

$$\left[\frac{1}{v - \omega e^{-(v-\omega)t}}\right]^{k+1}\left[v\left(1 - e^{(v-\omega)t}\right)\right]^{k-1}\left[\left(ve^{-(v-\omega)t} - \omega\right)\left(v - \omega e^{-(v-\omega)t}\right)\right] \quad (12.96)$$

that, after some algebra simplifies to

$$G_t(s) \triangleright \left\{ \left[e^{-(\nu-\omega)t} (\nu-\omega)^2 \right] \frac{\left[\nu\left(1-e^{(\nu-\omega)t}\right) \right]^{k-1}}{\left[\nu-\omega e^{-(\nu-\omega)t} \right]^{k+1}} \right\} \tag{12.97}$$

The inversion of $G_t(s)$ provides the probability of k patients with the disease at time t, given there is only one case in the system at time t. The solution is somewhat more complicated when there are a_0 cases at time t. To solve this problem, recall from Chapter 2, the result

$$\left[\frac{1}{1-s} \right]^n \triangleright \binom{n+k-1}{k} \tag{12.98}$$

will be helpful here since we recognize that G(s) is of the form $\left[\frac{as+b}{c-ds} \right]^n$ where

$$\begin{aligned} a &= \nu e^{-(\nu-\omega)t} - \omega; \qquad b = \omega\left(1-e^{-(\nu-\omega)t}\right) \\ c &= \nu - \omega e^{-(\nu-\omega)t}; \qquad d = \nu\left(1-e^{-(\nu-\omega)t}\right) \end{aligned} \tag{12.99}$$

and $a_0 = n > 1$. From Chapter 2, we know that

$$\left[\frac{a+bs}{c-ds} \right]^n \triangleright \left\{ \left[\frac{1}{c} \right]^n \sum_{j=0}^{Min(n,k)} \binom{n+k-j-1}{n-1} \left[\frac{d}{c} \right]^{k-j} \binom{n}{j} a^j b^{n-j} \right\} \tag{12.100}$$

we can now write the final solution:

$$P_k(t) = \left[\frac{1}{\nu-\omega e^{-(\nu-\omega)t}} \right]^{a_0} \text{x}$$

$$\sum_{j=0}^{Min(a_0,k)} \binom{a_0+k-j-1}{a_0-1} \binom{a_0}{j} \left[\frac{\nu\left(1-e^{-(\nu-\omega)t}\right)}{\nu-\omega e^{-(\nu-\omega)t}} \right]^{k-j} \left[\nu e^{-(\nu-\omega)t} - \omega \right]^j \left[\omega\left(1-e^{-(\nu-\omega)t}\right) \right]^{a_0-j} \tag{12.101}$$

12.7 Immigration–Birth–Death Model

12.7.1 Introduction

The next model we will consider is a combination of the models developed in the previous sections. This model will be a complicated representation of the spread of a contagious disease. Patients are allowed to arrive, and can spread the disease to other patients. In addition, patients are removed either through death or through cure with subsequent immunity.

In the derivation of $G_t(s)$ we will identify a partial differential equation in both t and s. Its solution will be straightforward, but there may be some complexities in the final simplification of $G_t(s)$.

12.7.2 Motivation

In this circumstance, we have three processes proceeding simultaneously. The first is the immigration force, bringing new cases. The second force is the birth force, spreading the disease. The final force is the death force, that acts to reduce the number of cases. Each of these three forces acts independently of the other two.

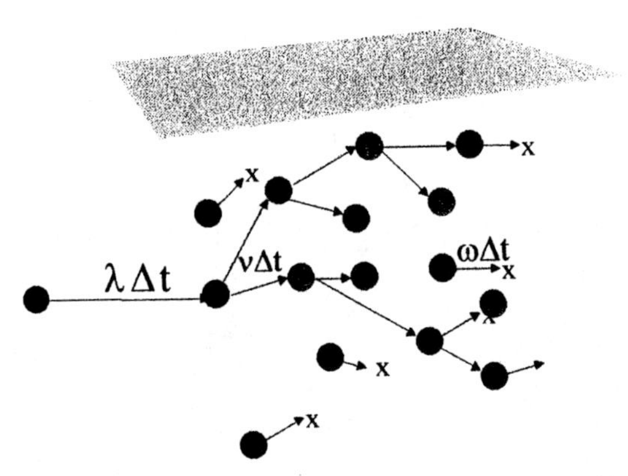

Figure 12.4 Immigration-birth-death model: Patients with disease enter the system at a constant rate, spreading the disease throughout the population, while some of the diseased patients die.

12.7.3 Chapman–Kolmogorov forward equations

Begin the development of the family of difference equations that govern this process using the above parameterization. If we slow time down sufficiently, how can there be n cases at time t + Δt, allowing only one event and one event only to occur in the time interval (t, t + Δt) ? This can happen one of three ways.

The first way this may happen is if there are n - 1 patients at time t, and an arrival occurs with probability $\lambda\Delta t$. A second possibility could be that there are n – 1 patients in the system at time t and a "birth" could occur, with probability $(n-1)\nu\Delta t$. Third, with n + 1 patients at time t, there could be a death, an event that occurs with probability $(n+1)\,\omega t$. Finally, there could be n patients in the system with neither an arrival, a birth, or a death occurring in time Δt. The Chapman–Kolmogorov equations for the immigration–birth–death process may now be written as

$$\begin{aligned} P_n(t+\Delta t) &= \lambda\Delta t\, P_{n-1}(t)+\nu\Delta t P_{n-1}(t)(n-1)+(n+1)\omega\Delta t P_n(t) \\ &\quad +\left(1-\lambda\Delta t-n\nu\Delta t-n\omega\Delta t\right)P_{n+1}(t) \end{aligned} \tag{12.102}$$

for n = 0 to ∞, $P_0(t) = 0$, and $P_{a_0}(0)=1$. Proceeding with the development of the differential equation that represents this process by consolidating the terms involving Δt

$$\begin{aligned} P_n(t+\Delta t) &= \lambda\Delta t P_{n-1}(t)+(n-1)\nu\Delta t P_{n-1}(t)+(n+1)\omega\Delta t P_{n+1}(t) \\ &+P_n(t)\left(1-\lambda\Delta t-n\nu\Delta t-n\omega\Delta t\right) \end{aligned} \tag{12.103}$$

$$\begin{aligned} P_n(t+\Delta t) - P_n(t) &= \lambda\Delta t P_{n-1}(t)+(n-1)\nu\Delta t P_{n-1}(t)+(n+1)\omega\Delta t P_{n+1}(t) \\ &\quad -\lambda\Delta t P_n(t)-n\nu\Delta t P_n(t)-n\omega\Delta t P_n(t) \end{aligned} \tag{12.104}$$

Consolidating this expression, bringing all terms involving the expression Δt to the left side of equation (12.104) reveals

$$\begin{aligned} \frac{P_n(t+\Delta t) - P_n(t)}{\Delta t} &= \lambda P_{n-1}(t)+(n-1)\nu P_{n-1}(t)+(n+1)\omega P_{n+1}(t) \\ &\quad -\lambda P_n(t)-n\nu P_n(t)-n\omega P_n(t) \end{aligned} \tag{12.105}$$

The next step requires taking a limit as t goes to infinity.

$$\begin{aligned} \lim_{\Delta t\to 0}\frac{P_n(t+\Delta t) - P_n(t)}{\Delta t} &= \lambda P_{n-1}(t)+(n-1)\nu P_{n-1}(t)+(n+1)\omega P_{n+1}(t) \\ &\quad -\lambda P_n(t)-n\nu P_n(t)-n\omega P_n(t) \end{aligned} \tag{12.106}$$

or

$$\frac{dP_n(t)}{dt} = \lambda P_{n-1}(t) + \nu P_{n-1}(t)(n-1) + P_{n+1}(t)(n+1)\omega \qquad (12.107)$$
$$-\lambda P_n(t) - n\nu P_n(t) - n\omega P_n(t)$$

12.7.4 Application of the generating function argument

Begin as always by first defining the generating function

$$G_t(s) = \sum_{n=0}^{\infty} s^n P_n(t) \qquad (12.108)$$

and move forward with the conversion and consolidation of equations (12.107)

$$\frac{dP_n(t)}{dt} = \lambda P_{n-1}(t) + \nu P_{n-1}(t)(n-1) + \omega P_{n+1}(t)(n+1)$$
$$-\lambda P_n(t) - n\nu P_n(t) - n\omega P_n(t)$$
$$s^n \frac{dP_n(t)}{dt} = \lambda s^n P_{n-1}(t) + \nu s^n P_{n-1}(t)(n-1) + \omega s^n P_{n+1}(t)(n+1)$$
$$-\lambda s^n P_n(t) \quad - \nu n s^n P_n(t) - \omega n s^n P_n(t)$$

(12.109)

Taking the sum over the range of the index of the generating function reveals

$$\sum_{n=o}^{\infty} s^n \frac{dP_n(t)}{dt} = \lambda \sum_{n=o}^{\infty} s^n P_{n-1}(t) + \nu \sum_{n=o}^{\infty} (n-1) s^n P_{n-1}(t)$$
$$+\omega \sum_{n=o}^{\infty} (n+1) s^n P_{n+1}(t) - \lambda \sum_{n=o}^{\infty} s^n P_n(t) - \nu \sum_{n=o}^{\infty} n s^n P_n(t) \qquad (12.110)$$
$$-\omega \sum_{n=o}^{\infty} n s^n P_n(t)$$

$$\frac{d\left[\sum_{n=o}^{\infty} s^n P_n(t)\right]}{dt} = \lambda s \sum_{n=o}^{\infty} s^{n-1} P_{n-1}(t) + vs^2 \sum_{n=o}^{\infty} (n-1)s^{n-2} P_{n-1}(t)$$

$$+\omega \sum_{n=o}^{\infty} (n+1)s^n P_{n+1}(t) - \lambda \sum_{n=o}^{\infty} s^n P_n(t) - vs \sum_{n=o}^{\infty} ns^{n-1} P_n(t) - \omega s \sum_{n=o}^{\infty} ns^{n-1} P_n(t)$$

(12.111)

Recognize these summands as functions of $G_t(s)$ and write

$$\begin{aligned} \frac{\partial G_t(s)}{\partial t} &= \lambda s G_t(s) + vs^2 \frac{\partial G_t(s)}{\partial s} + \omega \frac{\partial G_t(s)}{\partial s} \\ &-\lambda G_s(t) - vs \frac{\partial G_t(s)}{\partial s} - \omega s \frac{\partial G_t(s)}{\partial s} \end{aligned} \quad (12.112)$$

that upon simplification becomes

$$\begin{aligned} \frac{\partial G_t(s)}{\partial t} &= \lambda(s-1)G_t(s) + vs(s-1)\frac{\partial G_t(s)}{\partial s} + \omega(1-s)\frac{\partial G_t(s)}{\partial s} \\ \frac{\partial G_t(s)}{\partial t} &= (vs-\omega)(s-1)\frac{\partial G_t(s)}{\partial s} + \lambda(s-1)G_t(s) \end{aligned} \quad (12.113)$$

that may finally be written as

$$\frac{\partial G_t(s)}{\partial t} - (vs-\omega)(s-1)\frac{\partial G_t(s)}{\partial s} = \lambda(s-1)G_t(s) \quad (12.114)$$

12.7.5 Solution of the partial differential equation

Recognize the last line of equation (12.114) as a partial differential equation in t and in s, but of the form that will allow us to find a general solution using a subsidiary set of equations. These equations are written as

$$\frac{dt}{1} = \frac{ds}{-(vs-\omega)(s-1)} = \frac{dG_t(s)}{\lambda(s-1)G_t(s)} \quad (12.115)$$

From the second and third terms of equation (12.115)

$$\frac{ds}{-(\nu s-\omega)(s-1)}=\frac{dG_t(s)}{\lambda(s-1)G_t(s)} \tag{12.116}$$

$$\frac{-\lambda ds}{(\nu s-\omega)}=\frac{dG_t(s)}{G_t(s)} \tag{12.117}$$

Integrating both sides

$$\begin{aligned}
&\int\frac{-\lambda ds}{(\nu s-\omega)}=\int\frac{dG_t(s)}{G_t(s)}\\
&-\frac{\lambda}{\nu}\ln(\nu s-\omega)+C_1=\ln G_t(s)\\
&G_t(s)=(\nu s-\omega)^{-\frac{\lambda}{\nu}}C_1=(\nu s-\omega)^{-\frac{\lambda}{\nu}}\Phi(C_2)
\end{aligned} \tag{12.118}$$

Continue now with the first two terms of equation (12.115),

$$\frac{dt}{1}=\frac{ds}{-((\nu s-\omega)(s-1))} \tag{12.119}$$

This differential equation may be easy to solve, with the variables t and s already separated across the equal sign. We may carry out the integration directly. Using partial fractions, write

$$\frac{1}{(\nu s-\omega)(s-1)}=\frac{1/(\nu-\omega)}{s-1}-\frac{\nu/(\nu-\omega)}{\nu s-\omega} \tag{12.120}$$

Equation (12.119) may be written as

$$-dt = \frac{\frac{1}{(\nu-\omega)}}{s-1} - \frac{\frac{\nu}{(\nu-\omega)}}{\nu s-\omega}$$
$$dt = \frac{\frac{\nu}{(\nu-\omega)}}{\nu s-\omega} - \frac{\frac{1}{(\nu-\omega)}}{s-1} \tag{12.121}$$

and the integration can now proceed

$$\int dt = \int\left[\frac{\frac{\nu}{(\nu-\omega)}}{\nu s-\omega} - \frac{\frac{1}{(\nu-\omega)}}{s-1}\right]ds \tag{12.122}$$

$$t+C = \frac{1}{\nu-\omega}\ln(\nu s-\omega) - \frac{1}{\nu-\omega}\ln(s-1) = \frac{1}{\nu-\omega}\ln\left(\frac{\nu s-\omega}{s-1}\right) \tag{12.123}$$

What remains is to solve for the constant of integration

$$(\nu-\omega)t + C_2 = \ln\left(\frac{\nu s-\omega}{s-1}\right)$$
$$C_2 = -(\nu-\omega)t + \ln\left(\frac{\nu s-\omega}{s-1}\right) \tag{12.124}$$
$$C_2 = e^{-(\nu-\omega)t}\left(\frac{\nu s-\omega}{s-1}\right)$$

Combine $G_t(s) = (\nu s-\omega)^{-\frac{\lambda}{\nu}}\Phi(C_2)$ with $C_2 = e^{-(\nu-\omega)t}\left(\frac{\nu s-\omega}{s-1}\right)$:

$$G_t(s) = (\nu s-\omega)^{-\frac{\lambda}{\nu}}\Phi\left[e^{-(\nu-\omega)t}\left(\frac{\nu s-\omega}{s-1}\right)\right] \tag{12.125}$$

We are now ready to identify the specific solution to the immigration-birth-death model.

12.7.6 Specific solution for partial differential equation

Having identified $G_t(s) = (\nu s - \omega)^{-\frac{\lambda}{\nu}} \Phi\left[e^{-(\nu-\omega)t}\left(\frac{\nu s - \omega}{s-1}\right)\right]$ we can now pursue a specific solution, beginning with the boundary conditions. For $t = 0$, there are a_0 patients with the disease and find

$$G_0(s) = s^{a_0} = (\nu s - \omega)^{-\frac{\lambda}{\nu}} \Phi\left[\frac{\nu s - \omega}{s-1}\right] \tag{12.126}$$

Now, let $z = \frac{\nu s - \omega}{s-1}$ that gives $s = \frac{z-\omega}{z-\nu}$. Substituting this result into equation (12.126) reveals

$$\left(\frac{z-\omega}{z-\nu}\right)^{a_0} = \left(\nu\frac{z-\omega}{z-\nu} - \omega\right)^{-\frac{\lambda}{\nu}} \Phi[z] \tag{12.127}$$

or

$$\Phi[z] = \left(\nu\frac{z-\omega}{z-\nu} - \omega\right)^{\frac{\lambda}{\nu}} \left(\frac{z-\omega}{z-\nu}\right)^{a_0} \tag{12.128}$$

Acknowledging $G_t(s) = (\nu s - \omega)^{-\frac{\lambda}{\nu}} \Phi\left[e^{-(\nu-\omega)t}\frac{\nu s - \omega}{s-1}\right]$ $G_t(s)$ may be expressed as

$$G_t(s) = (\nu s - \omega)^{-\frac{\lambda}{\nu}} \left(\nu\frac{e^{-(\nu-\omega)t}\frac{\nu s-\omega}{s-1} - \omega}{e^{-(\nu-\omega)t}\frac{\nu s-\omega}{s-1} - \nu} - \omega\right)^{\frac{\lambda}{\nu}} \left(\frac{e^{-(\nu-\omega)t}\frac{\nu s-\omega}{s-1} - \omega}{e^{-(\nu-\omega)t}\frac{\nu s-\omega}{s-1} - \nu}\right)^{a_0} \tag{12.129}$$

This represents a simple convolution. The second term in equation (12.129) can be evaluated with some algebraic manipulation as

$$\left[v \frac{e^{-(v-\omega)t} \frac{vs-\omega}{s-1} - \omega}{e^{-(v-\omega)t} \frac{vs-\omega}{s-1} - v} - \omega \right]^{\frac{\lambda}{v}} = \left[\frac{e^{-(v-\omega)t}(v-\omega)(vs-\omega)}{v-\omega e^{-(v-\omega)t} - v\left(1-e^{-(v-\omega)t}\right)s} \right]^{\frac{\lambda}{v}}$$

$$\left[(v-\omega)e^{-(v-\omega)t} \left[\frac{vs-\omega}{v-\omega e^{-(v-\omega)t} - v\left(1-e^{-(v-\omega)t}\right)s} \right] \right]^{\frac{\lambda}{v}} \tag{12.130}$$

that is of the form $\left[\frac{as-b}{c-ds}\right]^n$; that using the tools from Chapter 2, may be inverted as

$$\left[\frac{as-b}{c-ds}\right]^n \triangleright \left\{ \left[\frac{1}{c}\right]^n \sum_{j=0}^{Min(n,k)} \binom{n}{j} a^j (-b)^{n-j} \binom{n+k-j-1}{k-j-1} \left[\frac{d}{c}\right]^{k-j} \right\} \tag{12.131}$$

Thus proceeding with inversion of this term

$$\left[(v-\omega)e^{-(v-\omega)t} \left[\frac{vs-\omega}{v-\omega e^{-(v-\omega)t} - v\left(1-e^{-(v-\omega)t}\right)s} \right] \right]^{\frac{\lambda}{v}}$$

$$= \left((v-\omega)e^{-(v-\omega)t}\right)^{\frac{\lambda}{v}} \left[\frac{vs-\omega}{v-\omega e^{-(v-\omega)t} - v\left(1-e^{-(v-\omega)t}\right)s} \right]^{\frac{\lambda}{v}}$$

$$\triangleright \left\{ \begin{matrix} \left[\frac{1}{v-\omega e^{-(v-\omega)t}}\right]^{\frac{\lambda}{v}} \left[\frac{1}{v-\omega e^{-(v-\omega)t}}\right]^{\frac{\lambda}{v}} \\ X \sum_{j=0}^{Min(\frac{\lambda}{v},k)} \binom{\frac{\lambda}{v}}{j} v^j (-\omega)^{n-j} \binom{\frac{\lambda}{v}+k-j-1}{k-j-1} \left[\frac{v\left(1-e^{-(v-\omega)t}\right)}{v-\omega e^{-(v-\omega)t}}\right]^{k-j} \end{matrix} \right\} \tag{12.132}$$

$$= \{\alpha_k\}$$

if $\frac{\lambda}{\nu}$ is assumed to be an integer.

The last expression in equation (12.129) can be expressed as

$$\frac{e^{-(\nu-\omega)t}\frac{\nu s-\omega}{s-1}-\omega}{e^{-(\nu-\omega)t}\frac{\nu s-\omega}{s-1}-\nu}=\frac{\left(\nu e^{-(\nu-\omega)t}-\omega\right)s+\omega\left(1-e^{-(\nu-\omega)t}\right)}{\nu-\omega e^{-(\nu-\omega)t}-\nu\left[1-e^{-(\nu-\omega)t}\right]s} \tag{12.133}$$

that is of the form $\frac{as+bs}{c-ds}$. Its inversion is as follows

$$\left[\frac{\left(\nu e^{-(\nu-\omega)t}-\omega\right)s+\omega\left(1-e^{-(\nu-\omega)t}\right)}{\nu-\omega e^{-(\nu-\omega)t}-\nu\left[1-e^{-(\nu-\omega)t}\right]s}\right]^{a_0}$$

$$\triangleright\left\{\begin{array}{l}\left[\frac{1}{\nu-\omega e^{-(\nu-\omega)t}}\right]^{a_0}X\\ \sum_{j=0}^{\mathrm{Min}(a_0,k)}\left[\begin{array}{l}\binom{a_0}{j}\left(\nu e^{-(\nu-\omega)t}-\omega\right)^j\left(\omega\left(1-e^{-(\nu-\omega)t}\right)\right)^{a_0-j}\\ X\binom{a_0+k-j-1}{k-j-1}\left[\frac{\nu\left[1-e^{-(\nu-\omega)t}\right]}{\nu-\omega e^{-(\nu-\omega)t}}\right]^{k-j}\end{array}\right]\end{array}\right\}$$

$$=\{\beta_k\} \tag{12.134}$$

Recall that the first term for G(s) in equation (12.129) is $(\nu s-\omega)^{-\frac{\lambda}{\nu}}$. If $\frac{\lambda}{\nu}$ is an integer then write

$$\frac{1}{(\nu s-\omega)^{\frac{\lambda}{\nu}}}\triangleright\left\{\left[\frac{-1}{\omega}\right]^{\frac{\lambda}{\nu}}\binom{\frac{\lambda}{\nu}+k-1}{k-1}\left[\frac{\nu}{\omega}\right]^k\right\}=\{\gamma_k\} \tag{12.135}$$

This preliminary work allows the completion of the inversion of G(s), that is the convolution of three series, $\{\alpha_k\},\{\beta_k\}$, and $\{\gamma_k\}$. From Chapter 2, the inversion may be written as

$$G(s) \triangleright \left\{ \sum_{i=0}^{k} \sum_{j=0}^{k-i} \alpha_i \beta_j \gamma_{k-i-j} \right\} \tag{12.136}$$

where the terms for $\{\alpha_k\}, \{\beta_k\}$, and $\{\gamma_k\}$ are defined as in equations (12.132), (12.134), and (12.135), respectively.

Some simplification of this inversion is available when $a_0 = 0$. In this case the inversion of $G_t(s)$ can be shown to yield $P_n(t)$ that follows a negative binomial distribution. In this case $G_t(s)$ is

$$(\nu s - \omega)^{-\frac{\lambda}{\nu}} \left((\nu - \omega) e^{-(\nu-\omega)t} \right)^{\frac{\lambda}{\nu}} \left[\frac{\nu - \omega e^{-(\nu-\omega)t} - \nu\left(1 - e^{-(\nu-\omega)t}\right)s}{\nu s - \omega} \right]^{\frac{\lambda}{\nu}}$$

$$= \left[\frac{\dfrac{(\nu - \omega) e^{-(\nu-\omega)t}}{\nu - \omega e^{-(\nu-\omega)t}}}{1 - \dfrac{\nu\left(1 - e^{-(\nu-\omega)t}\right)}{\nu - \omega e^{-(\nu-\omega)t}} s} \right]^{\frac{\lambda}{\nu}} \tag{12.137}$$

that is the generating function for the negative binomial distribution, where

$$p = \frac{(\nu - \omega) e^{-(\nu-\omega)t}}{\nu - \omega e^{-(\nu-\omega)t}} : \; q = \frac{\nu\left(1 - e^{-(\nu-\omega)t}\right)}{\nu - \omega e^{-(\nu-\omega)t}} \tag{12.138}$$

We can compute the average number of patients with the disease as

$$\frac{\lambda\left(1 - e^{-(\nu-\omega)t}\right)}{(\nu - \omega) e^{-(\nu-\omega)t}} \tag{12.139}$$

and the variance as

$$\frac{\lambda\left(e^{-(\nu-\omega)t} - 1\right)\left[\nu - \omega e^{-(\nu-\omega)t}\right]}{\left[(\nu - \omega) e^{-(\nu-\omega)t}\right]^2} \tag{12.140}$$

See Bailey [2] for a discussion of this case.

12.8 Immigration–Birth–Emigration Model

12.8.1 Introduction

This additional three-force model is a combination of a birth process, immigration process, and emigration process. Patients are allowed to arrive, and can spread the disease to other patients. In addition, patients are removed by moving away at a constant rate. In the derivation of $G_t(s)$ we will identify a partial differential equation in both t and s.

12.8.2 Chapman–Kolmogorov forward equations

Begin the development of the family of difference equations that govern this process by using the above parameterization. If time is slowed down sufficiently, how can we compute n cases at time t + Δt so that only one event can occur in the time interval (t, t + Δt) ? This can happen one of four ways. The first way is if there are n - 1 patients at time t, and an arrival occurs with probability λΔt. A second possibility could be that, in addition, their could be n – 1 patients in the system at time t and a "birth" could occur, with probability (n – 1)νΔt. Also, with n + 1 patients at time t, a patient could leave the system with probability μΔt. Finally, there could be n patients in the system with neither an arrival nor a birth, or emigration occurring in time Δt. As in previous sections, we will assume that $P_0(t)$ = 0, and $P_{a_0}(t)=1$. It is important to distinguish between an immigration and a birth in these processes. Certainly both the immigration and birth processes lead to an increase in the number of diseased patients. However, the increase from immigration is constant over time, independent of the number of diseased patients. The birth process force is proportional to the number of diseased patients. The Chapman–Kolmogorov equation for the immigration–birth–emigration process for n ≥ 0 are

$$\begin{aligned} P_n(t+\Delta t) &= \lambda\Delta t P_{n-1}(t) + (n-1)\nu\Delta t P_{n-1}(t) + \\ &+ \mu\Delta t P_{n+1}(t) + P_n(t)\left(1-\lambda\Delta t - n\nu\Delta t - \mu\Delta t\right) \end{aligned} \tag{12.141}$$

We proceed with the development of the differential equation that represents this process by consolidating the terms involving Δt

$$\frac{P_n(t+\Delta t) - P_n(t)}{\Delta t} = \lambda P_{n-1}(t) + (n-1)\nu P_{n-1}(t) + \\ +\mu P_{n+1}(t) - \lambda P_n(t) - n\nu P_n(t) - \mu P_n(t) \tag{12.142}$$

Then take a limit, observing

$$\lim_{\Delta t \to 0} \frac{P_n(t+\Delta t) - P_n(t)}{\Delta t} = \lambda P_{n-1}(t) + (n-1)\nu P_{n-1}(t) \\ + \mu P_{n+1}(t) - \lambda P_n(t) - n\nu P_n(t) - \mu P_n(t) \tag{12.143}$$

Complete this process

$$\frac{dP_n(t)}{dt} = \lambda P_{n-1}(t) + \nu P_{n-1}(t)(n-1) + \mu P_{n+1}(t) \\ - \lambda P_n(t) - n\nu P_n(t) - \mu P_n(t) \tag{12.144}$$

12.8.4 Application of the generating function argument

Begin as always by first defining the generating function

$$G_t(s) = \sum_{n=0}^{\infty} s^n P_n(t) \tag{12.145}$$

and move forward with the conversion and consolidation of equation (12.144). Multiplying each side of this equation by s^n to find

$$s^n \frac{dP_n(t)}{dt} = \lambda s^n P_{n-1}(t) + \nu s^n P_{n-1}(t)(n-1) + \mu s^n P_{n+1}(t) - \lambda s^n P_n(t) \\ - \nu n s^n P_n(t) - \mu s^n P_n(t) \tag{12.146}$$

Taking a sum provides

$$\sum_{n=0}^{\infty} s^n \frac{dP_n(t)}{dt} = \lambda\sum_{n=0}^{\infty} s^n P_{n-1}(t) + v\sum_{n=0}^{\infty}(n-1)s^n P_{n-1}(t)$$
$$+\mu\sum_{n=0}^{\infty} s^n P_{n+1} - \lambda\sum_{n=0}^{\infty} s^n P_n(t) - v\sum_{n=0}^{\infty} ns^n P_n(t) - \mu\sum_{n=0}^{\infty} s^n P_n(t) \tag{12.147}$$

and interchanging the summation and derivative procedures reveals

$$\frac{d\sum_{n=o}^{\infty} s^n P_n(t)}{dt} = \lambda s\sum_{n=0}^{\infty} s^{n-1} P_{n-1}(t) + vs^2\sum_{n=0}^{\infty}(n-1)s^{n-2} P_{n-1}(t)$$
$$+\mu\sum_{n=0}^{\infty} s^n P_{n+1}(t) - \lambda\sum_{n=0}^{\infty} s^n P_n(t) - vs\sum_{n=0}^{\infty} ns^{n-1} P_n(t) - \mu\sum_{n=0}^{\infty} s^n P_n(t) \tag{12.148}$$

We can now recognize these summands as functions of $G_t(s)$ and write

$$\frac{\partial G_t(s)}{\partial t} = \lambda s G_t(s) + vs^2\frac{\partial G_t(s)}{\partial s} + \mu s^{-1} G_t(s) - \lambda G_t(s)$$
$$- vs\frac{\partial G_t(s)}{\partial s} - \mu G_t(s) \tag{12.149}$$

that can be simplified as follows

$$\frac{\partial G_t(s)}{\partial t} = \left[\lambda(s-1) + \mu(s^{-1}-1)\right] G_t(s) + vs(s-1)\frac{\partial G_t(s)}{\partial s} \tag{12.150}$$

leading to

$$\frac{\partial G_t(s)}{\partial t} - vs(s-1)\frac{\partial G_t(s)}{\partial s} = \left[\lambda(s-1) + \mu(s^{-1}-1)\right] G_t(s) \tag{12.151}$$

12.8.5 Solution of the partial differential equation

Recognize equation (12.151) as a partial differential equation in t and in s, but of the form which will allow us to find a general solution using a subsidiary set of equations. Write these equations as

$$\frac{dt}{1}=\frac{ds}{-\nu s(s-1)}=\frac{dG_t(s)}{\left[\lambda(s-1)+\mu\left(s^{-1}-1\right)\right]G_t(s)} \tag{12.152}$$

Continue with the first two terms of equation (12.152), evaluate these equalities in two combinations to provide information on the form of the generating function $G_t(s)$. From the second and third terms of equation (12.152) we have

$$\begin{aligned}
&\frac{ds}{-\nu s(s-1)}=\frac{dG_t(s)}{\left[\lambda(s-1)+\mu\left(s^{-1}-1\right)\right]G_t(s)}\\
&\frac{-\left[\lambda(s-1)+\mu\left(s^{-1}-1\right)\right]ds}{\nu s(s-1)}=\frac{dG_t(s)}{G_t(s)}\\
&\frac{-\lambda ds}{\nu s}-\frac{\mu\left(s^{-1}-1\right)ds}{\nu s(s-1)}=\frac{dG_t(s)}{G_t(s)}\\
&\frac{-\lambda ds}{\nu s}+\frac{\mu ds}{\nu s^2}=\frac{dG_t(s)}{G_t(s)}
\end{aligned} \tag{12.153}$$

Integrating both sides of the last equation in (12.153) reveals

$$\begin{aligned}
&\int\frac{-\lambda ds}{\nu s}+\int\frac{\mu ds}{\nu s^2}=\int\frac{dG_t(s)}{G_t(s)}\\
&\frac{-\lambda}{\nu}\ln(s)+\frac{\mu}{\nu s}=\ln G_t(s)\\
&\ln s^{-\frac{\lambda}{\nu}}+\frac{\mu}{\nu s}+C=\ln G_t(s)\\
&s^{-\frac{\lambda}{\nu}}e^{\frac{\mu}{\nu s}}C_1=G_t(s)
\end{aligned} \tag{12.154}$$

This development provides the form of $G_t(s) = s^{-\frac{\lambda}{\nu}} e^{\frac{\mu}{\nu s}} C$ that can be written in the form

$$G_t(s) = s^{-\frac{\lambda}{\nu}} e^{\frac{\mu}{\nu s}} \Phi(C_2) \tag{12.155}$$

The remaining task is to identify the form $\Phi(C_2)$. Using the first and second terms from equation (12.152) $\frac{dt}{1} = \frac{ds}{-\nu s(s-1)}$ the development of the constant term C_2 proceeds as follows:

$$\begin{aligned} \frac{dt}{1} &= \frac{ds}{-\nu s(s-1)} \\ -\nu dt &= \frac{ds}{s(s-1)} \\ \int -\nu dt &= \int \frac{ds}{s(s-1)} \end{aligned} \tag{12.156}$$

Integration of the left side gives

$$-\nu t = \int \frac{ds}{s(s-1)} \tag{12.157}$$

The term on the right side of equation (12.157) is

$$\begin{aligned} \int \frac{ds}{s(s-1)} &= \ln(s-1) - \ln s + C \\ &= \ln\left(\frac{s-1}{s}\right) + C \end{aligned} \tag{12.158}$$

Combining equations (12.157) and (12.158)

$$-\nu t = \ln\left(\frac{s-1}{s}\right) + C$$

$$C = \nu t + \ln\left(\frac{s-1}{s}\right) = -\nu t + \ln\left(\frac{s}{s-1}\right) \tag{12.159}$$

$$C_2 = \frac{s}{s-1}e^{-\nu t}$$

Combining $G_t(s) = s^{-\frac{\lambda}{\nu}} e^{\frac{\mu}{\nu s}} \Phi(C_2)$ with $C_2 = e^{-\nu t}\frac{s}{s-1}$ reveals

$$G_t(s) = s^{-\frac{\lambda}{\nu}} e^{\frac{\mu}{\nu s}} \Phi\left(e^{-\nu t}\frac{s}{s-1}\right) \tag{12.160}$$

allowing identification of the specific solution to the immigration-birth-emigration model.

12.8.6 Specific solution for partial differential equation

Having identified $G_t(s) = s^{-\frac{\lambda}{\nu}} e^{\frac{\mu}{\nu s}} \Phi\left(e^{-\nu t}\frac{s}{s-1}\right)$ we can now pursue a specific solution, beginning with the boundary conditions. At $t = 0$, there are a_0 patients with the disease therefore

$$G_0(s) = s^{a_0} = s^{-\frac{\lambda}{\nu}} e^{\frac{\mu}{\nu s}} \Phi\left(\frac{s}{s-1}\right) \tag{12.161}$$

Now, let $z = \frac{s}{s-1}$, then $s = \frac{z}{z-1}$ and substituting this result into equation (12.161)

$$\left(\frac{z}{z-1}\right)^{a_0} = \left(\frac{z}{z-1}\right)^{-\frac{\lambda}{\nu}} e^{\frac{\mu}{\nu\frac{z}{z-1}}} \Phi(z) \tag{12.162}$$

or

$$\Phi(z)=\left(\frac{z}{z-1}\right)^{a_0+\frac{\lambda}{v}} e^{-\frac{\mu}{v}} e^{\frac{\mu}{vz}} \tag{12.163}$$

allowing us to write

$$\begin{aligned} G_t(s) &= s^{-\frac{\lambda}{v}} e^{-\frac{\mu}{vs}} \Phi\left(e^{-vt}\frac{s}{s-1}\right) \\ &= s^{-\frac{\lambda}{v}} e^{\frac{\mu}{vs}} e^{-\frac{\mu}{v}} \left(\frac{e^{-vt}\frac{s}{s-1}}{e^{-vt}\frac{s}{s-1}-1}\right)^{a_0+\frac{\lambda}{v}} e^{\frac{\mu}{ve^{-vt}\frac{s}{s-1}}} \end{aligned} \tag{12.164}$$

and further simplification reveals

$$G_t(s) = s^{-\frac{\lambda}{v}} \left(\frac{se^{-vt}}{1-\left(1-e^{-vt}\right)s}\right)^{a_0+\frac{\lambda}{v}} e^{\frac{\mu}{v}\left(s^{-1}-1\right)} e^{\frac{\mu(s-1)}{vse^{-vt}}} \tag{12.165}$$

12.8.7 The Inversion of $G_t(s)$

The inversion of $G_t(s)$ as expressed in equation (12.165) is straightforward, requiring only some algebra to complete. The second expression on the right side of equation (12.165) is recognized as the genrating function for a negative binomial random variable.

$$\left(\frac{se^{-vt}}{1-\left(1-e^{-vt}\right)s}\right)^{a_0+\frac{\lambda}{v}} \triangleright \left\{\binom{a_0+\frac{\lambda}{v}-1}{k-1} e^{-kvt}\left(1-e^{-vt}\right)^{a_0+\frac{\lambda}{v}-k}\right\} \tag{12.166}$$

The expression $e^{\frac{\mu}{v}\left(s^{-1}-1\right)}$ may be inverted as follows:

$$e^{\frac{\mu}{\nu}\left(s^{-1}-1\right)} = e^{-\frac{\mu}{\nu}} e^{\frac{\mu}{\nu s}} = e^{-\frac{\mu}{\nu}} \left(1 + \frac{\left(\frac{\mu}{\nu}\right)}{1!} s^{-1} + \frac{\left(\frac{\mu}{\nu}\right)^2}{2!} s^{-2} + \frac{\left(\frac{\mu}{\nu}\right)^3}{3!} s^{-3} + \ldots + \right) \quad (12.167)$$

Finally, the rightmost expression on the right side of equation (12.165) is inverted as follows

$$e^{\frac{\mu(s-1)}{\nu s e^{-\nu t}}} = e^{\frac{\mu s}{\nu s e^{-\nu t}}} e^{\frac{-\mu}{\nu s e^{-\nu t}}} = e^{\frac{\mu}{\nu e^{-\nu t}}} e^{\frac{-\mu}{\nu s e^{-\nu t}}} \quad (12.168)$$

that may be expanded as

$$e^{\frac{\mu}{\nu e^{-\nu t}}} e^{\frac{-\mu}{\nu s e^{-\nu t}}} = e^{\frac{\mu}{\nu e^{-\nu t}}} \left(1 - \frac{\frac{\mu}{\nu e^{-\nu t}} s^{-1}}{1!} + \frac{\left(\frac{\mu}{\nu e^{-\nu t}}\right)^2 s^{-2}}{2!} - \frac{\left(\frac{\mu}{\nu e^{-\nu t}}\right)^3 s^{-3}}{3!} + \ldots + \right) \quad (12.169)$$

It remains to assemble this piecewise inversion into the final solution.

$$\left(\frac{s e^{-\nu t}}{1-\left(1-e^{-\nu t}\right)s} \right)^{a_0+\frac{\lambda}{\nu}} e^{\frac{\mu}{\nu}\left(1-s^{-1}\right)} e^{-\frac{\mu(s-1)}{\nu s e^{-\nu t}}}$$

$$\triangleright \left\{ e^{\frac{\mu}{\nu}\left(e^{\nu t}-1\right)} \sum_{j=0}^{\infty} \left[\binom{a_0+\frac{\lambda}{\nu}-1}{k+j-1} e^{-(k+j)\nu t} \left(1-e^{-\nu t}\right)^{a_0+\frac{\lambda}{\nu}-k-j} \right] \frac{\left[\frac{\mu}{\nu}\left(e^{\nu t}-1\right)\right]^j}{j!} \right\} \quad (12.170)$$

and the final application of the sliding rule reveals

$$P_n(t) = e^{\frac{\mu}{\nu}\left(e^{\nu t}-1\right)} \sum_{j=0}^{\infty} \left[\binom{a_0+\frac{\lambda}{\nu}-1}{k+j+\frac{\lambda}{\nu}-1} e^{-\left(k+j+\frac{\lambda}{\nu}\right)\nu t} \left(1-e^{-\nu t}\right)^{a_0-k-j} \right] \frac{\left[\frac{\mu}{\nu}\left(e^{\nu t}-1\right)\right]^j}{j!} \quad (12.171)$$

12.9 Immigration–Death–Emigration Model

12.9.1 Introduction

In this section, the model considers new arrivals, deaths, and emigrants. Patients are allowed to arrive, but cannot spread the disease. The patients with the disease can either die from (or be cured of) the disease. However, they can leave the system as well. This could describe the spread and eventual removal of a noncontagious disease or condition such as amyotrophic lateral sclerosis (Lou Gehrig's disease). We might expect in this derivation of G(s) a partial differential equation in both t and s. Its solution will be straightforward.

12.9.2 Motivation

In this circumstance, there are three processes working simultaneously (Figure 12.5). The first is the immigration force, bringing new cases. The second is the death force, that extinquishes the disease, either through the death of the patient, or the cure of the patient. The final force is the emigration force, that removes patients with the disease. Recall that the distinction between a death process and the emigration process is that the death process produces deaths with a force that is proportional to the number of patients with the disease, while the emigration process is a force that is independent of the number of diseased patients. Each of these three forces acts independently of the other two.

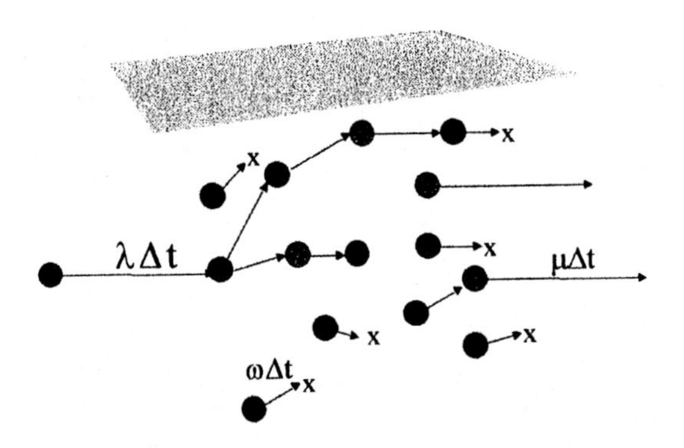

Figure 12.5 Immigration-death-emigration model: Patients with disease enter the population at a constant rate. Once in the population, patients can die or emigrate out.

12.9.3 Chapman–Kolmogorov forward equations

We begin the development of the family of difference equations that govern this process using the above parameterization. If time is slowed down sufficiently, how can we compute n cases at time $t + \Delta t$ allowing only one event and one event only to occur in the time interval $(t, t + \Delta t)$? This can happen one of four ways. The first way this may happen is if there are $n - 1$ patients at time t, and an arrival occurs with probability $\lambda\Delta t$. This is the only way the number of patients with the disease can increase. However, there are two ways by which the number of patients with the disease could decrease. The first is that there are $n + 1$ patients at time t, and a death occurs, an event that occurs with probability $(n + 1)\omega\Delta t$. In addition, again with $n + 1$ patients at time t, a patient could leave the system with probability $\mu\Delta t$. Finally, there could be n patients in the system with neither an arrival, a death, or an emigration occurring in time Δt. Assume that $P_0(t) = 0$, and $P_{a_0}(0) = 1$. For $n \geq 0$, the Chapman–Kolmogorov equations for this system are

$$\begin{aligned} P_n(t+\Delta t) &= P_{n-1}(t)\lambda\Delta t + P_{n+1}(t)(n+1)\omega\Delta t \\ &\quad + P_{n+1}(t)\mu\Delta t + P_n(t)\left(1 - \lambda\Delta t - n\omega\Delta t - \mu\Delta t\right) \end{aligned} \tag{12.172}$$

Collect all of the terms involving Δt on the left side of the equation and take a limit as Δt decreases in size to zero.

$$\begin{aligned} \frac{P_n(t+\Delta t) - P_n(t)}{\Delta t} &= \lambda P_{n-1}(t) + (n+1)\omega P_{n+1}(t) + \mu P_{n+1}(t) \\ &\quad - \lambda P_n(t) - n\omega t P_n(t) - \mu P_n(t) \\ \lim_{\Delta t \to 0} \frac{P_n(t+\Delta t) - P_n(t)}{\Delta t} &= \lambda P_{n-1}(t) + (n+1)\omega P_{n+1}(t) + \mu P_{n+1}(t) \\ &\quad - \lambda P_n(t) - n\omega P_n(t) - \mu P_n(t) \end{aligned} \tag{12.173}$$

leading to

$$\frac{dP_n(t)}{dt} = \lambda P_{n-1}(t) + (n+1)\omega P_{n+1}(t) + \mu P_{n+1}(t) - \lambda P_n(t) - n\omega P_n(t) - \mu P_n(t) \quad (12.174)$$

12.9.4 Application of the generating function argument

Begin as always by first defining the generating function

$$G_t(s) = \sum_{n=0}^{\infty} s^n P_n(t) \quad (12.175)$$

and move forward with the conversion and consolidation of equation (12.174)

$$\frac{dP_n(t)}{dt} = \lambda P_{n-1}(t) + \omega(n+1)P_{n+1}(t) + \mu P_{n+1}(t) - \lambda P_n(t) - \omega n P_n(t) - \mu P_n(t)$$

$$s^n \frac{dP_n(t)}{dt} = \lambda s^n P_{n-1}(t) + \omega(n+1) s^n P_{n+1}(t) + \mu s^n P_{n+1}(t) - \lambda s^n P_n(t) - \omega n s^n P_n(t) - \mu s^n P_n(t) \quad (12.176)$$

Summing both the sides

$$\sum_{n=o}^{\infty} s^n \frac{dP_n(t)}{dt} = \lambda \sum_{n=o}^{\infty} s^n P_{n-1}(t) + \omega \sum_{n=o}^{\infty} (n+1) s^n P_{n+1}(t) + \mu \sum_{n=0}^{\infty} s^n P_{n+1}(t) - \lambda \sum_{n=o}^{\infty} s^n P_n(t) - \omega \sum_{n=o}^{\infty} n s^n P_n(t) - \mu \sum_{n=0}^{\infty} s^n P_n(t)$$

$$\frac{d\sum_{n=0}^{\infty} s^n P_n(t)}{dt} = \lambda s \sum_{n=o}^{\infty} s^{n-1} P_{n-1}(t) + \omega \sum_{n=o}^{\infty} (n+1) s^n P_{n+1}(t) + \mu \sum_{n=o}^{\infty} s^n P_{n+1}(t) - \lambda \sum_{n=o}^{\infty} s^n P_n(t) - \omega s \sum_{n=o}^{\infty} n s^{n-1} P_n(t) - \mu \sum_{n=0}^{\infty} s^n P_n(t) \quad (12.177)$$

Recognize these summands as functions of $G_t(s)$ and write

$$
\begin{aligned}
\frac{\partial G_t(s)}{\partial t} &= \lambda s G_t(s) + \omega \frac{\partial G_t(s)}{\partial s} + \mu s^{-1} G_t(s) - \lambda G_t(s) \\
&\quad - \omega s \frac{\partial G_t(s)}{\partial s} - \mu G_t(s) \\
\frac{\partial G_t(s)}{\partial t} &= \left[\lambda(s-1) + \mu\left(s^{-1}-1\right)\right] G_t(s) + \omega(1-s)\frac{\partial G_t(s)}{\partial s} \\
\frac{\partial G_t(s)}{\partial t} &= \omega(1-s)\frac{\partial G_t(s)}{\partial s} + \left[\lambda(s-1) + \mu\left(s^{-1}-1\right)\right] G_t(s)
\end{aligned}
\tag{12.178}
$$

that becomes

$$
\frac{\partial G_t(s)}{\partial t} - \omega(1-s)\frac{\partial G_t(s)}{\partial s} - \left[\lambda(s-1) + \mu\left(s^{-1}-1\right)\right] G_t(s) = 0 \tag{12.179}
$$

12.9.5 Solution of the partial differential equation

Recognizing the last line of equation (12.179) as a partial differential equation in t and in s, but of the form that will allow us to find a general solution using a subsidiary set of equations. Write these equations as

$$
\frac{dt}{1} = \frac{ds}{-\omega(1-s)} = \frac{dG_t(s)}{\left[\lambda(s-1) + \mu\left(s^{-1}-1\right)\right] G_t(s)} \tag{12.180}
$$

Continuing with the first two terms of equation (12.180), one can evaluate these equalities in two combinations to provide information on the form of the generating function $G_t(s)$. From the second and third terms of equation (12.180) we have

$$\frac{ds}{-\omega(1-s)}=\frac{dG_t(s)}{\left[\lambda(s-1)+\mu(s^{-1}-1)\right]G_t(s)}$$

$$\frac{\left[\lambda(s-1)+\mu(s^{-1}-1)\right]ds}{\omega(1-s)}=\frac{dG_t(s)}{G_t(s)}$$

$$\frac{\lambda ds}{\omega}-\frac{\mu(s^{-1}-1)ds}{\omega(1-s)}=\frac{dG_t(s)}{G_t(s)} \tag{12.181}$$

$$\frac{\lambda ds}{\omega}-\frac{\mu(1-s)ds}{\omega s(1-s)}=\frac{dG_t(s)}{G_t(s)}$$

$$\frac{\lambda ds}{\omega}-\frac{\mu ds}{\omega s}=\frac{dG_t(s)}{G_t(s)}$$

Proceeding as before

$$\frac{\lambda ds}{\omega}-\frac{\mu ds}{\omega s}=\frac{dG_t(s)}{G_t(s)}$$

$$\int\frac{\lambda ds}{\omega}-\int\frac{\mu ds}{\omega s}=\int\frac{dG_t(s)}{G_t(s)} \tag{12.182}$$

$$\frac{\lambda}{\omega}s-\frac{\mu}{\omega}\ln s+C=\ln G_t(s)$$

$$\frac{\lambda}{\omega}s+\ln s^{\frac{-\mu}{\omega}}+C=\ln G_t(s)$$

Exponentiating both sides gives

$$e^{\frac{\lambda}{\omega}s}s^{\frac{-\mu}{\omega}}C=G_t(s) \tag{12.183}$$

that can be written in the form $G_t(s)=e^{\frac{\lambda}{\omega}s}s^{\frac{-\mu}{\omega}}\Phi(C_2)$. The remaining test is to identify the form $\Phi(C_2)$. Using the first and second terms from equation (12.180) write

$$\frac{dt}{1}=\frac{ds}{-\omega(1-s)} \tag{12.184}$$

The above equation may be solved as follows.

$$\frac{dt}{1} = \frac{ds}{-\omega(1-s)}$$

$$\omega dt = \frac{-ds}{(1-s)}$$

$$\int \omega dt = \int \frac{-ds}{(1-s)} \tag{12.185}$$

$$\omega t + C = \ln(1-s)$$

$$C = -\omega t + \ln(1-s)$$

From which find

$$C_2 = e^{-\omega t}(1-s) \tag{12.186}$$

Combining $G_t(s) = e^{-\frac{\lambda}{\omega}s} s^{\frac{\mu}{\omega}} \Phi(C_2)$ with $C_2 = e^{-\omega t}(1-s)$ gives

$$G_t(s) = e^{\frac{\lambda}{\omega}s} s^{\frac{-\mu}{\omega}} \Phi\left(e^{-\omega t}(1-s)\right) \tag{12.187}$$

and we are now ready to identify the specific solution to the immigration-death-emigration model.

12.9.6 Specific solution for partial differential equation

Having identified $G_t(s) = e^{\frac{\lambda}{\omega}s} s^{\frac{-\mu}{\omega}} \Phi\left(e^{-\omega t}(1-s)\right)$ now pursue a specific solution, beginning with the boundary conditions. At t = 0, if there are a_0 patients with the illness, therefore

$$G_0(s) = s^{a_0} = e^{\frac{\lambda}{\omega}s} s^{-\frac{\mu}{\omega}} \Phi(1-s) \tag{12.188}$$

Now, let z = 1−s, then $s = 1 - z$ and substituting this result into equation (12.188) reveals

$$(1-z)^{a_0} = e^{\frac{\lambda}{\omega}(1-z)} (1-z)^{-\frac{\mu}{\omega}} \Phi(z) \tag{12.189}$$

or

$$\Phi(z) = (1-z)^{a_0+\frac{\mu}{\omega}} e^{-\frac{\lambda}{\omega}(1-z)} \tag{12.190}$$

and write

$$G_t(s) = s^{-\frac{\mu}{\omega}}\left[1-e^{-\omega t}(1-s)\right]^{a_0+\frac{\mu}{\omega}} e^{\frac{\lambda}{\omega}\left(1-e^{-\omega t}\right)[s-1]} \tag{12.191}$$

12.9.7 The Inversion of $G_t(s)$

The inversion of $G_t(s)$ can be carried out in a straightforward manner. The first term on the right side of equation (12.191) will require use of only the sliding tool. Then the inversion of the second term of this equation we can recognize as

$$\left(1-e^{-\omega t}(1-s)\right)^{a_0+\frac{\mu}{\omega}} = \left[\left(1-e^{-\omega t}\right)+se^{-\omega t}\right]^{a_0+\frac{\mu}{\omega}} \tag{12.192}$$

that will be recalled from Chapter 2 is the probability generating function of the binomial distribution with $p = e^{-\omega t}$. Thus

$$\left[\left(1-e^{-\omega t}\right)+se^{-\omega t}\right]^{a_0+\frac{\mu}{\omega}} \rhd \left\{\binom{a_0+\frac{\mu}{\omega}}{k} e^{-\omega t k}\left(1-e^{-\omega t}\right)^{a_0+\frac{\mu}{\omega}-k}\right\} \tag{12.193}$$

The remaining term is inverted as follows

$$e^{\frac{\lambda}{\omega}\left(1-e^{-\omega t}\right)[s-1]} \rhd \left\{\frac{\left[\frac{\lambda}{\omega}\left(1-e^{-\omega t}\right)\right]^k}{k!} e^{-\frac{\lambda}{\omega}\left(1-e^{-\omega t}\right)}\right\} \tag{12.194}$$

The inversion may now proceed as the product of the convolution of a series with a finite number of terms and those with an infinite number of terms. Recall from Chapter 2 that if and $G_2(s) \rhd \{b_k\}$ $k = 0,1,\ldots,\infty$ then

$$G_1(s)G_2(s) \triangleright \left\{ \sum_{j=0}^{\min(n,k)} a_j b_{k-j} \right\} \qquad (12.195)$$

This observation leads to the inversion of $G_t(s)$ as

$$G_t(s)$$

$$\triangleright \left\{ e^{-\frac{\lambda}{\omega}\left(1-e^{-\omega t}\right)} \sum_{j=0}^{\min\left(a_0+\frac{\mu}{\omega},k+\frac{\mu}{\omega}\right)} \binom{a_0+\frac{\mu}{\omega}}{j} e^{-j\omega t}\left(1-e^{-\omega t}\right)^{a_0+\frac{\mu}{\omega}-j} \frac{\left(\frac{\lambda}{\omega}\left[1-e^{-\omega t}\right]\right)^{k+\frac{\mu}{\omega}-j}}{\left(k+\frac{\mu}{\omega}-j\right)!} \right\}$$

(12.196)

that is simply the convolution of a binomial and Poisson process.

12.10 Immigration–Birth–Death–Emigration Model

12.10.1 Introduction

The last model to be considered is a combination of the models developed in the previous section. This model will be the most complicated representation of the spread of a contagious disease that will be considered. Patients are allowed to arrive and spread the disease to other patients. In addition, patients are removed either through death or through cure with subsequent immunity. In the derivation of $G_t(s)$ we will identify a partial differential equation in both t and s. Its solution will be straightforward, but there may be some complications in the final simplification of $G_t(s)$.

12.10.2 Motivation

In this circumstance, four processes are proceeding simultaneously (Figure 12.6) In this set of circumstances, there are four forces to contend with. The first is the immigration force, bringing new cases. The second force is the birth force, spreading the disease. The final force is the death force, that acts to reduce the number of cases. Each of these three forces acts independently of the other two.

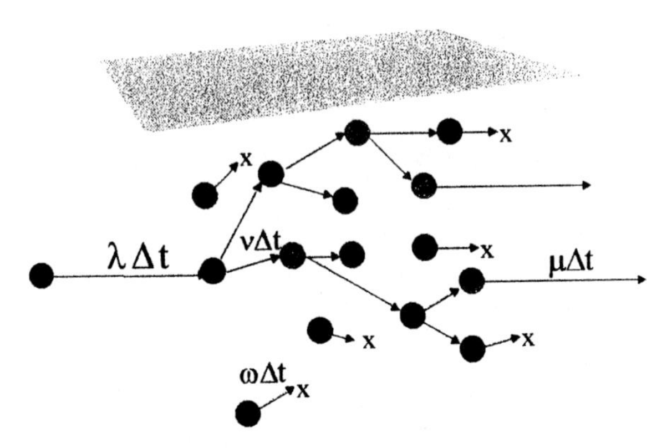

Figure 12.6 Immigration-birth-death-emigration model: Patients with disease enter the system at a constant rate, leading to the spread of the disease in the population. Deaths occur in the population, while other patients emigrate away.

12.10.3 Chapman–Kolmogorov forward equations

Begin the development of the family of difference equations that govern this process by using the some useful notation. If time is slowed down sufficiently, how can we compute n cases at time $t + \Delta t$ allowing only one event and one event only to occur in the time interval $(t, t + \Delta t)$? This can happen one of five ways. The first way this may happen is if there are n - 1 patients at time t, and an arrival occurs with probability $\lambda\Delta t$. A second possibility could be that, in addition, there could be n – 1 patients in the system at time t and a "birth" could occur, with probability $(n - 1)v\Delta t$. Also, with n + 1 patients at time t, there could be a death, an event that occurs with probability $(n + 1)\omega\Delta t$. With n + 1 patients at time t, a patient could leave the system with probability $\mu\Delta t$. Finally, there could be n patients in the system with neither an arrival, nor a birth, nor a death, nor emigration occurring in time Δt. As in previous sections, we will assume that $P_0(t) = 0$, and $P_{a_0}(0) = 1$.

It is important to distinguish between the immigration and emigration from the birth and death processes. Certainly both the immigration and birth processes lead to an increase in the number of diseased patients. However, the increase from immigration is constant over time, independent of the number of diseased patients. The birth process force is proportional to the number of patients with the disease. Similarly, people emigrate independent of the number of ill patients. Chapman–Kolmogorov equation for the birth–death–emigration process for $n \geq 0$ are

$$
\begin{aligned}
P_n(t+\Delta t) &= \lambda\Delta t P_{n-1}(t) + (n-1)\nu\Delta t P_{n-1}(t) + (n+1)\omega\Delta t P_{n+1}(t) \\
&+ \mu\Delta t P_{n+1}(t) + P_n(t)\left(1-\lambda\Delta t - n\nu\Delta t - n\omega\Delta t - \mu\Delta t\right)
\end{aligned} \tag{12.197}
$$

We proceed with the development of the differential equation that represents this process by consolidating the terms involving Δt.

$$
\begin{aligned}
\frac{P_n(t+\Delta t) - P_n(t)}{\Delta t} &= \lambda P_{n-1}(t) + (n-1)\nu P_{n-1}(t) + (n+1)\omega P_{n+1}(t) \\
+\mu P_{n+1}(t) &- \lambda P_n(t) - n\nu P_n(t) - n\omega P_n(t) - \mu P_n(t)
\end{aligned} \tag{12.198}
$$

Then take a limit, observing

$$
\begin{aligned}
\lim_{\Delta t \to 0} \frac{P_n(t+\Delta t) - P_n(t)}{\Delta t} &= \lambda P_{n-1}(t) + (n-1)\nu P_{n-1}(t) + (n+1)\omega P_{n+1}(t) \\
&- \mu P_{n+1}(t) - \lambda P_n(t) - n\nu P_n(t) - n\omega P_n(t) - \mu P_n(t)
\end{aligned} \tag{12.199}
$$

Complete this process:

$$
\begin{aligned}
\frac{dP_n(t)}{dt} &= \lambda P_{n-1}(t) + \nu P_{n-1}(t)(n-1) + \omega(n+1)P_{n+1}(t) - \mu P_{n+1}(t) \\
&- \lambda P_n(t) - n\nu P_n(t) - n\omega P_n(t) - \mu P_n(t)
\end{aligned} \tag{12.200}
$$

12.10.4 Application of the generating function argument

Begin as always by first defining the generating function

$$
G_t(s) = \sum_{n=0}^{\infty} s^n P_n(t) \tag{12.201}
$$

and move forward with the conversion and consolidation of equation (12.200). Multiply each side this equation by s^n to find

$$s^n \frac{dP_n(t)}{dt} = \lambda s^n P_{n-1}(t) + \nu s^n P_{n-1}(t)(n-1) + \omega s^n P_{n+1}(t)(n+1)$$
$$+ \mu s^n P_{n+1}(t) - \lambda s^n P_n(t) - \nu n s^n P_n(t) - \omega n s^n P_n(t) - \mu s^n P_n(t) \tag{12.202}$$

Taking a sum provides

$$\sum_{n=o}^{\infty} s^n \frac{dP_n(t)}{dt} = \lambda \sum_{n=o}^{\infty} s^n P_{n-1}(t) + \nu \sum_{n=o}^{\infty} (n-1) s^n P_{n-1}(t) + \omega \sum_{n=o}^{\infty} (n+1) s^n P_{n+1}(t)$$
$$+\mu \sum_{n=0}^{\infty} s^n P_{n+1} - \lambda \sum_{n=o}^{\infty} s^n P_n(t) - \nu \sum_{n=o}^{\infty} n s^n P_n(t) - \omega \sum_{n=o}^{\infty} n s^n P_n(t) - \mu \sum_{n=0}^{\infty} s^n P_n(t) \tag{12.203}$$

and interchanging the summation and derivative procedures gives

$$\frac{d\sum_{n=o}^{\infty} s^n P_n(t)}{dt} = \lambda s \sum_{n=o}^{\infty} s^{n-1} P_{n-1}(t) + \nu s^2 \sum_{n=o}^{\infty} (n-1) s^{n-2} P_{n-1}(t) + \omega \sum_{n=o}^{\infty} (n+1) s^n P_{n+1}(t)$$
$$+\mu \sum_{n=o}^{\infty} s^n P_{n+1}(t) - \lambda \sum_{n=o}^{\infty} s^n P_n(t) - \nu s \sum_{n=o}^{\infty} n s^{n-1} P_n(t) - \omega s \sum_{n=o}^{\infty} n s^{n-1} P_n(t) - \mu \sum_{n=0}^{\infty} s^n P_n(t) \tag{12.204}$$

We can now recognize these summands as functions of $G_t(s)$ and write

$$\frac{\partial G_t(s)}{\partial t} = \lambda s G_t(s) + \nu s^2 \frac{\partial G_t(s)}{\partial s} + \omega \frac{\partial G_t(s)}{\partial s} + \mu s^{-1} G_t(s) - \lambda G_t(s)$$
$$- \nu s \frac{\partial G_t(s)}{\partial s} - \omega s \frac{\partial G_t(s)}{\partial s} - \mu G_t(s) \tag{12.205}$$

that can be simplified as follows

$$\frac{\partial G_t(s)}{\partial t} = \left[\lambda(s-1) + \mu(s^{-1}-1)\right] G_t(s) + \nu s(s-1) \frac{\partial G_t(s)}{\partial s} + \omega(1-s) \frac{\partial G_t(s)}{\partial s}$$
$$\frac{\partial G_t(s)}{\partial t} = (\nu s - \omega)(s-1) \frac{\partial G_t(s)}{\partial s} + \left[\lambda(s-1) + \mu(s^{-1}-1)\right] G_t(s)$$

(12.206)

leading to

$$\frac{\partial G_t(s)}{\partial t} - (\nu s-\omega)(s-1)\frac{\partial G_t(s)}{\partial s} - \left[\lambda(s-1)+\mu\left(s^{-1}-1\right)\right]G_t(s) = 0 \quad (12.207)$$

12.10.5 Solution of the partial differential equation

Recognize the last line of equation (12.207) as a partial differential equation in t and in s, but of the form that will allow us to find a general solution using a subsidiary set of equations. Write these equations as

$$\frac{dt}{1} = \frac{ds}{-(\nu s-\omega)(s-1)} = \frac{dG_t(s)}{\left[\lambda(s-1)+\mu\left(s^{-1}-1\right)\right]G_t(s)} \quad (12.208)$$

Continuing with the first two terms of equation (12.208), evaluate these equalities in two combinations to provide information on the form of the generating function $G_t(s)$. From the second and third terms of equation (12.208) we have

$$\begin{aligned}
&\frac{ds}{-(\nu s-\omega)(s-1)} = \frac{dG_t(s)}{\left[\lambda(s-1)+\mu\left(s^{-1}-1\right)\right]G_t(s)} \\
&\frac{\left[\lambda(s-1)+\mu\left(s^{-1}-1\right)\right]ds}{-(\nu s-\omega)(s-1)} = \frac{dG_t(s)}{G_t(s)} \\
&\frac{-\lambda ds}{(\nu s-\omega)} - \frac{\mu\left(s^{-1}-1\right)ds}{(\nu s-\omega)(s-1)} = \frac{dG_t(s)}{G_t(s)} \\
&-\frac{\lambda ds}{(\nu s-\omega)} - \frac{\mu(1-s)ds}{s(\nu s-\omega)(s-1)} = \frac{dG_t(s)}{G_t(s)} \\
&\frac{-\lambda ds}{(\nu s-\omega)} + \frac{\mu ds}{s(\nu s-\omega)} = \frac{dG_t(s)}{G_t(s)}
\end{aligned} \quad (12.209)$$

Using partial fractions,

$$\frac{\mu}{s(\nu s-\omega)}=\frac{\mu\nu}{\omega}\left[\frac{1}{(\nu s-\omega)}\right]-\frac{\mu}{\omega}\left[\frac{1}{s}\right] \tag{12.210}$$

Continue

$$\left(\frac{-\lambda}{(\nu s-\omega)}+\frac{\mu\nu}{\omega}\left[\frac{1}{(\nu s-\omega)}\right]-\frac{\mu}{\omega}\left[\frac{1}{s}\right]\right)ds = \frac{dG_t(s)}{G_t(s)}$$

$$\left(\frac{-\lambda+\frac{\mu\nu}{\omega}}{(\nu s-\omega)}-\frac{\mu}{\omega}\left[\frac{1}{s}\right]\right)ds = \frac{dG_t(s)}{G_t(s)} \tag{12.211}$$

$$\int\left(\frac{-\lambda+\frac{\mu\nu}{\omega}}{(\nu s-\omega)}-\frac{\mu}{\omega}\left[\frac{1}{s}\right]\right)ds = \int\frac{dG_t(s)}{G_t(s)}$$

Continuing

$$\left(\frac{-\lambda}{\nu}+\frac{\mu}{\omega}\right)\ln(\nu s-\omega)-\frac{\mu}{\omega}\ln(s)+C=\ln G_t(s)$$

$$\frac{-\lambda}{\nu}\ln(\nu s-\omega)+\frac{\mu}{\omega}\ln\left[\frac{(\nu s-\omega)}{s}\right]+C=\ln G_t(s) \tag{12.212}$$

$$\ln(\nu s-\omega)^{-\frac{\lambda}{\nu}}+\ln\left[\frac{(\nu s-\omega)}{s}\right]^{\frac{\mu}{\omega}}+C=\ln G_t(s)$$

$$(\nu s-\omega)^{-\frac{\lambda}{\nu}}\left[\frac{(\nu s-\omega)}{s}\right]^{\frac{\mu}{\omega}}C=G_t(s) \tag{12.213}$$

This development provides the form of $G_t(s)=(\nu s-\omega)^{-\frac{\lambda}{\nu}+\frac{\mu}{\omega}}s^{-\frac{\mu}{\omega}}C$ that can be written in the form

$$G_t(s)=(\nu s-\omega)^{\frac{-\lambda}{\nu}+\frac{\mu}{\omega}}s^{-\frac{\mu}{\omega}}\Phi(C_2) \tag{12.214}$$

The remaining task is to identify the form $\Phi(C_2)$. Using the first and second terms from equation (12.208) write

$$\frac{dt}{1} = \frac{ds}{-(\nu s-\omega)(s-1)} \tag{12.215}$$

Using partial fractions,

$$\frac{1}{(\nu s-\omega)(s-1)} = \frac{1/(\nu-\omega)}{s-1} + \frac{-\nu/(\nu-\omega)}{\nu s-\omega} \tag{12.216}$$

So equation (12.215) may be written as

$$\begin{aligned} -dt &= \frac{1/(\nu-\omega)}{s-1} - \frac{\nu/(\nu-\omega)}{\nu s-\omega} \\ dt &= \frac{\nu/(\nu-\omega)}{\nu s-\omega} - \frac{1/(\nu-\omega)}{s-1} \\ \int dt &= \int\left[\frac{\nu/(\nu-\omega)}{\nu s-\omega} - \frac{1/(\nu-\omega)}{s-1}\right] ds \end{aligned} \tag{12.217}$$

Continuing

$$\begin{aligned} t+C &= \frac{1}{\nu-\omega}\ln(\nu s-\omega) - \frac{1}{\nu-\omega}\ln(s-1) = \frac{1}{\nu-\omega}\ln\left(\frac{\nu s-\omega}{s-1}\right) \\ (\nu-\omega)t+C_2 &= \ln\left(\frac{\nu s-\omega}{s-1}\right) \\ C_2 &= -(\nu-\omega)t+\ln\left(\frac{\nu s-\omega}{s-1}\right) \\ C_2 &= e^{-(\nu-\omega)t}\left(\frac{\nu s-\omega}{s-1}\right) \end{aligned} \tag{12.218}$$

From which

$$C_1 = \Phi(C_2) = \Phi\left[e^{-(\nu-\omega)t}\left(\frac{\nu s-\omega}{s-1}\right)\right] \tag{12.219}$$

Combining $G_t(s) = (\nu s-\omega)^{-\frac{\lambda}{\nu}+\frac{\mu}{\omega}} s^{-\frac{\mu}{\omega}}\Phi(C_2)$ with $C_2 = e^{-(\nu-\omega)t}\left(\frac{\nu s-\omega}{s-1}\right)$ reveals

$$G_t(s) = (\nu s-\omega)^{-\frac{\lambda}{\nu}+\frac{\mu}{\omega}} s^{-\frac{\mu}{\omega}}\Phi\left(e^{-(\nu-\omega)t}\left(\frac{\nu s-\omega}{s-1}\right)\right) \tag{12.220}$$

allowing identification of the specific solution to the immigration-birth-death-emigration model.

12.10.6 Specific solution for partial differential equation

Having identified $G_t(s) = (\nu s-\omega)^{-\frac{\lambda}{\nu}+\frac{\mu}{\omega}} s^{-\frac{\mu}{\omega}}\Phi\left(e^{-(\nu-\omega)t}\left(\frac{\nu s-\omega}{s-1}\right)\right)$ we can now pursue a specific solution, beginning with the boundary conditions. At t = 0, there are a_0 patients with the disease, therefore

$$G_0(s) = s^{a_0} = (\nu s-\omega)^{-\frac{\lambda}{\nu}+\frac{\mu}{\omega}} s^{-\frac{\mu}{\omega}}\Phi\left(\frac{\nu s-\omega}{s-1}\right) \tag{12.221}$$

Now, let $z = \frac{\nu s-\omega}{s-1}$, then $s = \frac{z-\omega}{z-\nu}$ and substituting this result into equation (12.221), and

$$\left(\frac{z-\omega}{z-\nu}\right)^{a_0} = \left(\nu\frac{z-\omega}{z-\nu}-\omega\right)^{-\frac{\lambda}{\nu}+\frac{\mu}{\omega}}\left[\frac{z-\omega}{z-\nu}\right]^{-\frac{\mu}{\omega}}\Phi(z) \tag{12.222}$$

or

$$\Phi(z) = \left(\frac{z-\omega}{z-\nu}\right)^{a_0}\left(\nu\frac{z-\omega}{z-\nu}-\omega\right)^{\frac{\lambda}{\nu}-\frac{\mu}{\omega}}\left[\frac{z-\omega}{z-\nu}\right]^{\frac{\mu}{\omega}} \tag{12.223}$$

allowing us to write

$$G_t(s) = (\nu s - \omega)^{\frac{-\lambda}{\nu}+\frac{\mu}{\omega}} s^{-\frac{\mu}{\omega}} \Phi\left(e^{-(\nu-\omega)t}\left(\frac{\nu s - \omega}{s-1}\right)\right)$$

$$= (\nu s - \omega)^{-\frac{\lambda}{\nu}+\frac{\mu}{\omega}} s^{-\frac{\mu}{\omega}} \left[\frac{\left(\nu e^{-(\nu-\omega)t} - \omega\right)s + \omega\left(1 - e^{-(\nu-\omega)t}\right)}{\nu\left(e^{-(\nu-\omega)t} - 1\right)s + \left(\nu - \omega e^{-(\nu-\omega)t}\right)}\right]^{a_0+\frac{\mu}{\omega}} \tag{12.224}$$

$$\text{x}\left(\nu\frac{e^{-(\nu-\omega)t}\left(\frac{\nu s-\omega}{s-1}\right)-\omega}{e^{-(\nu-\omega)t}\left(\frac{\nu s-\omega}{s-1}\right)-\nu}-\omega\right)^{\frac{\lambda}{\nu}-\frac{\mu}{\omega}}$$

12.10.7 The Inversion of $G_t(s)$

The inversion of $G_t(s)$ is straightforward, requiring only some algebra to complete. Begin by rewriting the last term on the right side of equation (12.224) as

$$\left(\nu\frac{e^{-(\nu-\omega)t}\left(\frac{\nu s-\omega}{s-1}\right)-\omega}{e^{-(\nu-\omega)t}\left(\frac{\nu s-\omega}{s-1}\right)-\nu}-\omega\right)^{\frac{\lambda}{\nu}-\frac{\mu}{\omega}} = \left[\frac{e^{-(\nu-\omega)t}(\nu-\omega)(\nu s-\omega)}{\nu\left(e^{-(\nu-\omega)t}-1\right)s+\left(\nu-\omega e^{-(\nu-\omega)t}\right)}\right]^{\frac{\lambda}{\nu}-\frac{\mu}{\omega}} \tag{12.225}$$

Substituting equation (12.225) into equation (12.224) and simplifying reveals

$$G_t(s) = s^{-\frac{\mu}{\omega}}\left[e^{-(\nu-\omega)t}(\nu-\omega)\right]^{\frac{\lambda}{\nu}-\frac{\mu}{\omega}} \frac{\left[\left(\nu e^{-(\nu-\omega)t}-\omega\right)s+\omega\left(1-e^{-(\nu-\omega)t}\right)\right]^{a_0+\frac{\mu}{\omega}}}{\left[\nu\left(e^{-(\nu-\omega)t}-1\right)s+\left(\nu-\omega e^{-(\nu-\omega)t}\right)\right]^{a_0+\frac{\lambda}{\nu}}} \tag{12.226}$$

This expression, despite its complicated appearance, may be inverted in a straightforward manner. As has been our approach in each of Chapter 11 and this chapter, we will take each expression in the final equation for $G_t(s)$ one at a time, inverting each term. After completing this procedure, we will be in the position to use a convolution argument from Chapter 2 to complete the inversion.

The first term in on the right side of equation will involve only the sliding tool. The expression $\left[e^{-(\nu-\omega)t}(\nu-\omega)\right]^{\frac{\lambda}{\nu}-\frac{\mu}{\omega}}$ is a constant with respect to s, and the final term $\frac{\left[\left(\nu e^{-(\nu-\omega)t}-\omega\right)s+\omega\left(1-e^{-(\nu-\omega)t}\right)\right]^{a_0+\frac{\mu}{\omega}}}{\left[\nu\left(e^{-(\nu-\omega)t}-1\right)s+\left(\nu-\omega e^{-(\nu-\omega)t}\right)\right]^{a_0+\frac{\lambda}{\nu}}}$ is of the form $\frac{[as+b]^{n_1}}{[c+ds]^{n_2}}$.

From the work in Chapter 2 using the binomial theorem, the application of the derivative process to the denominator, and a convolution, it can be shown that

$$\frac{[as+b]^{n_1}}{[c+ds]^{n_2}} \triangleright \left[\frac{1}{c}\right]^{n_2} \sum_{j=0}^{\min(k,n_1)} \binom{n_2+j-1}{j} \left[\frac{d}{c}\right]^j \binom{n_1}{k-j} b^{k-j} a^{n_1-k+j} \tag{12.227}$$

Applying this result to the final expression in equation (12.226) reveals

$$\frac{\left[\left(\nu e^{-(\nu-\omega)t}-\omega\right)s+\omega\left(1-e^{-(\nu-\omega)t}\right)\right]^{a_0+\frac{\mu}{\omega}}}{\left[\left(\nu-\omega e^{-(\nu-\omega)t}\right)+\nu\left(e^{-(\nu-\omega)t}-1\right)s+\right]^{a_0+\frac{\lambda}{\nu}}}$$

$$\triangleright \left[\frac{1}{\nu-\omega e^{-(\nu-\omega)t}}\right]^{a_0+\frac{\lambda}{\nu}} \sum_{j=0}^{\min\left(k,a_0+\frac{\mu}{\omega}\right)} \binom{a_0+\frac{\lambda}{\nu}+j-1}{j} \left[\frac{\left(\nu-\omega e^{-(\nu-\omega)t}\right)}{\nu\left(e^{-(\nu-\omega)t}-1\right)}\right]^j \binom{a_0+\frac{\mu}{\omega}}{k-j} \frac{\left[\omega\left(1-e^{-(\nu-\omega)t}\right)\right]^{k-j}}{\left[\left(\nu e^{-(\nu-\omega)t}-\omega\right)\right]^{-\left[a_0+\frac{\mu}{\omega}-k+j\right]}}$$

(12.228)

and the final solution follows immediately from an application of the scaling and sliding tool

$$P_k(t) = \left[e^{-(\nu-\omega)t}(\nu-\omega)\right]^{\frac{\lambda}{\nu}-\frac{\mu}{\omega}} \left[\frac{1}{\nu-\omega e^{-(\nu-\omega)t}}\right]^{a_0+\frac{\lambda}{\nu}}$$

$$\sum_{j=0}^{\min\left(k+\frac{\mu}{\omega},\, a_0+\frac{\mu}{\omega}\right)} \binom{a_0+\frac{\lambda}{\nu}+j-1}{j} \left[\frac{\left(\nu-\omega e^{-(\nu-\omega)t}\right)}{\nu\left(e^{-(\nu-\omega)t}-1\right)}\right]^j \binom{a_0+\frac{\mu}{\omega}}{k+\frac{\mu}{\omega}-j} \frac{\left[\omega\left(1-e^{-(\nu-\omega)t}\right)\right]^{k+\frac{\mu}{\omega}-j}}{\left[\left(\nu e^{-(\nu-\omega)t}-\omega\right)\right]^{[a_0-k+j]}} \tag{12.229}$$

Problems

1. By modifying the Chapman–Kolmogorov forward equations for the immigration process, find the distribution of the number of patients with disease given that there are three sources of immigrants, one source arrives with arrival rate λ_1, the second source submits arrivals with arrival rate λ_2, and the third arrives with arrival rate λ_3.
2. Obtain $P_k(t)$ for $t \to \infty$ in equation (12.26). What probability model is this?
3. Simplify equation (12.26) for the case of $\lambda = \nu$.
4. Derive the Chapman–Kolmogorov equations for the immigration–death process when $\lambda = \omega$.
5. Obtain the steady state solution by letting $t \to \infty$ and $\lambda = \omega$ in equation (12.47)
6. For the immigration–death model, obtain the expected number in the system at time t.
7. For the birth–death model, obtain $G_t(s)$ for the case of $\nu = \omega$ and use it to obtain $P_k(t)$.
8. Obtain the limit of $P_k(t)$ as $t \to \infty$ in problem 6.
9. Using equation (12.101) obtain the expected number in the system at time t when $\nu = \omega$ for the birth–death model.
10. In equation (12.196) let $\lambda = 0$. How does $G_t(s)$ when $\lambda = 0$ compare with $G_t(s)$ for the birth–death model?
11. Obtain $P_k(t)$ in equation (12.229) by inversion.
12. Obtain $\lim_{t\to\infty} P_k(t)$ in problem 10 above.
13. Show that the generating function for the immigration–emigration–birth–death process reduces to the generating function for the immigration–birth–death process. Hint: set $\mu = 0$ in equation (12.224).

References

1. Bailey N.T. The elements of stochastic processes with applications to the natural sciences. New York. John Wiley and Sons. 1964.

Index